ENVIRONMENTAL AND SOCIAL IMPACT ASSESSMENT

An Introduction

C. J. Barrow

Centre for Development Studies,
University of Wales Swansea

A member of the Hodder Headline Group
LONDON • NEW YORK • SYDNEY • AUCKLAND

First published in Great Britain in 1997 by
Arnold, a member of the Hodder Headline Group
338 Euston Road, London NW1 3BH

Co-published in the US, Central and South America by
John Wiley & Sons, Inc., 605 Third Avenue,
New York, NY 10158–0012

British Library Cataloguing in Publication Data
A catalogue record for this book is available from the British Library

Library of Congress Cataloging-in-Publication Data
A catalog record for this book is available from the Library of Congress

ISBN 0 340 66271 9
ISBN 0 470 23528 4 (Wiley)

Typeset in 10/12pt Palatino by
J&L Composition Ltd, Filey, North Yorkshire
Printed and bound by
JW Arrowsmith, Bristol

CONTENTS

PREFACE

The procedures, methods and techniques examined in this book gather, assess and present information that can potentially aid in planning and decision-making, and act as research tools to improve understanding of the interaction between, first, development and the environment and, second, development and its socio-economic effects. Impact assessment and some related approaches offer a bridge between environmental management theory and real-life development, and should play a vital part in the quest for sustainable development.

Having edited a book on the theory and practice of environmental impact assessment, *Environmental impact assessment: theory and practice* (London, Unwin Hyman, 1988), P. Wathern noted the difficulty that any one person has in giving a competent, comprehensive overview. I have tried to give an introduction which is accessible to those who are interested in, or who might commission or make use of, impact assessment and related activities. I do not seek to provide a handbook or to train impact assessors.

There are two pitfalls that it is all too easy to fall into: advocacy and encyclopaedic coverage. The first error is easy to make, for it is less of a challenge to say what *ought* to be done than to give precise guidance as to how it could actually be achieved. The second has to be countered by careful selection of information. I hope that I have managed to avoid these traps.

C.J. Barrow
University of Wales, Swansea

ACKNOWLEDGEMENTS

I would like to thank the staff of the University of Wales Swansea Library. I am also grateful for the sharp eyes of Peter Harrison, for the cheerful encouragement and support from Laura McKelvie and Wendy Rooke and other members of the production team at Arnold – somehow they managed to decipher my handwriting!

Chris Barrow
Swansea
November 1996

The author and publishers would like to thank the following for permission to reproduce the following copyright material:

Garden City Press, New York for verse 6 of Boulding's 'Ballad of Ecological Awareness' from Farvar, M.T. and Milton, J.P. (eds) 1972, *The Careless Technology*; McGraw-Hill, Inc. for the Adkins-Burke checklist table from Canter, L.W. (ed.) 1977, *Environmental Impact Assessment*, p. 204; D. Reidel for figures 1 and 2 from Clark, B.D., Gilad, A., Bisset, R. and Tomlinson, P. (eds) 1984, *Perspectives on Environmental Impact Assessment*; John Wiley & Sons, Inc., for the box material from Westman, W.E. 1985, *Ecology, Impact Assessment and Environmental Planning*.

INTRODUCTION

> Impact assessment should be designed as a bridge that integrates the science of environmental analysis with the policies of resource management. (Graham Smith, 1993: 12)

The ethos that has predominated in 'modern' times has been one of 'develop now, minimize associated costs and, if forced to, clean up later'. Since the 1960s there has been a shift toward more environmentally and socially appropriate development, and the tendency to damage the environment and societies in the name of 'progress' is being questioned more often. There is increased awareness that technology and biotechnology can pose huge threats, and there is growing interest in sustainable development. To minimize problems, maximize benefits and, increasingly, to involve the public and win their support, planners and decision-makers need to assess how activities have affected or might affect the environment – including social conditions.

WHAT IS IMPACT ASSESSMENT?

Impact assessment has been evolving for over a quarter of a century, is still far from perfect, and is often misapplied or misused. Impact assessment has been dominated by environmental impact assessment, the origins and spread of which are discussed in Chapter 6. A selection of guidebooks, handbooks and bibliographies on this topic are provided at the end of the chapter, together with a list of journals which contain articles on the subject. Selected guidebooks, handbooks and a list of journals which contain articles on social impact assessment, the social and cultural consequences to human populations of development, are provided at the end of Chapter 8.

There is no commonly agreed definition of what exactly environmental impact assessment is, so it is best treated as a generic term for a process that seeks to blend administration, planning, analysis and public involvement in assessment prior to the taking of a decision. A shorter explanation might be 'an approach which seeks to improve development by a-priori assessment'. Kozlowski (1989: 6) noted, 'The EIA concept is rooted in the common sense wisdom that it is better to prevent a problem than to cure it.' Identifying future consequences of a current or proposed activity is hardly a revolutionary idea; however, for much of history it has not been the development approach adopted.

Impact assessment offers much more than simply a common-sense approach to development. It can be a policy instrument, a planning tool, a means of public involvement. Some claim that it should be part of a broad framework, crucial to environmental management and the drive for sustainable development, as advocated by Heer and Hagerty (1977: 18) before sustainable development was a fashionable concept. Graham Smith (1993: 1) viewed impact assessment as a philosophy, rather than just a technique, and argued for it to be treated as such. An increasing number of people take Graham Smith's line, and see impact assessment as something more than a technocratic tool – as, in fact, a process that goes beyond prediction to improve decision-making and environmental management in the face of political and other realities.

The arguments for impact assessment being a major innovation for policy-making were covered in a text edited by Bartlett (1989), who felt that it had a long way to go to shift from being a 'tool' to becoming 'part of policy-making'. Bartlett (1989: ix) observed: 'Impact assessment failed to excite large numbers of policy scholars because, like engineers, scientists, and lawyers, they too saw it as a technique or non-substantive legal procedure or, at most, as a simplistic decision-making reform.'

Environmental and social impact assessment are effectively opposite ends of the same spectrum, but they nevertheless overlap. Some recognize a third subdivision of impact assessment: cultural impact assessment (*see* Chapter 8). Running parallel to environmental and social impact assessment, with broadly similar goals, frequently exchanging information and methods and sometimes overlapping with them, are hazard and risk assessment, technology assessment, monitoring, futures forecasting and eco-auditing. These approaches are dealt with in Chapter 2. Hyman and Stiftel (1988: 17) argued that environmental impact assessment brought environmental and social impact assessment brought social information into planning. Environmental and social impact assessment have stronger regulatory frameworks than most of the other approaches considered, although eco-auditing is rapidly becoming more formalized.

'Environment' is often poorly defined. As there is so much interaction between physical, biological, social, cultural and economic components it makes sense to use the word in its broadest sense, which is what I do. However, there are a number of 'schools' of impact assessment which adopt different approaches and methods and vary in how they interpret 'environment'. There are increasingly strong arguments for impact assessment to adopt a holistic view if it is to achieve successful environmental management in practice (Rossini and Porter, 1983) (*see* Figs 1.1 and 1.2). In spite of

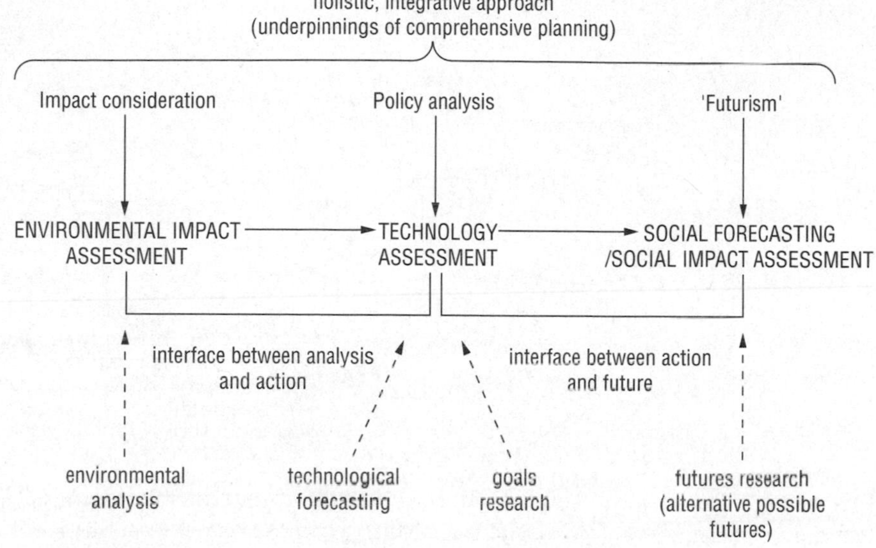

FIGURE 1.1 Relationship of environmental impact assessment, technology assessment, social forecasting/social impact assessment
Source: Based on a figure on p. 54 in Vlachos (1985)

BOX 1.1 CHARACTERISTICS OF ENVIRONMENTAL IMPACT ASSESSMENT (EIA)

- Proactive assessment should be initiated pre-project/programme/policy before development decisions are made. Unfortunately, in-project/programme/policy and post-project/programme/policy assessment are very common. Even so, while these may not allow as much problem avoidance, they can advise on problem mitigation, gather data, feed into future impact assessment, improve damage control and enable the exploitation of unexpected benefits).
- Initial screening and scoping should determine what is subjected to EIA, and decide what form the assessment should take.
- Systematic, interdisciplinary study is needed to improve planning/decision-making.
- Independent, objective review of EIA results is needed, as an ideal at least.
- An account of likely impacts, with an indication of their significance, should be published.
- Declaration should be made of possible alternative development options, including nil development, and their likely impacts.
- Public disclosure of EIA is desirable, so that planners/administrators are more accountable. (Often, disclosure is made only to a restricted audience.)
- Public participation in EIA is desirable; often, participation is partial or avoided altogether.
- EIA should be integrated into the planning/legal process; often, integration is weak.

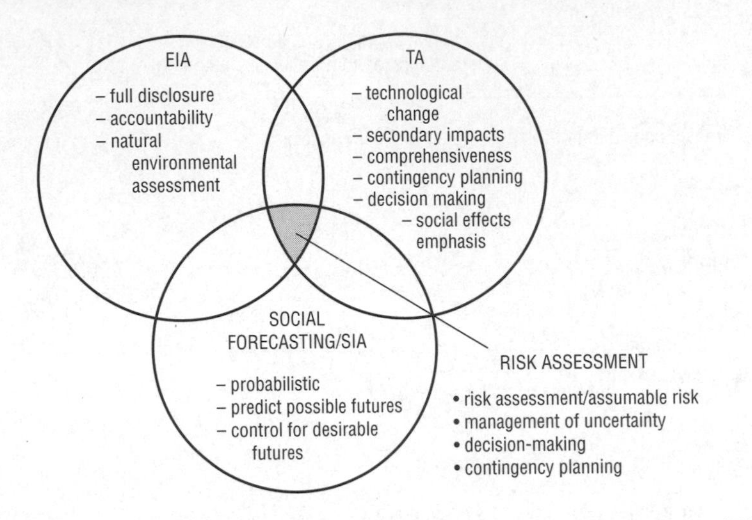

FIGURE 1.2 Interrelationship of environmental impact assessment (EIA), technology assessment (TA), social forecasting and risk assessment. SIA = social impact assessment
Source: Based on Covello *et al.* (1985: 5)

differences in the respective literatures and in attitudes and backgrounds of practitioners, environmental impact assessment, social impact assessment and many related activities have much in common, especially with respect to aims, procedures and processes (Westman, 1985; Graham Smith, 1993: 14). For example:

- They focus on effects.
- They adopt, or should adopt, a future (proactive) orientation.
- Their approach is systematic, focused, interdisciplinary and comprehensive, and generally iterative (*see* Box 1.1).

Given the similarities between environmental impact assessment, social impact assessment and related fields *impact assessment* is used in this book as a convenient shorthand, unless specific fields are being discussed. 'Assessment' is deliberately chosen, rather than 'analysis', 'appraisal' or 'evaluation', although these latter terms are sometimes used in the literature. 'Analysis' is rejected because it implies scientific determination of research results by a fully objective process, which is clearly not the case with most impact assessment. 'Assessment' is the most appropriate term because it implies evaluation of results with some degree of subjectivity, and in recent years has been more often used than 'appraisal'.

Impact assessment is an art rather than a science (Matthews, 1975), although Wathern (1988: 5) felt that the links were such that the distinction was not crucial. Assessment also hints at a guiding role. The 'unscientific' qualities of impact assessment can pose problems: reporting on a two-year study to determine the extent to which ecology could contribute to the design and conduct of environmental impact assessment, Beanlands and Duinker (1983: 13) noted that some scientists could be 'reluctant to become directly involved in impact assessment since they feel it is not an acceptable forum in which to apply their scientific method'.

Although it is difficult to give a precise definition of environmental impact assessment, the following points convey its character:

- study of the effects of a proposed action intended to protect the environment (which can include biophysical, social, economic, cultural and aesthetic aspects) and the public from the consequences of reckless or inadequately informed actions;
- systematic evaluation of all significant environmental (including social and economic) consequences an action (or policy, programme, project or plan) is likely to have upon the environment, ideally undertaken before the decision to take the action is made;
- a process leading to a statement to aid and guide decision-makers;
- a structured, comprehensive approach to controlling development;
- a learning process and means to find the optimum development path;
- a process by which information is collected and assessed to determine whether it is wise to proceed with a proposed development;
- an activity designed to identify and predict the impacts (good and bad) of an action on the biogeophysical environment and the impacts on human health and well being of legislative proposals, policies, programmes, projects and operational procedures, and to interpret and communicate information about the impacts (Munn, 1979: 1);
- a process which forces (or should force) developers to reconsider proposals and which seeks to cope with decision making under conditions of uncertainty (Taylor, 1984);
- a process that has the potential to increase developers' accountability to the public;
- prediction of future changes in environmental quality and the valuation of these changes.

There is thus a 'spectrum' of attitudes towards impact assessment, ranging from the view that it is just a required procedure – a 'rubber-stamping' activity to ensure that development meets government requirements – or that it determines optimal development, to the idea that it has a vital role to play in improving environmental management and planning and in achieving sustainable development. The assessment of impacts depends on human perception, something that is complex and not static. O'Riordan and Rayner (1991: 92) graphically represented human perspectives along time and spatial scales: most concern is focused at family (spatial) and short-term (time) levels, much less at national or global, long-term levels – the levels of future generations. A long-term focus is vital, however, if sustainable development is a goal. Assessment is also affected by whether an impact is reversible and obvious to observers.

WHAT ARE THE GOALS OF IMPACT ASSESSMENT?

Until quite recently, resource management mainly asked:

- Is it technically feasible?
- Is it financially viable?
- Is it legally permissible?

Gradually, resource managers began to enquire about the balance between benefits and costs and public acceptance, and then to enquire whether proposed courses of action were 'environmentally sound'. By considering goals, realities and available alternative developments it should be possible to identify the *best* options, rather than proposals that are merely acceptable (Wood, 1995: 10; Morrisey, 1993). Impact assessment is therefore concerned with the continued welfare of people and the stewardship of nature. Impact assessment has tended to flag negative impacts; however, it should also point to potentially beneficial opportunities. The emphasis tends to be on the avoidance of problems, but impact assessment can also help ensure that benefits are fully realized. It is important to stress that impact assessment should consider all options, including the option to undertake no development, to make no change. Often it is initiated much too late to do this and focuses on an already chosen plan of development. Environmental impact assessment is not just of passive use for getting development at least environmental (and, it is to be hoped, socio-economic) cost; it can, by improving understanding of relationships between development and environment and by prompting studies, actively lead to improvement of the environment.

SEEKING SUSTAINABLE DEVELOPMENT: THE VALUE OF IMPACT ASSESSMENT

If impact assessment is to become an integral part of planning it must be applied before development decisions are made. However, a large portion of the literature consists of retrospective impact assessments. These are still of value because they can help clarify problems and add to hindsight knowledge, even though they do not immediately assist the cause of sustainable development or ensure optimum planning decisions. Communities and governments should use impact assessment because it offers a chance of avoiding the public sector having to pay to rectify environmental and socio-economic impacts often caused by the private sector.

Impact assessment is seen by many as a promising tool in the quest for sustainable development, particularly the strategic environmental assessment approach (Thérivel *et al.*, 1992: 22; and *see* Chapter 3), as are eco-auditing (*see* Chapter 2) and social impact assessment. Many attempts at developments have been 'Faustian bargains' – they sacrifice long-term well-being (in the case of Marlow's Dr Faustus, his soul) for short-term gains often obtained at the expense of damage to the environment or to people's physical and mental welfare. A single precise definition of sustainable development is impossible (*see* Barrow, 1995: 66–72), but Box 1.2 and Fig. 1.3 give some idea of the concept's meaning. A useful 'rough definition' is: development that maintains and, if possible, improves the long-term condition of the environment and people's quality of life.

There is much confusion as to whether the development being 'sustained' is ongoing and unrestricted ('sustainable growth') or steady-state and limited; 'sustainable development' is a term used carelessly and to mean different things. 'Mainstream' sustainable development demands the consideration of existing and future, potential hazards or risks and limits, and the adoption of measures to counter poverty.

People tend to react toward environmental and social threats in one of three general ways (O'Riordan and Rayner, 1991: 99):

- They adopt a preventive approach – 'nature is fragile, it is morally wrong to abuse nature'.
- They adopt an adaptive approach – 'nature is robust, it is morally wrong to curtail development'.

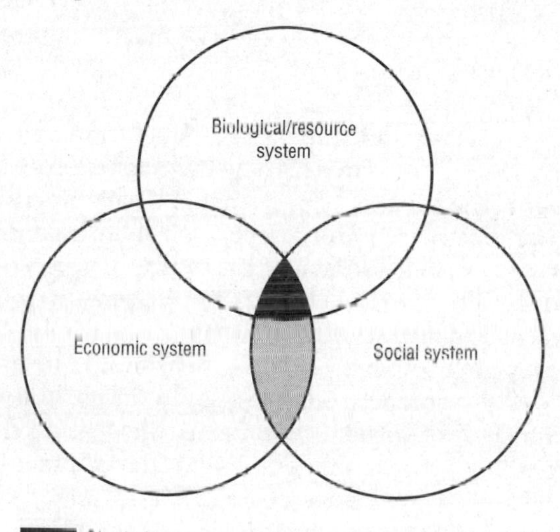

Shaded area = maximization of goals across the three systems and, in effect, represents sustainable development.
= Marxist economics.

Goals of biological system: keep genetic diversity, maximize productivity.
Goals of economic system: supply basic needs, improve equity, improve goods and services.
Goals of social system: sustain institutions, improve social justice, improve participation.
Conventional economics maximizes the economic system and the social system.

FIGURE 1.3 Sustainable development: a diagrammatic explanation of the concept
Source: Barrow (1995: 67, Fig. 4.6)

Box 1.2 Sustainable Development

A commonly offered definition (derived from WCED, 1987) of sustainable development would run: development that seeks to satisfy the needs of present without compromising the ability of future generations to meet their own needs.

Most definitions incorporate three main aims:

- Maintenance of the essential ecological processes ('life-support systems') and human well-being has priority.
- It is important to reserve genetic diversity (we should try to avoid extinction of wild species and domesticated varieties).
- We should aim for sustainable use of genetic diversity and environment.

Support of basic human needs (reduction of poverty) is generally part of the concept of sustainable development. There is also likely to be acceptance that environment and development issues should not be considered in separation and that attitudes and ethics will almost certainly have to change.

- They adopt a sustainable development approach – 'nature is robust, within limits, it is morally imperative to preserve nature enough not to reduce choice of options for the future'.

Environmental protection and impact assessment are still often seen as secondary objectives within resource management, or even as hurdles in the path of development. The concept of sustainable development offers opportunities for bringing together environmental, social and economic planning and management, and impact assessment is a means of achieving such a synthesis (Jacobs and Sadler, 1989; Pritchard, 1993; Gilpin, 1995: 35). Graham Smith (1993: 4) warned that the world is facing the possibility of potentially irreparable and damaging impacts and that there should be efforts to avoid them. A narrow, *ad hoc* approach, Graham Smith felt, is not enough: problems are often a function of social, economic, political and physical factors, demanding integration of environmental and development planning, a process that can be helped by the use of impact assessment. Graham Smith (1993: 1) argued that impact assessment is crucial for sustainable development, because it is, or must become 'a process for resource management and environmental planning that provides for the achievement of the goal of sustainability.' James (1994) stressed the value of impact assessment that considers environmental and economic aspects in efforts to achieve sustainable development, and Dalal-Clayton (1992) saw modified environmental impact asessment as a promising route to such development.

Impact assessment should emphasize a preventive, systematic, holistic, multidisciplinary approach to environmental protection and sustainable development. It can also be organized to help ensure intergenerational equity (that is, to ensure that future generations are not denied opportunities by today's developments) and to ensure that decision-makers acknowledge environmental limits (Rees, 1988; Glasson *et al.*, 1994: 3; Morris and Thérivel, 1995: 2; Trzyna, 1995). Interest in impact assessment was stimulated and supported by two of the most influential publications of the 1980s: the *World conservation strategy* (IUCN, WWF and UNEP, 1980) and the 'Brundtland Report' (WCED, 1987).

Decision-makers aiming for sustainable development frequently face uncertainty, a fact that has led to widespread acceptance in Western societies of a sort of moral injunction to err on the side of caution known as the 'precautionary principle'. This may be broadly defined as follows: where there are threats of serious or irreversible environmental changes, lack of full scientific certainty should not be used as a reason for postponing measures to prevent environmental degradation (Dovers and Handmer, 1995: 92). The precautionary principle encourages careful evaluation of available development options. Such evaluation can be through impact assessment and eco-auditing, but, as it is impossible to be certain about uncertainties, these are not perfect 'arts', nor are they easy to abide by.

The 1992 UN Conference on Environment and Development (UNCED – the Rio 'Earth Summit') stressed the value of impact assessment. Of its 27 principal declarations, 17 deal with some aspect of environmental impact assessment. A publication released by UNCED, *Agenda 21*, called for 'integration of environment and development in decision making'. This call has been heeded by many governments, and the value of impact assessment has thus had a boost. Since 1992 the European Union has stressed the value of impact assessment as a route to sustainable development. Some forms of impact assessment, particularly strategic environmental assessment (*see* Chapter 3), offer ways of viewing and coordinating development from policy and plan levels down to project level (Sadler, 1994).

Impact assessment seems to have great potential for supporting sustainable development, but this will not be realized unless it is better applied and, above all, better integrated into policy-making, planning and administration (Kozlowski, 1990; Nay Htun, 1990; Hare, 1991; Jenkins, 1991; Pearce, 1992; Pritchard, 1993; Van Pelt, 1993: 99; Sadler, 1994; Ortolano and Shepherd, 1995: 16). Impact assessment should deal with transboundary issues and global environmental change as well as local and regional or national issues, adopt a more flexible and longer-term outlook, and be much better integrated with environmental management and development planning (Slocombe, 1993; Bowyer, 1994). It must be used much earlier in planning, and effectively consider indirect and cumulative impacts (Gardiner, 1989; Jacobs and Sadler, 1989; Anon., 1990; Jenkins, 1991; Wallington, *et al.*, 1994). All this involves consideration of more 'unknowns', such as: 'how will today's developments affect the future?' and 'what will be needed in future?' (Gilpin, 1995: 9–10). One problem at present is that sustainable development is a rather vague concept, making it difficult to focus impact assessment on it.

FORECASTING, FUTURES EVALUATION AND IMPACT ASSESSMENT

Impact assessment (and some related approaches) should be predictive and feed into planning so that beneficial modification is possible. In practice, the application is often 'postdictive' – retrospective and stock-taking. Forecasting is an essential part of planning and policy formulation. It has been tried from prehistoric times in hunting, settlement siting, agriculture, decisions to embark on migrations or warfare, and many other things. Some past civilizations, such as the Mayan civilization, seem to have devoted great effort to calendrical observations to try to foretell and record natural catastrophes. (Closer examination of some of these records might be worth while today.) Forecasting or prophesy has been accomplished by reference to accumulated knowledge and tradition, often with a strong seasoning of magic, religion and superstition. For example, in the seventeenth century BC, the Old Testament prophet Joseph's interpretations of the Pharaoh Khyan's dreams reputedly warned Egypt of drought and pestilence (Genesis 40, verse 9).

Faith in astrology, necromancy and other mystical or semi-mystical futurology is far from dead, even in largely secular Western societies, and elsewhere, it is often heeded very closely. It would be rash to reject all of this mystical forecasting, because geomancers, astrologers and the like, may read

BOX 1.3 FENG SHUI

Feng shui ('vital design') is a Chinese philosophy, perhaps three thousand years old, which seeks to understand and control 'good' and 'bad' invisible life forces or energy, symbolized by the yang/yin symbol:

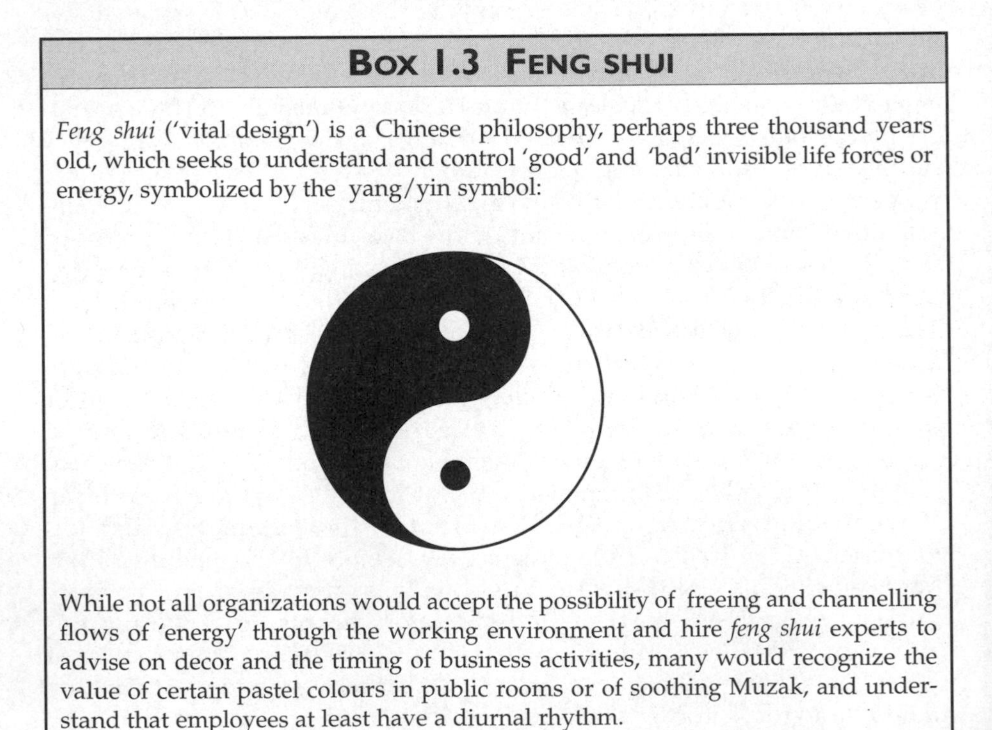

While not all organizations would accept the possibility of freeing and channelling flows of 'energy' through the working environment and hire *feng shui* experts to advise on decor and the timing of business activities, many would recognize the value of certain pastel colours in public rooms or of soothing Muzak, and understand that employees at least have a diurnal rhythm.

social and environmental conditions to usefully suggest ways in which people may become less stressed and more confident, thereby increasing successes (see Box 1.3).

EVOLUTION OF IMPACT ASSESSMENT AND RELATED ACTIVITIES

Since roughly the mid-eighteenth century Western societies have based forecasting on rational observation, projection of trends and hindsight knowledge. Here and there, rational stocktaking occurred even earlier. An early example, not called an impact assessment but in many ways similar, was the Royal Commission in England charged in 1548 to examine the effect of the iron mills and furnaces of the Weald (an early industrial development area in south-east England which suffered pollution, labour in-migration and deforestation). This commission's investigators produced a thorough situation report on the physical, social and economic impacts as rapidly as their modern equivalent (Fortlage, 1990: 1). By the 1930s Europe, the USSR and the USA were conducting post- and in-project assessments of development problems, and cautionary guidebooks, checklists, procedural manuals and planning regulations (and, in the UK, occasional public inquiries) were in use to guide decision-making (Caldwell, 1989).

The banking, investment and insurance industries developed hazard and risk assessment methods by the 1940s, and military tacticians were trying to predict war scenarios during the Second World War and the 'Cold War'. Cost–benefit analysis was well-established by the 1960s as a means of assessing development proposals, but it often ignored or misvalued environmental and social factors. (Futures research, modelling and cost–benefit analysis are discussed in Chapter 2.) Legislation on food and drugs safety, public sanitation, workplace conditions and pollution control were tightened between the 1930s and 1960s, especially in Europe and North America. However, these measures were largely applied to specific problems. In North America, there was need for coordination of development, environmental management and national environmental protection legislation. Also missing were adequate consideration and dissemination of information on development alternatives. The development of impact assessment has been influenced by the evolution of related fields, especially rational planning theory, technology assessment, risk assessment and cost–benefit analysis (Mayda, 1993).

In 1969 the US National Environmental Policy Act, which sowed the seeds of environmental impact assessment, was put before Congress to try to address these problems. Passed in 1970, the National Environmental Policy Act, was a seminal piece of legislation that triggered interest in impact assessment, especially environmental impact assessement around the world. It is discussed in Chapter 6. Impact assessment came to offer a structured comprehensive approach that is also more than just precept in that it can compel action.

FURTHER READING

Journals

The following journals regularly contain articles on environmental impact assessment and related activities:
Impact Assessment Bulletin (in 1994 became *Impact Assessment*)
Environmental Impact Assessment Review
Environmental Monitoring and Assessment
Risk Analysis
Journal of Risk Analysis
Environmental Monitoring and Assessment
Environmental Management
Journal of Environmental Management
Science of the Total Environment
Project Appraisal
Environmental Impact Assessment Worldletter
Global Environmental Change
Environmental Assessment (Institute of Environmental Assessment and the Environmental Auditors Registration Association, UK)
Risk, Decision and Policy
EIA Trainer's Newsletter (EIA Centre, University of Manchester, Manchester M13 9PL, UK)

Organizations

International Association of Impact Assessment (IAIA). Established 1980 (USA). Voluntary body of researchers and practitioners concerned with environmental impact assessment, social impact assessment, eco-auditing, risk assessment, technology assessment, etc. Seeks to promote impact assessment (publishes *Impact Assessment*). For background information on IAIA, *see*: *Impact Assessment Bulletin* 1989: **7(4)**, 5–15 (e-mail address: RHAMM@ndsu.ext.nodak.edu).

EIA Centre, University of Manchester, Manchester, UK. Publishes a regularly updated series of leaflets on a range of topics related to environmental impact assessment.
Institute of Environmental Assessment. Established 1990 (UK). Non-profit-making organization established to promote best practices in environmental assessment and eco-auditing. Provides a registration scheme (accreditation) for UK environmental auditors, assists with standards and guidelines, and can provide independent non-governmental review of EISs (telephone: UK 01522 540069).

Environmental Auditors Registration Association (EARA). UK non-profit-making non-governmental organization that registers environmental auditors.

Guidebooks, handbooks, reviews and bibliographies on environmental impact assessment and related approaches

Guidebooks and Handbooks

Becker, H.A. and **Porter, A.L.** (eds) 1986: *Methods and experiences in impact assessment*. New York: D. Reidel.

Black, P. 1981: *Environmental impact assessment*. New York: Praeger.

Brown, G.A. 1994: *Environmental audit guidebook*. Melbourne: Law Book Group.

Buckley, R.C. 1991: *A handbook for environmental audit*. Canberra: Australian International Development Assistance Bureau.

Burchell, R.W. and **Listokin, D.** 1975: *The environmental impact handbook*. New Brunswick, NJ: Rutgers University, Center for Urban Policy Research.

Canter, L.W. 1996: *Environmental impact assessment*, 2nd edn. New York: McGraw-Hill. (1st edn 1977)

Cheremisinoff, P.N. and **Morresi, A.O.** 1977: *Environmental assessment and impact statement handbook*. Chichester: Wiley.

Department of the Environment 1981: *A manual for the assessment of major development proposals*. London: HMSO.

Department of the Environment 1989: *Environmental assessment: a guide to the procedures*. London: HMSO.

Erickson, P.A. 1979 *Environmental impact assessment: principles and appliances*. New York: Academic Press.

Fortlage, C.A. 1990: *Environmental assessment: a practical guide*. Aldershot: Gower Technical.

Gilpin, A. 1995: *Environmental impact assessment (EIA): cutting edge for the twenty-first century*. Cambridge: Cambridge University Press.

Glasson, J., **Thérivel, R.** and **Chadwick, A.** 1994: *Introduction to environmental impact assessment*. London: University College London Press.

Golden, J., **Ouellette, R.P.**, **Saari, S** and **Cheremisinoff, P.N.** 1980: *Environmental impact data book*, 2nd edn. Ann Arbor, MI: Ann Arbor Science.

Heer, J.E. and **Haggerty, D.J.** 1977: *Environmental impact assessments and statements*. New York: Van Nostrand Reinhold.

ICSU 1987: *Environmental impact assessment*. Washington DC: International Council for Scientific Unions.

IES 1996: *Environmental assessment: a guide for the non-specialist*. London: Institution of Environmental Sciences (14 Princes Gate, London SW7 1PU).

Landy, M. 1981: *Environmental impact statement directory*. New York: Plenum.

Morris, P. and **Thérivel, R.** (eds) 1995: *Methods of environmental impact assessment*. London: University College London Press.

Munn, R.E. (ed.) 1979: *Environmental impact assessment: principles and procedures*, 2nd edn. SCOPE Report 5. Chichester: Wiley.

Porter, A.L., **Rossini, F.A.**, **Carpenter, R.A.** and **Roper, A.T.** 1980: *A guidebook for technology assessment and impact analysis*. New York: North-Holland.

Porter, C.F. 1985: *Environmental impact assessment: a practical guide*. St Lucia, Brisbane: University of Queensland Press.

Pritchard, D. 1993: Towards sustainability in the planning process: the role of EIA. *Ecos* **14(3-4)**, 10–15.

Rau, J.G. and **Wooten, D.C.** (eds) 1980: *Environmental impact analysis handbook*. New York: McGraw-Hill.

Roe, D., Dalal-Clayton, B. and **Hughes, R.** 1995: *A directory of impact assessment guidelines*. London: International Institute for Environment and Development for the International Union for the Conservation of Nature and Natural Resources.

Rosen, S.J. 1977: *Manual for environmental impact evaluation*. Englewood Cliffs, NJ: Prentice-Hall.

Vanclay, F. and **Bronstein, D.A.** (eds) 1995: *Environmental and social impact assessment*. Chichester: Wiley.

Wathern, P. (ed.) 1988: *Environmental impact assessment: theory and practice*. London: Unwin Hyman.

Wood, C. 1995: *Environmental impact assessment: a comparative review*. Harlow: Longman.

Wood, C. and **Gazidellis, V.** 1985: *A guide to training materials for environmental impact assessment*. Occasional Paper 14. Manchester: University of Manchester, Department of Town and Country Planning.

World Bank 1992: *Environmental assessment sourcebook* (2 vols). World Bank Technical Papers 139 and 140. Washington DC: World Bank.

Bibliographies

Clark, B.D., Bisset, R. and **Wathern, P.** 1980: *Environmental impact assessment: a bibliography with abstracts*. London: Mansell.

Duinker, P.N. and **Beanlands, G.E.** 1983: *Ecology and environmental impact assessment: an annotated bibliography*. Halifax, Nova Scotia: Institute for Resource and Environmental Studies, Dalhousie University.

Landy, M. (ed.) 1979: *Environmental impact statement glossary: a reference source for EIS writers, reviewers and citizens*. New York: IFI/Plenum.

Lawson, W.J. 1989: *Environmental impact assessment: a selected bibliography*. Brisbane: Griffith University Library.

Meagher, J. and Pickett, R. undated: *Bibliography on environmental impact assessment methodologies*. Washington DC: Office of Environmental Review, US Environmental Protection Agency.

Williams, J. 1982: *Environmental assessment methods, guidelines and policies: a selected, annotated bibliography*. East–West Center, Environment Policy Institute Working Paper. Honolulu: East–West Center.

REFERENCES

Anon. 1990: Draft convention on environmental impact assessment in a transboundary context. *Environmental Policy and Law* **20(4–5)**, 181–6.

Barrow, C.J. 1995: *Developing the environment: problems and management*. Harlow: Longman.

Bartlett, R.V. (ed.) 1989: *Policy through impact assessment: institutional analysis as a policy strategy*. New York: Greenwood Press.

Beanlands, G.E. and **Duinker, P.N.** 1983: *An ecological framework for environmental impact assessment in Canada*. Halifax, Nova Scotia: Institute for Resource and Environmental Studies, Dalhousie University.

Bowyer, J.L. 1994: Needed: a global environmental impact assessment. *Journal of Forestry* **92(6)**, 6.

Caldwell, L.K. 1989: A constitutional law for the environment: 20 years with NEPA indicates the need. *Environment* **31(10)**, 6–11, 25–8.

Covello, V.T., Mumpower, J.L., Stallen, P.J.M. and **Uppuluri, V.R.R.** (eds) 1985: *Environmental impact assessment, technology assessment, and risk analysis: contributions from the psychological and decision sciences*. Berlin: Springer-Verlag.

Dalal–Clayton, B. 1992: *Modified EIA and indicators of sustainability: first steps towards sustainability analysis*. Environmental Planning Issues 1. London: Earthscan.

Dovers, S.R. and Handmer, J.W. 1995: Ignorance, the precautionary principle, and sustainability. *Ambio* **XXIV(2)**, 92–7.

Fortlage, C.A. 1990: *Environmental assessment: a practical guide.* Aldershot: Gower.

Gardiner, J.E. 1989: Decision making for sustainable development: selected approaches to environmental assessment and management. *Environmental Impact Assessment Review* **9(4)**, 337–66.

Gilpin, A. 1995: *Environmental impact assessment: cutting edge for the twenty first century*. Cambridge: Cambridge University Press.

Glasson, J., Thérivel, R. and **Chadwick, A.** 1994: *Introduction to environmental impact assessment: principles and procedures, process, practice and prospects*. London: University College London Press.

Graham Smith, L. 1993: *Impact assessment and sustainable resource management*. Harlow: Longman.

Hare, B. 1991: Environmental impact assessment: broadening the framework. *Science of the Total Environment* **108(1–2)**, 17–32.

Heer, J.E. and **Hagerty, D.J.** 1977: *Environmental assessment and statement*. New York: Van Nostrand Reinhold.

Hyman, E.L. and **Stiftel, B.** 1988: *Combining facts and values in environmental impact assessment: theories and techniques*. Boulder, CO: Westview.

IUCN, WWF and **UNEP** 1980: *The world conservation strategy: living resources for sustainable development*. Gland: International Union for the Conservation of Nature and Natural Resources, Worldwide Fund for Nature and United Nations Environment Programme.

Jacobs, P and **Sadler, B.** (eds) 1989: *Sustainable development and environmental assessment: perspectives on planning for a common future*. Hull, Quebec: Canadian Environmental Assessment Research Council.

James, D. (ed.) 1994: *The application of economic techniques in environmental impact assessment*. Dordrecht: Kluwer.

Jenkins, B.R. 1991: Changing Australian monitoring and policy practice to achieve sustainable development. *Science and the Total Environment* **108(1)**, 33–50.

Kozlowski, J.M. 1989: Integrating ecological thinking into the planning process: a comparison of EIA and UET concepts. WZB Paper FS-II-89-404. Berlin: Wissenschaftszentrum Berlin für Sozialforschung.

Kozlowski, J.M. 1990: Sustainable development in professional planning: a potential contribution of the EIA and UET concepts. *Landscape and Urban Planning* **9**, 307–32.

Matthews, W.H. 1975: Objective and subjective judgements in environmental impact analysis. *Environmental Conservation* **2(2)**, 121–31.

Mayda, J. 1993: Historical roots of EIA? *Impact Assessment Bulletin* **11(4)**, 411–15.

Morris, P. and **Thérivel, R.** (eds) 1995: *Methods of environmental impact assessment*. London: University College London Press.

Morrisey, D.J. 1993: Environmental impact assessment: a review of its aims and recent developments. *Marine Pollution Bulletin* **26(10)**, 540–5.

Munn, R.E. (ed.) 1979: *Environmental impact assessment: principles and procedures*, 2nd edn SCOPE Report 5. Chichester: Wiley.

Nay Htun 1990: EIA and sustainable development. *Impact Assessment Bulletin* **8(1–2)**, 16–23.

O'Riordan, T. and **Rayner, S.** 1991: Risk management for global environmental change. *Global Environmental Change: Human and Policy Dimensions* **1(2)**, 91–108.

Ortolano, L. and **Shepherd, A.** 1995: Environmental impact assessment. In Vanclay, F. and Bronstein, D.A. (eds), *Environmental and social impact assessment*. Chichester: Wiley, 3–30.

Pearce, D. 1992: Sustainable development and environmental impact appraisal. In *Forum Valuazione: semestrale a cura del Comitato Internazionale per lo Sveluppo dei Popoli* (CISP)(IT) no. 4, 13–22.

Pritchard, D. 1993: Towards sustainability in the planning process: the role of EIA. *Ecos* **14(3/4)**, 10–15.

Rees, W.E. 1988: A role for environmental assessment in achieving sustainable development. *Environmental Impact Assessment Review* **8(4)**, 273–91.

Rossini, F.A. and **Porter, A.** (eds) 1983: *Integrated impact assessment*. Boulder, CO: Westview.

Sadler, B. 1994: Environmental assessment and sustainability at the project and programme level. In Goodland, R.G. and Edmundson, V. (eds.) *Environmental assessment and development*. Washington DC: World Bank, 3–19.

Slocombe, D.S. 1993: Environmental planning, ecosystem science, and ecosystem approaches for integrating environment and development. *Environmental Management* **17(3)**, 289–303.

Taylor, S. 1984: *Making bureaucracies think: the environmental impact statement strategy of administrative reform*. Stanford, CA: Stanford University Press.

Thérivel, R., Wilson, E., Thompson, S., Heaney, D. and **Pritchard, D.** 1992: *Strategic environmental assessment*. London: Earthscan.

Trzyna, T.C. (ed.) 1995: *A sustainable world: defining and measuring sustainable development*. Sacramento , CA: International Center for the Environment and Public Policy, published for the IUCN (also London: Earthscan).

Van Pelt, M.J.F. 1993: *Ecological sustainability and project appraisal*. Aldershot: Avebury.

Vlachos, E. 1985: Assessing long-range cumulative impacts. In Covello, V.T., Munpower, J.L., Stallen, P.J.M. and Uppuluri, V.R.R. (eds), *Environmental impact assessment, technology assessment and risk analysis: contributions from the psychological and decision sciences*. NATO ASI series G, Ecological Science, vol. 4. Berlin: Springer-Verlag, 49–80.

Wallington, T.J., Schneider, W.F., Worsnop, D.R., Neilsen, O.J., Sehested, J., Debruyn, W.J. and **Shorter, J.A.** 1994: The environmental impact of CFC replacements – HFCs and HCFCs. *Environmental Science and Technology* **28(7)**, 320A–326A.

Wathern, P. (ed.) 1988: *Environmental impact assessment: theory and practice*. London: Unwin Hyman.

WCED 1987: *Our common future*. Report of the World Commission on Environment and Development – the 'Brundtland Report'. Oxford: Oxford University Press.

Westman, W. 1985: *Ecology, impact assessment and environmental planning*. Chichester: Wiley.

Wood, C.M. 1995: *Environmental impact assessment: a comparative approach*. Harlow: Longman.

2

TECHNIQUES THAT SUPPORT IMPACT ASSESSMENT

INTRODUCTION

Techniques are ways of working that provide, process or present data that can be used by *methods*. A method may use one or more techniques at a given point in time and space to achieve a goal; at some other time or in some other place that same method might modify the techniques, or use different ones. A conceptual approach or activity, like environmental impact assessment, social impact assessment, eco-auditing or risk assessment, may use one or more methods, these methods possibly changing over time as new challenges appear and new techniques and concepts evolve.

Authorities generally seek to establish procedures and processes to guide or regulate impact assessment, risk assessment, eco-auditing, etc., and ensure more effective results. Nevertheless, there is still variation in methods and techniques. If two case-studies of similar conceptual approach were selected at random it is highly unlikely that they would share more than general similarities of method. As basic 'building-blocks', techniques tend to vary less from usage to usage than do methods and procedures.

This chapter reviews techniques that pre-date, feed into, support, run parallel to or overlap with impact assessment. Examples are cost–benefit analysis, hazard and risk assessment, and eco-auditing. *Approaches* to impact assessment are considered in the second half of Chapter 3. The environmental impact assessment process is examined in Chapter 4 whereas Chapter 5 outlines environmental impact assessment methods.

REVIEW OF PAST DEVELOPMENTS

Study of past developments similar to one that is being proposed or is under way often feeds into impact assessment, risk assessment and eco-auditing.

The study may take the form of direct observation or be limited to 'desk-studies' (examination of available literature). Sometimes this sort of input is the main component of impact assessment, and certainly, with the availability of on-line literature searching, it is becoming easier, faster and cheaper. Caution should be exercised, as apparently similar developments under what seem to be matching environmental and socio-economic conditions can generate different impacts. Hindsight knowledge is valuable but it must be treated with caution and should be supplemented with other approaches. There is no substitute for adequate first-hand familiarity with the environment and society involved and form of development proposed, although study of similar developments or situations is the next best thing.

Pilot studies

A common-sense, simple and valuable approach to assessing development impact is to conduct a pilot study (a smaller-scale or simplified forerunner of the main development). However, pilot studies are unlikely to get under way unless a development has come close to implementation or has already begun, whereas ideally, impact assessment and risk assessment should influence planning at an earlier stage. Pilot studies are less useful for small-scale developments and those that take place rapidly. A reason why they are often neglected is that planning horizons tend to be too short to allow enough time for pre-development study. The assessor must also be aware that what happens on a small scale in a pilot study may not accurately represent developments on a larger scale.

Delphi technique

The Delphi technique (some people deem it a method) is a way of seeking consensus among a panel of evaluators on questions that involve judgements of relative worth (Munn, 1979: 29). It was developed in California by the RAND Corporation in the late 1940s (well before environmental impact assessment appeared). It is a means of obtaining a reliable consensus of opinion about future developments from a group of experts – it is a panel evaluation technique (Stouth *et al.*, 1993). Because the Delphi technique was used for Cold War purposes, little was published for a decade until a report by Gordon and Helmer (1964) appeared. It has since been applied to forecasting and impact assessments and has been described as a 'cornerstone of futures research'. It relies on group communication (a Gestalt approach in which the group is perceived as an organized whole that is more than the sum of its parts) for systematically eliciting and 'pooling' judgements, having ensured anonymity for the expert assessors in order to prevent peer pressure or intimidation from influencing the results, controlled feedback between assessors, and a statistical group response (Ament, 1970; Pill, 1971;

Linstone and Turoff, 1975; Anon., 1978; Jain *et al.*, 1981: 65–70; Linstone, 1985; Richey *et al.*, 1985a, 1985b; Miller and Cuff, 1986). In summary, it gives a group viewpoint and an aggregate judgement. The expert assessors are asked to give their views without communicating with each other; these views are pooled, and evaluated, and the assessors are allowed to see the result as a form of controlled feedback and given the chance to modify their views; the feedback–pooled response process may be repeated three or more times to produce the final conclusions.

The Delphi technique has been used in health care predictions, gambling, tourism, marketing, management studies, resources allocation, technology innovation studies, war games, environmental impact assessment and social impact assessment. The Environmental Evaluation Systems approach to environmental impact assessment (*see* Chapter 5) uses the Delphi technique, and the cross-impact matrix has also been developed from it (Soderstrom, 1981: 20).

Judged a rapid technique, it is useful for short-range and good for longer-range (over 15 years into the future) forecasting, especially if high degrees of uncertainty are involved and where there is a need to predict impacts on culture (Ono and Wedemeyer, 1994). The results are, of course, subjective and qualitative. Impact assessment asks what impacts may occur, whereas the Delphi technique asks about the likelihood that some impact will happen and its possible date. It can thus complement impact assessment and cost–benefit analysis (Richey *et al.*, 1985a, 1985b; Green *et al.*, 1990; Glasson *et al.*, 1994: 113). It has become much easier to run with the aid of modern computers and may be done through a communications network or even by mail without the need to gather expert assessors in one place. Assessments of the technique suggest that it is valuable, but has often been poorly applied. Careful selection of the experts is crucial to avoid 'gaps' or bias, and it is also important to ensure that the questions they are asked are not too limited, otherwise their expertise could be constrained and lost. Bias can be introduced if assessors are allowed to suggest other assessors.

ASSESSING ECONOMIC IMPACTS

Although cost–benefit analysis is well-established, there have been repeated attempts to develop better approaches to economic impact assessment (Cooper, 1981; Hufschmidt and Hyman, 1982; Dixon *et al.*, 1986; James, 1994). For a review of the application of economics methods and techniques, other than cost–benefit analysis, to environmental impact assessment, *see* James (1994).

Cost–benefit analysis (CBA)

Designed to help decision-makers select one of a set of development alternatives, cost–benefit analysis (sometimes termed benefit–cost analysis, especially in the USA) is a method for systematically evaluating alternative developments and weighing relevant social costs with social benefits (social cost–benefit analysis has similarities to cost–benefit analysis – *see* Chapter 8). It is applied to plans, projects, programmes or policies, with results expressed in monetary terms. If need be, difficult-to-value things can be costed by contingent valuation, reference to property values, opportunity costs or shadow prices (for a discussion of how difficult-to-value things may be dealt with, *see* James, 1994: 78–92; Gilpin, 1995). Cost–benefit analysis pre-dates environmental impact assessment and is still widely used, either alone or as part of impact assessment or risk assessment. The underlying theory of cost–benefit analysis originated in France in the 1860s but practical development took place in the USA after 1936, mainly in connection with flood control and water resources exploitation, interest having been stimulated by welfare economics and works like those of Krutilla and Eckstein (1958), Hundloe *et al.* (1990), Leistritz (1994), James (1994) and James and Morris (1995).

Applied to aid, health care, social development, environmental issues, large development projects (e.g. the third London airport siting: HMSO, 1971), cost–benefit analysis generates what *seems* to be an objective single value expressing net benefits of development or proposed development. It is thus attractive to administrators seeking benchmarks; a development that gives a cost–benefit analysis result showing the greatest net social benefits is considered preferable. It has been popular with engineers and decision-makers because it appears to give an objective and tangible measure of social utility. Cost–benefit analysis tries to mimic the decisions that would have been reached by a perfectly competitive market mechanism, which may not reflect reality. Whereas cost–benefit analysis seeks – and sometimes struggles unsatisfactorily – to express diverse and difficult-to-value impacts as common (monetary) units, impact assessment uses subjective judgement to interpret and compare impacts. Costs and benefits which cannot be easily valued tend to be deemed 'intangibles', and thus cost–benefit analysis has often overlooked environmental and social values. Kenneth Boulding made this point in the last verse of his 'Ballad of ecological awareness' (Farvar and Milton, 1972):

> There are benefits, of course, which may be countable, but which
> Have a tendency to fall into the pockets of the rich,
> While the costs are apt to fall upon the shoulders of the poor.
> So cost–benefit analysis is nearly always sure
> To justify the building of a solid concrete fact,
> While Ecologic Truth is left behind in the Abstract.

There are a number of variants of cost–benefit analysis, and efforts have been made to adapt economics and cost–benefit analysis to cope better with environmental and social issues (Pigou, 1920; Kapp, 1950; Pearce, 1971; Abelson, 1979; Bohm and Henry, 1979; Cooper, 1981; Biswas and Qu Geping, 1987: 204; Winpenny, 1991) and use in developing countries (Little and Mirrlees, 1974).

Cost–benefit analysis can focus on indirect costs and benefits (indirect impacts), and, like impact assessment, depends on value judgements (e.g. for shadow prices). It is best done before development but is often undertaken later and, like impact assessment, is frequently poorly applied. The approach has been criticized (*see* O'Riordan and Turner, 1983; Westman, 1985: 168–97; Nijkamp, 1986: 17) for:

- emphasizing short-term issues;
- adopting monetary units, which are not good measures of utility (a resource can vary in value over time and various sections of a society may value a thing differently);
- difficulty in incorporating uncertainties;
- missing equity issues, despite its, recognition of efficiency;
- difficulties in allocating monetary values to some things, notably environmental features and environmental impacts (often the environment is not fully understood, which makes it difficult to value);
- encouraging rigid auditing that fails to encourage improvement of approach to problems;
- failing to look at 'trade offs'.

Debate over the value of cost–benefit analysis continues: in a recent editorial, Adams (1996) put forward strong criticisms of it, and O'Neill (1996), not convinced that decision-makers really needed a common unit of value, was thus sceptical of its worth.

Cost-effectiveness analysis

Cost-effectiveness analysis is used to select development alternatives with the lowest monetary cost; that is, those that represent the best 'value-for-money' (James, 1994: 90; Save the Children, 1995: 193–7 – the latter gives a short list of further reading). Project planners and aid donors often make use of it. A goal is set in the form of a physical target, social indicator, health standard, or environmental quality measure, and the assessor then seeks the least-cost way of achieving it. One possibility is to compare impact mitigation costs with benefits expected.

Fiscal impact assessment

Fiscal impact assessment should not be confused with cost–benefit analysis, as it has a different focus, namely on changes in revenues and costs. It has attracted a number of supporters, mainly policy-makers (for an introduction, *see* Burchell *et al.*, 1985; Leistritz, 1994, 1995).

Logical framework analysis

Logical framework analysis (also known as the project framework approach) is a tool introduced to development planning from management studies and developed by the US Agency for International Development in 1969–70. It is a way of establishing a structure to describe a project. It is an aid to logical thinking useful in the early stages of planning to help develop management skills and to establish communication between various people involved in the development. It is also an aid to assessing the relationships between activities and objectives (Save the Children, 1995: 192). It tests the logic of a plan of action in terms of means and ends. It can help clarify how objectives will be achieved; show up implications of developments; ensure that strategies, objectives and aims are linked; force planners to state the assumptions they have made (a quality shared with environmental impact assessment and social impact assessment); and stimulate discussion of feasibility and expectations (also like environmental impact assessment and social impact assessment). Logical framework analysis is quite widely used, mainly at the project level (Save the Children, 1995: 178–92). A good overview is provided by Wiggins and Shields (1995).

Input–output analysis

A number of econometric techniques have been used in environmental impact assessment, such as shift-share analysis, linear programming and input–output analysis. Input–output analysis is probably the most commonly used, and is a form of modelling used to assess the secondary, or indirect, impacts of development. It normally has a project or a regional focus (Bruckner *et al.*, 1987). Much of the early development was carried out by Leontief in the mid-1930s to early 1940s (for a relatively recent edition of his work, *see* Leontief, 1986). This approach assumes linear relationships, which in practice is often not the case, to account for linkages between economic and environmental systems. Input–output models have been used in strategic environmental assessment (*see* the discussion of strategic environmental assessment in Chapter 3) and integrated environmental management; for example, Whitney (1985) examined input–output analysis and an approach for integrated environmental impact assessment (i.e. assessment that considers environmental and economic factors). Input–output and linear programming approaches are demanding of computer facilities and mathematical skills (Solomon, 1985; James, 1994: 97).

Multiple-criteria analysis

Multiple-criteria (decision) analysis seeks to make a multi-dimensional evaluation of expected impacts – in some respects it is similar to cost–benefit analysis (Nijkamp, 1986; Tamura *et al.*, 1994).

Environmental assessment and environmental appraisal

Environmental assessment, also called environmental appraisal or evaluation, can refer to a stage in impact assessment, but it can also mean assessment of environment with developments in mind, although it is not as structured in approach as environmental impact assessment. In the UK, however, environmental assessment has been used by the authorities to mean broadly the same as environmental impact assessment. In some cases environmental assessment is a desk-research exercise. Aid agencies such as the Overseas Development Administration and the World Bank have published environmental appraisal guidelines (ODA, 1989; World Bank, 1991a, 1991b).

Ecologists and biologists have long been providing data, concepts and techniques that are used for impact assessment (Eberhardt, 1976; Ward, 1978). However, more caution is needed; some of the concepts provide tempting indices or benchmarks for planners and administrators, but their limitations tend to be overlooked (Holling, 1973; Westman, 1978; Pielou, 1981). Geomorphologists and other natural scientists are frequently called upon to contribute to environmental assessments, identifying natural hazards and opportunities for development (Stocking, 1984; Gonzales *et al.*, 1995).

Ecological impact assessment

Ecological impact assessment is impact assessment that considers how organisms in general, not merely people, will be affected (Doremus *et al.*, 1978; Westman, 1985: 86; Duinker, 1989). Recently the expression has been applied to the description and evaluation of the ecological baseline (*see* the Glossary for an explanation of the term 'baseline') used by environmental impact assessment (Institute of Environmental Assessment, 1995). Ecological assessment is concerned with establishing the 'state of the environment' (*see* 'environmental inventory' in the Glossary), whereas impact assessment focuses on predicted and actual effects of change. Treweek (1995a, 1995b) reviews ecological impact assessment and regards it as a valuable support for environmental impact assessment. An aspect of ecological impact assessment that is growing in importance is its application to the loss of biodiversity (Hirsch, 1993).

Ecological impact assessment may rely on selected ecosystem components as indicators. Ecosystem function can be complex and often is poorly understood, making assessment difficult.

Habitat evaluation

The goal of habitat evaluation is to assess the suitability of an ecosystem for a species (Suter, 1993: 8) or the impact of development on a habitat. More than one habitat may be affected by a development, in which case each is dealt with separately (Hyman and Stiftel, 1988: 212–18). This approach has been used by the US Fish and Wildlife Service in assessments of the impacts of federal water resource development projects, and the US Army Corps of Engineers (Canter, 1996: 390).

ECOSYSTEM APPROACHES

Ecosystem approaches to environmental quality assessment, including *ecosystem assessment*, *ecosystem analysis*, and *ecosystem impact assessment*, use a systems approach to assess environment and development interactions. An ecosystem is a functional system (ecological system) that includes organisms together with their physical environment. A systems approach seeks to study the interacting parts of an ecosystem. A systems diagram may be used, typically showing energy flows between components. Within an ecosystem, change of one component may promote changes in one or more other components. The focus is primarily on the quality of the whole ecosystem, rather than one or a few specific impacts. There are various approaches reflecting differing levels of knowledge about various ecosystems' function, and at some point there is reference to benchmarks or standards to permit an assessment of quality, stability, etc. (Auerbach, 1978; Holling, 1986; Nip and Udo deHaes, 1995). McHarg's work (1969) adopted a form of environmental systems analysis to assess ecosystems impacts, pre-dating environmental impact assessment. Ecosystem study and systems approaches are not universally supported (Cooper, 1976; Hillborn, 1979). Human ecologists have used a similar approach – *ecological analysis* – since the 1920s.

Ecosystem simulation modelling (*see* Ahmad, 1993) generally deals with a single or limited range of impacts in a given ecosystem; an example might be the effects of development on salmon in a river. A more complex development of environmental simulation modelling is the Holling-type adaptive environmental impact assessment and the related approach of adaptive environmental assessment and management (*see* Chapter 3), which involves constructing ecosystem(s) models to assess likely impacts in the context of environmental change (Holling, 1978).

An ecosystem approach can be valuable when there is a conservation or sustainable development goal. Unfortunately, ecosystem impact assessments are often *ad hoc*, general reviews, rather than adopting a consistent methodology and ongoing approach. There have been attempts to assess impacts in complex environmental systems, with possible application to the taxation of pollution, and the control of pollution and technology for making this possible (Vizayakumar and Mohapatra, 1989).

ENVIRONMENTAL MODELLING, FUTURES MODELLING AND FUTURES RESEARCH

Models are used to study and assess the impacts of a wide range of developments, such as altered land use, effluent discharges, global climatic change, modification of river channels, changes in estuarine conditions, coastal erosion, the use of agricultural chemicals, catchment management, and acid deposition (Whitehead, 1992). Social modelling is considered in

Chapter 8; modelling techniques are discussed in Chapter 5). Depending on the circumstances, one can use physical models (e.g. laboratory tests, scale models of estuaries or catchments) or statistical models (e.g. principal components analysis), computer models, systems models, world models (Forrester, 1971), etc.

Futures research makes use of modelling, including trend extrapolation and informed speculation (e.g. the Delphi technique). Futures modelling, futures research and 'futurology' attracted attention in the early 1970s following the publication of *The limits to growth* (Meadows *et al.*, 1974). This reported the results of systems modelling with a computer 'world model' that tried to assess population and industrial, technological and consumption trends, predicting a global crisis and suggesting strategies for avoiding or mitigating the expected scenarios. *The limits to growth* helped stimulate interest in the concept of sustainable development (which was less drastic than the strategy of 'zero growth' to which Meadows *et al.* drew public attention). A sequel to *The limits to growth*, reviewing how accurate the warnings had been, appeared at the time of the 1992 Rio 'Earth Summit' (Meadows *et al.*, 1992), and it seems likely there will be a revival of interest in what Westman (1985: 3) called 'the murky world of futurology' as the new millennium arrives.

Futures research is a difficult and imprecise art that has to allow for both gradual and sudden changes due to new inventions, attitude changes, environmental alterations, etc. (Linstone and Simmons, 1977). The further ahead one attempts to make predictions, the less accurate they are likely to be; Hyman and Stiftel (1988: 21) caution that when an 'assessment adopts a very long time horizon, it is likely to become an exercise in science fiction'. The results of futures research are useful, but must thus be treated with caution.

LAND USE PLANNING, LAND CLASSIFICATION AND RELATED APPROACHES

Land use planning is a process that may operate at local, regional or national scale; land capability assessment, land appraisal, land evaluation, land suitability assessment and terrain evaluation feed into that process. A *land use survey* indicates the situation at the time of study and is not the same as a *capability classification*, which looks to the future. There are various approaches and methods for land use classification; an example is the Holdridge life zones system (*see* Glossary; Holdridge, 1947, 1964; Davidson, 1980). The land use planning approach adopted largely depends on a country's politics.

It is widely felt that land use planning is a valuable ingredient of environmental impact assessment, and can help in the quest for sustainable development. Conversely, environmental impact assessment can feed into land use planning – it is a two-way process (Keyes, 1976; Marsh, 1978; Rivas *et al.*, 1994).

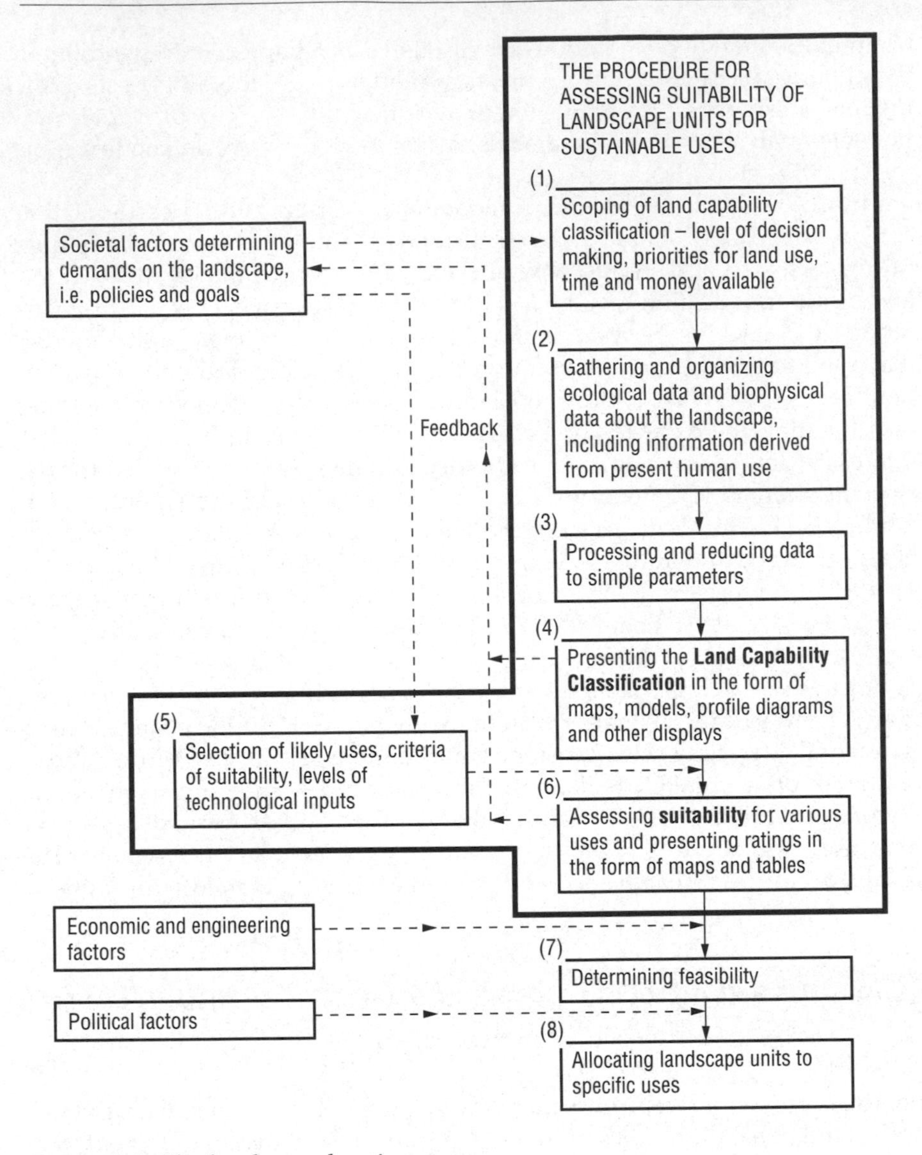
THE PROCEDURE FOR ASSESSING SUITABILITY OF LANDSCAPE UNITS FOR SUSTAINABLE USES
Societal factors determining demands on the landscape, i.e. policies and goals
(1) Scoping of land capability classification – level of decision making, priorities for land use, time and money available
(2) Gathering and organizing ecological data and biophysical data about the landscape, including information derived from present human use
Feedback
(3) Processing and reducing data to simple parameters
(4) Presenting the **Land Capability Classification** in the form of maps, models, profile diagrams and other displays
(5) Selection of likely uses, criteria of suitability, levels of technological inputs
(6) Assessing **suitability** for various uses and presenting ratings in the form of maps and tables
Economic and engineering factors
(7) Determining feasibility
Political factors
(8) Allocating landscape units to specific uses

FIGURE 2.1 The land use planning process

Land capability assessment, land evaluation and land appraisal generally follow a process similar to that followed for environmental impact assessment (proactive: scoping, data collection, evaluation, presentation of evaluation) in the production of a land capability classification or land evaluation (*see* Fig. 2.1; Beek, 1978; Patricos, 1986). Some approaches consider a range of factors, which might include the concept of carrying capacity (*see* the Glossary); others consider simply soil characteristics and slope. The end-product is a description of landscape units in terms of inherent capacity to produce a combination of plants, animals, etc.. It is also likely to reflect government development goals, market opportunities, labour availability and public demands (for example, terraced agriculture might be possible given the environment and available technology in a particular case, but labour might not be available).

Simple inventories of land use and, to a limited extent, capability were made in medieval times – notably the Domesday Book (eleventh-century England) or the work of Yahya ibn Mohammed in the twelfth century. Land capability classification was in part developed by the US Soil Conservation Service in the 1930s and 1940s following problems like the 'Dust Bowl'. (The Dust Bowl is a name given to a region of short-grass prairies of the southern Midwest of the USA settled in the mid-nineteenth century and ploughed from the 1880s. Poor husbandry and drought led to severe soil erosion and dust-storms in the early to mid-1930s. Huge areas of land were abandoned and many farmers were forced to relocate.) Linked to consideration of conservation and development aspects, land capability classification can lead to production of a *land suitability assessment*, a rating of landscape units showing what development they might best support. The approach adopted by McHarg (1969) is essentially a land suitability assessment approach (it considers constraints and opportunities of an environment in relation to proposed developments), and depends on overlay maps of various landscape or development attributes. Alternatively a 'Gestalt' approach, which involves direct field observation of clues – for example, plants indicative of good soil, may be used to prepare a rapid suitability assessment.

A crucial aspect of all these approaches is that they should seek to develop dialogue between assessor and landusers.

Landscape planning is not discussed in this chapter, however, some of the techniques used in aesthetic impact assessment have been developed by landscape planners and those in related disciplines.

Universal Soil Loss Equation

The Universal Soil Loss Equation is a predictive tool that uses data on a wide range of parameters to estimate average annual soil loss. It was developed in the 1930s by the US Soil Conservation Service and was improved in 1954 and again in 1978 by the US Department of Agriculture. It is widely used by planners and consultants to check on existing and likely future soil loss, and to select agricultural practices and crops suitable for sustaining production. Developed in the midwestern USA, it has been modified to suit other environments (Hudson, 1981: 258; Landon, 1991: 309–16).

The Universal Soil Loss Equation should be used with caution. Problems arise when data are imprecise or unavailable, and it is best applied in situations where water rather than wind erosion occurs (although there is a modified version intended to cope with wind erosion). A typical form of the equation would be:

$$A = 0.224 \times RKLSCP$$

where: A is soil loss (in kg/m^2 per second); R is the rainfall erosivity factor (i.e. the degree to which rainfall can erode soil); K is the soil erodability factor (i.e. soil's vulnerability to erosion); L is the slope length factor; S is the slope gradient factor; C is the cropping management factor (what is grown and how); and P is the erosion control practice factor (i.e. the rate by which erosion is reduced by conservation measures from the worst possible case).

AGROECOSYSTEM ASSESSMENT AND FARMING SYSTEMS RESEARCH

The agroecosystem zones concept was promoted by the Food and Agriculture Organization of the United Nations (FAO, 1978) to provide a framework for considering a range of parameters over a limited planning term with the aim of promoting sustainable development. An agroecosystem is an ecological system modified by humans to produce food and commodities, which generally means a reduction in diversity of wildlife.

Agroecosystem assessment

Agroecosystem assessment, also called agroecosystem analysis, evolved in Thailand (Conway and Barbier, 1990: 162–93) and attempts rapid multidisciplinary diagnosis using ecological and socio-economic concepts and parameters (Fig. 2.2). It considers not only the farming system but also household characteristics and regional, national and even global factors likely to affect the farming community. The area under consideration is zoned – often with the use of a land use survey or land capability assessment.

It would be difficult to identify research priorities or to make improvements without an understanding of existing farming systems and the problems faced by farmers. A multidisciplinary team of researchers uses semi-structured procedures to study the agroecosystem and its parameters and hierarchies (i.e. the relationships from the plants and animals found at a site to the field level, the regional market level, the national market level and to the international trade level. Once the relationships are understood, descriptive diagrams are prepared for use in discussion workshops to aid communication between various disciplines so that problems, opportunities and research priorities can be agreed upon. Agroecosystem assessment needs to be approached with some caution because it can lead to over-simple interpretation.

Farming systems research

Farming systems research is an open-ended, iterative, multidisciplinary, holistic (*see* Glossary), continuous, farmer-centred and dynamic process applied to agricultural research and development (Gilbert *et al.*, 1980; Shaner *et al.*, 1982; Brush, 1986: 221). Some argue that it is just an extension of farm

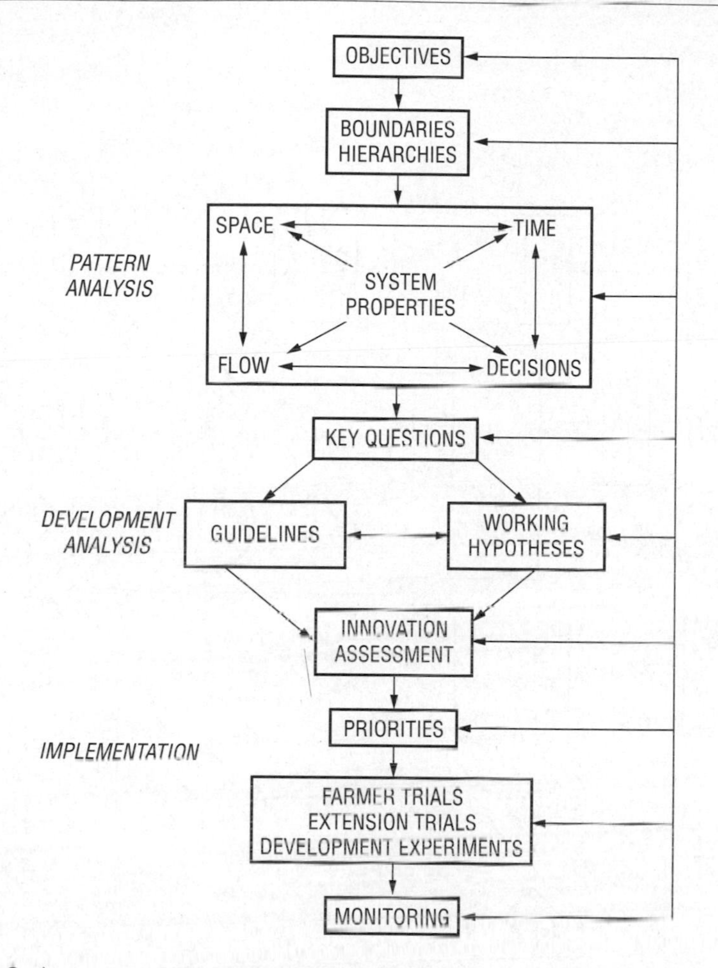

FIGURE 2.2 Agroecosystem assessment or analysis – procedure
Source: Based on Conway and Barbier (1990: 174, Fig. A.6)

management. There is no single method, but all approaches share five basic steps (*see* Maxwell, 1986):

1. classification – the identification of homogeneous groups ('target groups') of farmers;
2. diagnosis – identification of limiting factors, opportunities, threats, etc. for the target group;
3. generation of recommendations – which may require field experiments, pilot studies and/or research station work;
4. implementation – usually working with an agricultural extension service; and
5. evaluation – which may lead to revision of what is being done (Fig. 2.3).

Farming systems research views the whole farm as a system; it is a systems approach applied to on-farm research. It is promoted as a way to 'reach'

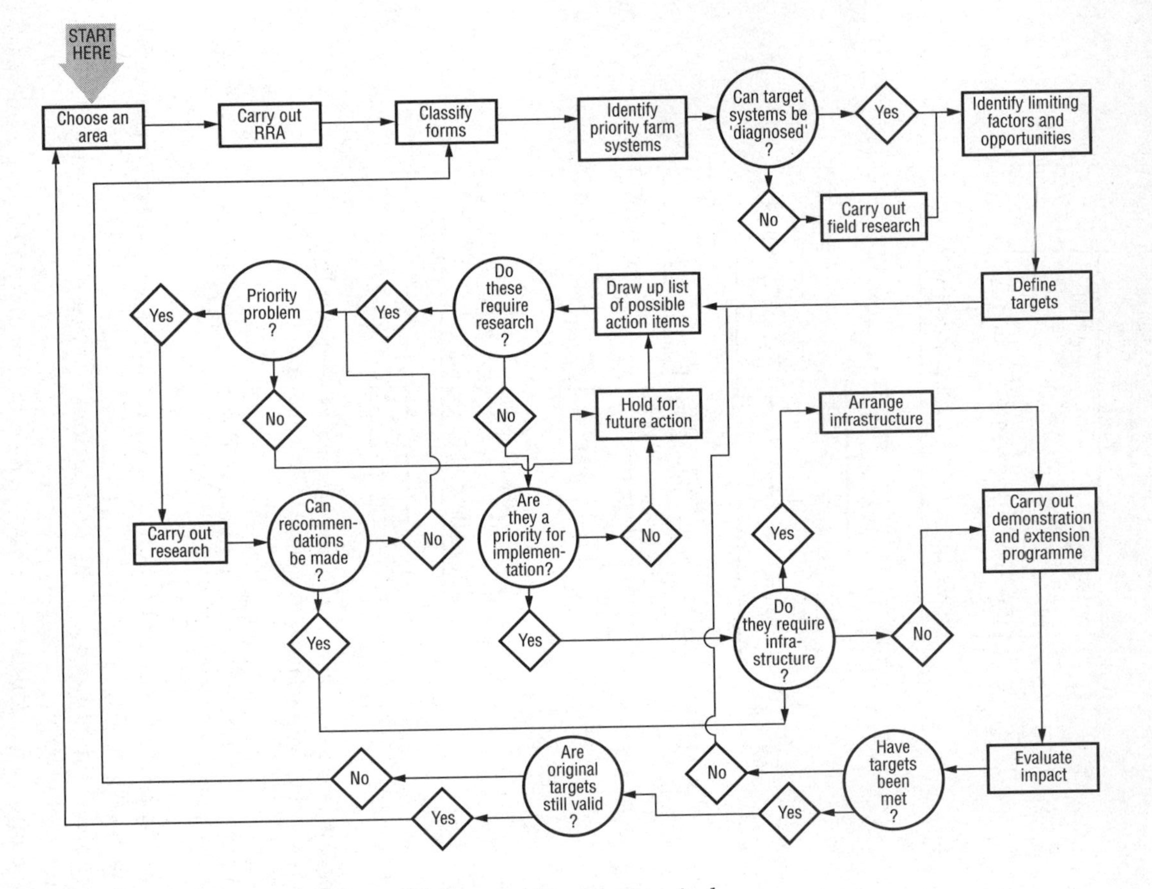

FIGURE 2.3 Farming system research methodology. RRA = rapid rural appraisal.
Source: Based on Maxwell (1986: 68, unnumbered figure)

small farmers (that is, to increase farmer participation in development) and to generate improved approaches and technology. It is seen by many as a way of generating research and technology that are appropriate – meaning, developed from the 'bottom up'). Technology and procedures developed from the bottom up are usually preferable to trying, perhaps with poor success, to impose on farmers what has been developed in research stations – a 'top-down' approach). Farming systems research includes study of factors that may be beyond the control of the farming community, such as world trade issues, or global warming, etc. Also, unless some 'off-the-shelf' input is available, it usually takes time – often a couple of years or more and sometimes from 5 to 15 or more years.

Participatory Assessment

There is considerable overlap between agroecosystem assessment and farming systems research, on the one hand, and the *participatory assessment* approaches outlined in this section – rapid rural appraisal, culturally adapted market research and participatory rural appraisal – on the other. The main reason for separating them is that more stress is placed on participation (by the local people or target group) in the latter three approaches. An increasing numbers of impact assessment experts are promoting forms of *participatory impact assessment and monitoring* (Yap, 1990). Participatory assessment can be defined as qualitative research or survey work that seeks to provide an in-depth understanding of a community or situation. It can be slow or rapid, but the emphasis is on participation (Save the Children, 1995: 144–77).

Rapid rural appraisal

Rapid rural appraisal is a family of approaches to land capability assessment that seek to incorporate (or involve) local people in assessment and to reduce the time and costs of preparation. It is a systematic, semi-structured activity carried out in the field by a multidisciplinary team and is designed to acquire quickly new information on, and new hypotheses about, rural life (Conway and Barbier, 1990: 177). Rapid rural appraisal has evolved considerably since the late 1970s, and there is no single standardized methodology. For an introduction, *see Agricultural Administration* 1981: **8(6)**, which is a special issue on rapid rural appraisal; *IDS Bulletin* 1991: **12(4)**, also a special issue on the topic; McCracken *et al.*, (1988); Conway and McCracken (1990); and Chambers (1992). A central thesis of the technique is 'optimal ignorance' – the idea that the amount of information required should be kept to the necessary minimum (something some environmental impact assessment practitioners should also bear in mind). Another central thesis is 'diversity of analysis' – the use of different sources of data or means of data-gathering, and the participation of a range of experts, if possible familiar with every aspect of rural life.

Rapid rural appraisal, according to Conway and Barbier (1990: 177–8) is iterative (i.e. processes and goals are not fixed and can be modified as an exercise progresses); innovative (it is adapted to suit needs); interactive (team members work in such a way as to gain interdisciplinary insight); informal (it often relies on opportunistic interviews); and is in contact with the community. It can vary in character. For example, it can be exploratory, seeking, like agroecosystem analysis, information on a new rural topic or agroecosystem. Or it can be topical in cases when a specific output is expected, often in the form of a hypothesis that can be a basis for research or development.

Rapid urban environmental assessment

Rapid urban environmental assessment has been reviewed by Leitmann (1993). Given the huge and very rapid growth of cities, especially in less developed countries, and the misery and environmental damage that such growth can cause, it is strange that environmental assessment and environmental impact assessment have been so little applied.

Participatory rural appraisal

Participatory rural appraisal approaches seek to enable local people to share, enhance and analyse their knowledge of life and conditions, to plan and to act. Participatory rural appraisal differs from rapid rural appraisal, in that the latter extracts information, whereas the former shares it and seeks rapport (Chambers, 1994a, 1994b, 1994c; Save the Children, 1995: 144–71).

Multidisciplinary-team studies and a stress on participatory public involvement offer possibilities for better conduct of impact assessment. However, there has been a tendency to emphasize the strengths of both rapid rural appraisal and participating rural appraisal, and understate the problems that may be encountered (Jiggins, 1995: 59). Sometimes 'rapid' seems to mean speed for assessors' fieldwork, rather than meaning an approach designed to give useful results fast.

Culturally adapted market research

Culturally adapted market research is a variant of rapid rural appraisal (Save the Children, 1995: 161–2). It draws upon market research techniques. (quantitative and qualitative surveys) to find out about attitudes, opinions and behaviour of people in relation to proposed development activities. Surveys are generally carried out by indigenous interviewers. The aim is to ensure that proposals are relevant by checking the needs and wishes of those involved. Culturally adapted market research is a way of ensuring that the views of those likely to be affected by development are available to assessors.

HAZARD AND RISK ASSESSMENT

The importance of hazard and risk assessment hardly needs stressing in the wake of accidents like those at Minamata, Seveso, Love Canal, Three Mile Island, Bhopal and Chernobyl or natural disasters like the Bangladesh floods or the earthquakes in Japan and California. They are activities that run parallel with and overlap impact assessment, although many regard hazard assessment as an essential component of environmental impact assessment (Gilpin, 1995).

Hazard assessment

A hazard is a *perceived* event or source of danger that threatens life or property or both. A disaster is the realization of a hazard. If the disaster is on a particularly serious scale it becomes a catastrophe. Hazard assessment may be said to seek to recognize things that give rise to concern. Clark *et al.* (1984: 501) attempted a definition: 'the determination of the existence of a hazard – either as a preliminary to other assessments – or as an alternative assessment method.' Human activities may initiate natural hazards and alter the vulnerability of the environment, wildlife or humans to them. Some people classify hazards as either natural, quasi-natural or man-made. Hazard assessment tends to deal with natural hazards, and is well-developed in the fields of flood, storm, tornado, tsunami, locust swarm warning, geomorphological hazards and avalanche warning (White and Haas, 1975; Bryant, 1991; Degg, 1992; Blaikie *et al.*, 1994; Cavallin *et al.*, 1994), but it also deals with hazards generated by technology, industry or human activity (e.g. crime)(Kates, 1978).

Various human groups, even within one society, can perceive and evaluate hazards and risks differently, and often vary in their vulnerability. Risk acceptability is fickle: the perception of risk may not be based on rational judgements, because people may have 'gut reactions' to certain risks and little awareness of other, perhaps more real, threats. Risk and hazard perception has generated a large and growing literature from behavioural psychologists, health and safety specialists, anthropologists and specialist risk or hazard assessors (see, for example, Lowrance, 1976; Fischhoff *et al.*, 1981; Douglas and Wildavsky, 1982, Carpenter, 1995a: 163).

One difficult problem faced by hazard or risk assessment, and to some extent by environmental impact assessment, is assessing what is 'acceptable'. Acceptability varies between people, and for any given group through time. For example, a youthful person may be more tolerant of risk than someone older; and commonly the poor face and have to accept more risks. In general, people are more concerned about short-term risks than long-term ones, and about 'concentrated' hazards – an air crash that kills 300 people is seen as greater cause for concern than 300 fatalities resulting from, say, household accidents dispersed over a country or in time. Unfamiliar risks tend to receive more attention: radiation hazards are taken more seriously than traffic accidents. If people think they are in control (as many a car driver, for example), they are less worried than as a passenger on a train or boat. Perception is not a rational process and can be greatly affected by media and myth. For example, airships that make use of modern materials for their envelope and have helium as a lifting gas have great potential, but there is public concern about them, mainly generated by the *Hindenburg* disaster and other airship accidents in the 1930s.

Faced with a hazard or risk, people's responses are diverse (*see* Fig. 2.4). The assessor can categorize hazard or risk according to criteria such as: minor/severe, infrequent/frequent, localized/widespread, and may resort to estimating the value of a life to weigh against risk probability and risk avoidance costs. In this connection, it is interesting to recall that the Bhopal tragedy in India raised the question of the higher 'life valuations' given to people in developed countries as compared to those in developing ones. Another problem is that assessment may involve industrial activities that need to keep some of their practices secret, perhaps to avoid giving away the fruits of their experience or research, or because they involve strategic information or some activity that the public or factions (e.g. terrorists) may attack or misinterpret. Hazard, risk and technology assessment, in common with impact assessment, have to address the question of how to deal with public involvement (Rakos, 1988).

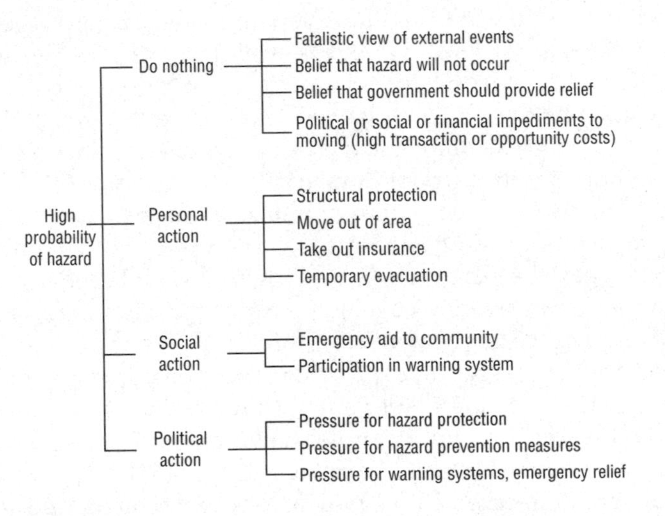

FIGURE 2.4 Response to hazard or risk

Risk assessment

Risk assessment, also called risk appraisal or analysis, is a loose term. It considers hazard and vulnerability: how people react to risk and their pattern of exposure. It has been defined as

> the process of assigning magnitudes and probabilities to the adverse effects of human activities (including technical innovation) or natural

> catastrophes (Horlick-Jones *et al.*, 1995; Taylor, 1993). This process involves identifying hazards . . . and using measurement, testing and mathematical or statistical methods to quantify the relationship between the initiating event and the effects . . . (Suter, 1993: 3).

Alternatively, one might define risk as the expression of the chance or probability of a danger or hazard taking place, and risk assessment, as going beyond predicting probability to objectively identify the frequency, causes, extent and severity of exposure of people or things or activities. Risk assessment may go on to identify coping strategies or establish what people will pay to avoid a risk (Carpenter, 1995a). Some researchers recognize risk appraisal as the assessment of communities' attitudes to risks. Risk assessment can be divided into that concerned with risks to the environment or biota and that concerned with risks to humans. Risk assessment typically consists of risk identification, risk estimation (establishment of nature and levels) and risk evaluation (assessment of probability of occurrence, consequences, etc.). It may study effects, pathways or factors involved; an example would be laboratory experiments into toxicity. Often, it involves weighing dangers against benefits – the threat, say, of asbestos-related illness as balanced against its value in fire protection.

Hazard and risk assessment are not precise arts: different risk assessments may assign hugely different predictions to a risk. Added to the fickle nature of public acceptance of threats, these variable assessments can pose problems for the planner and decision-maker. Risk assessment is an 'analytical tradition', not a legal definition. It has centuries-old roots in the actuarial, investment and insurance professions, and has spread to engineering, the development of new materials (especially chemicals and innovations in biotechnology), economics, health care, criminology and the pharmaceuticals industry (Furlong, 1995). Insurance companies and bankers have long realized the need to know risks before providing cover or loans (Schmid *et al.*, 1995). Administrators, public bodies and companies are increasingly using risk assessment to reduce the likelihood of their being accused of negligence if something goes wrong or for contingency planning. There are published tables, and various bodies hold records, of risks associated with most methods of transport and types of recreation, employment, industrial activity and places of residence (Whyte and Burton, 1980: 31).

Risk assessment can provide, according to Suter, 1993: 3,

- a quantitative basis from which to compare and prioritize risks;
- a systematic means of improving understanding of risks; and
- a means of making assessment more useful and credible by giving probabilities to predicted impacts.

One type of risk assessment is the screening of a new product or activity, under laboratory or test-bed conditions, to ensure that it is safe to user and environment and does not accumulate, before it is actually released for general use.

Risk and hazard assessment and impact assessment have to some extent evolved in parallel; they often overlap and are mutually supportive or have things to offer each other, but they emphasize different things (Andrews, 1988; Ratanachai, 1991; Graham Smith, 1993: 28). Some authors argue that unification of impact and risk assessment is possible and desirable (e.g. Canter, 1993; Carpenter, 1995b: 193). All three deal with uncertainty, adopt a multidisciplinary approach and are applied predictive assessment processes that seek to improve policy and plan implementation, although they have their roots in different disciplines.

Risk assessment has its ancestry mainly in engineering and decision sciences and has been prompted by legislation such as the US Toxic Substances Control Act 1976 (which requires regulation if there is a risk to human health or environment through use or release of a harmful chemical or biological agent), or the UK Environment Act 1995 (which requires local authorities to carry out risk assessment and maintain registers of contaminated land). Separation of 'natural' from 'man-made' industrial or technical hazard and risk assessment is mainly maintained by practitioners and the literature rather than reflecting different concepts.

Risk assessment tends not to address development alternatives, as does environmental impact assessment, is less integrated in approach, tends not to identify opportunities as well as threats, and is at present less likely to be required by government policy or law (which is also true for technology risk assessment). It is often better at estimating the magnitude, certainty and timing of impacts than environmental impact assessment (or at least makes an effort to be better). Risk and hazard assessment are often applied where there is more uncertainty than environmental impact assessment faces (Covello *et al.*, 1985: 16). Bodies such as the US Environmental Protection Agency use risk assessment to assist in the regulating of industry and other human activities. 'Mainstream' environmental impact assessment also differs from risk and hazard assessment in that it focuses on impacts caused by human actions (although a crime risk assessment also does so). There has been a little less incorporation of risk assessment into planning than there has of environmental impact assessment. Another difference, according to Andrews (1988: 93), is that environmental impact assessment, at least in the USA, increases planners' accountability to citizen groups; risk assessment is more concerned with internal management within a company or organization. Environmental impact assessment has been applied to aid risk assessment (Ratanachai, 1991), and risk assessment can input to impact assessment, usually in the scoping stage (*see* the Chapter 4 section on scoping).

Hazard assessment and risk assessment usually use a template (to help order the process) to generate a statistical estimate of probability of occurrence of a certain level of impact. Such an estimate is not a forecast, but a statistical recurrence interval; for example, it might specify a 1 in 100 year chance of a serious flood. The estimate can be used to generate a zoned map for the purpose of determining land use or building regulations, preparing a contingency plan or emergency procedure, or installing emergency equipment such as hurricane shelters or tsunami warning systems and protective walls. Insurance companies often use these techniques to determine insurance charges, for example; by mapping risks against postcodes or ZIP codes. Some threats appear suddenly, but others creep up, and in these latter cases it may be possible to spot when a threat is approaching a threshold and so give a warning forecast (Stromquist, 1992).

Well-developed areas to which risk analysis is applied include ecological risks, health risks and technological and industrial risks. Ecologists have researched factors involved in risks to ecosystems, and some of their concepts have been applied to social risk assessment. Examples are the assessment of vulnerability (if damaged, what will be the effect?), elasticity (the ability of an ecosystem or society to recover from damage), inertia (resistance to damage) and resilience (how often can the ecosystem or society recover?). For a review of this kind of social risk assessment, *see* Cairns and Dickson (1980). Like environmental impact assessment, risk assessment is mainly applied at project level (*see* Raferty, 1993) or to a particular process, although it is sometimes used on policies, plans and programmes.

Environmental risk assessment, a subfield of risk assessment, seeks to assess the risks to the environment resulting from industrial activity and other developments, including the use of various products. *Ecological risk assessment* seeks to define and quantify risks to non-human biota and determine the acceptability of those risks (Clark *et al.*, 1984: 498). *See* Whyte and Burton (1980), Bartel *et al.* (1992); Suter (1993) and Carpenter (1995a: 172–80).

TECHNOLOGY ASSESSMENT

'Technology assessment' is an expression often applied to technical evaluation – establishing whether equipment and techniques work. It can also mean (and is more often used in this context to mean) the assessment of technology risks and impacts in order to inform decision-making and clarify problems and opportunities. (*Impact Assessment Bulletin* 1987: **53** is a special issue entitled 'International impacts of technology'; and *see also* Clark, 1990; Porter, 1995a.) It follows a path broadly parallel to that of environmental impact assessment and may involve evaluation of indirect and cumulative impacts (*see* Chapter 4). It involves systematic study of the effects on environment and society that occur when a technology is introduced, extended or modified (for example, *see* Armstrong and Harmon, 1980). Increasingly, realization of the potential threat of technological developments is prompting proactive assessment. As well as performing this 'alerting function', technology assessment can, like environmental impact assessment, aid decision-making and planning. Technology assessment may be initiated by government, international bodies, non-governmental organizations or the industries or agencies that plan to innovate.

A risk may be posed by siting a plant, or operating a process, that is known to be potentially dangerous in a populated area. An example is the

Canvey Island petrochemical and liquid propane gas complex east of London, which was the subject of a risk assessment by the British Gas Corporation in 1976 (Ricci, 1981: 101). Other risks may arise from new and untested technology, and new chemicals and pharmaceuticals.

Technological innovation may relate to virtually any aspect of life: attempts to improve agriculture, industrial activities, transport, etc. (*see* ASCE, 1977; Conrad, 1980; Dierkes *et al.*, 1980; Porter *et al.*, 1980; Kates and Hohenemser, 1982; O'Brien and Marchand, 1982; Rakos, 1988; Coates and Coates, 1989). Industrial hazard and risk assessment mainly examines established manufacturing practices, and is less likely to deal with 'unknowns' arising from technical innovation than is technology assessment proper (World Bank, 1985).

The adoption of new technology in developing countries has generated difficulties in the past, with the result that there has been increasing interest in the application of technology assessment to minimize problems and to 'tune' the technology (UN Branch for Science and Technology Development, 1991; Wiesbecker and Porter, 1993; Porter, 1995b: 75–80). It is increasingly being applied to biotechnology development (Ginzberg, 1991), including 'genetic engineering' (Thakur *et al.*, 1991; Furlong, 1995).

Technology assessment was widely conducted in the USA by 1967, and so predates environmental impact assessment (Medford, 1973: 34). In 1973 the US Congress created the Office of Technology Assessment (OTA) to promote and oversee its use. An International Society for Technology Assessment operated from the USA in the mid-1970s, but wound up and developed into the International Association for Impact Assessment, a body that promotes environmental impact assessment, social impact assessment, technology assessment, hazard assessment, risk assessment and related activities. The National Science Foundation in the USA supports technology assessment, and bodies similar to the OTA had been set up in the European Community (now the European Union), Japan, Canada, Australia and elsewhere by the late-1980s.

Technology impacts can be a function of a variety of factors, including technology failure, operator failure, poor maintenance, poor design, faulty installation, natural or human accident, or adaptations prompted by the innovation. Not surprisingly, practitioners of technology assessment are often engineers. Technology assessment has in the past concentrated on threats relating to morbidity and mortality, but there is now also increasing interest in civil liberties, and in the social aspects of technology innovation such as the effects of innovations in television broadcasting. Social impacts of technology innovations are discussed in Chapter 8.

Technology assessment can help identify appropriate technology and practices as well as spot potential and actual problems. Where new technology and substances are likely to be quickly and widely adopted, it is important that the results of hazard and risk assessment are rapidly and accurately disseminated to all who need to know. Technology assessment also has an important part to play in the quest for sustainable development by identifying threats and promising development paths (Coates, 1971; Porter, 1995a: 137). It can also support policy-making (Coates and Coates, 1989): the European Union is applying it to long-term strategic policy-making and as a means of providing early warning of difficulties (Rakos, 1988).

Many practitioners feel that environmental impact assessment should integrate with social impact assessment, technology assessment, risk assessment, hazard assessment and related approaches (Porter and Rossini, 1980), and impact assessment's main professional society, the International Association for Impact Assessment, encourages such links. Hazard and risk assessment increasingly require international co-operation as the threat of global change and transboundary impacts from technology grow. Examples include nuclear accidents, storm tracking, locust monitoring, air-quality forecasting and global climate change. International co-operation is also necessary because there is a tendency for technological and industrial hazard to be 'exported' to less developed countries where laws, monitoring and enforcement may be less stringent and planners and regulators less well informed of risks. Large sums of money may be at stake in such 'exporting', making objective assessment a challenge.

LIFE CYCLE ASSESSMENT

When impact assessment, risk assessment or eco-auditing (to be considered later in this chapter) deal with a process, for example manufacturing, or establishing and running a power-station, there should be consideration of impacts throughout the life of the activity: at construction, during use, at closure and during reinstatement of the site. This process is known as life cycle assessment. Equipment is often subject to wear and tear, and so presents different risks as it ages and as management acquire experience (or become complacent). Sites of industrial activity, power generation, etc. often accumulate contamination, presenting a changing threat that may remain after the site activity has terminated.

MONITORING AND SURVEILLANCE

Monitoring, is an important activity that aims at establishing a system of continued observation, measurement and evaluation for defined purposes (Whyte and Burton, 1980: 44; Clark *et al.*, 1984: 365). It is dealt with more fully in Chapter 4. Monitoring may thus provide information at the start of an impact assessment, risk assessment, eco-auditing, etc., either during it or after completion. Monitoring should be operated to agreed schedules with comparable methods. Usually it focuses on specific elements or indicators.

Monitoring can:

- improve understanding of environmental, social or economic processes;
- provide early warning of possible difficulties;
- help optimize use of environment and resources;
- assist in regulating environmental and resources usage (for example, it may provide information for law courts).

The United Nations Environment Programme promoted global environmental monitoring at the 1972 UN Conference on the Human Environment at Stockholm. Since the 1970s there has been increasing interest, spurred by transboundary problems, in developing international or global monitoring systems. These seek to monitor at the global level and ideally offer wide access to their information. The bodies involved include the United Nations Environment Programme, the Organization for Economic Co-operation and Development, the European Union, the International Atomic Energy Commission and many others. The United Nations Environment Programme has established a Global Environmental Monitoring System (GEMS), which is a co-ordinated programme for gathering data for use in environmental management and for providing early warning of disasters. An independent international research unit, the Monitoring and Assessment Research Centre (MARC), was founded in 1975 to assist international organizations with monitoring. This concentrates on biological and ecological monitoring, particularly of pollution. The World Conservation Monitoring Centre was established in 1980 by the upgrading of a body formerly run by the International Union for the Conservation of Nature and Natural Resources. It monitors endangered plant and animal species. Bodies like the US Food and Drugs Administration monitor pharmaceuticals and foods, and the spread and use of weapons (especially nuclear, chemical and biological) are increasingly monitored. In a number of countries doctors, vets and other professionals report observed effects to a central monitoring body. In the USA the Environmental Protection Agency and in Europe the European Environmental Agency (which has slightly weaker powers than its US counterpart) promote and oversee monitoring.

Environmental, social and economic monitoring have each generated their own practitioners and literature, which may focus at local, regional, national or global level or study 'pathways' (for pollution, etc.). Monitoring or surveillance can be done at source (in other words, where something is being generated), at selected sample points, at random, along transects or by sampling some suitable material or organism. For example, pollution might be monitored by checking a smokestack, by a network of instruments, or by surveying the diversity and growth of lichen species. Regulatory monitoring checks its findings against set 'in-house', national or international standards or stated objectives. Monitoring can look back from the present to establish trends, for example by using dendrochronology, palynology, ice-core samples, tissue samples from burials, historical records, etc. This can be important when there is inadequate baseline information. (The term 'baseline' is an important one in the field of impact analysis; it means what things are or were like before the occurrence of an event or development. *See also* the Glossary.)

Surveillance, is *repetitive* measurement over a period of time but with a less clearly defined purpose than monitoring. It is more exploratory and can be undertaken to determine trends, calibrate or validate models, make short-term forecasts, ensure optimal development, or warn of the unexpected (Suter, 1993: 377; Furlong, 1995). Surveillance, like monitoring, can focus on the environment, people or an economy, and may:

- check whether statutory regulations are being complied with (without monitoring and surveillance, the setting of standards and rules is of little value);
- provide information for systems control or management;
- assess environmental quality to see if it remains satisfactory;
- detect unexpected changes.

Monitoring, surveillance and screening (which in this context refers to checking for a specific thing, such as a particular disease in a population, and is not to be confused with the impact assessment 'screening' referred to in Chapter 4) are valuable development aids, but they cost money, may delay new developments, and generate problems over who should administer, enforce and pay for them.

Ultimate environmental thresholds assessment

Ultimate environmental thresholds assessment is derived from threshold analysis, and is based on the assumption that there are environmental thresholds which, if exceeded, may mean serious, perhaps difficult-to-rectify, changes, possibly on a large, even global, scale. Ultimate environmental thresholds are thus final boundaries that may be reached and even broken by direct or indirect (including cumulative) impacts. Kozlowski (1986: 146) defined these thresholds as 'stress limits beyond which a given ecosystem becomes incapable of returning to its original condition and balance.' It is possible to recognize temporal, quantitative, qualitative and spatial dimensions of these thresholds and to seek to assess their present and future status.

Ultimate environmental threshold assessment was developed by Polish national park planners in an attempt to define absolute ecological limits. The approach has many things in common with environmental impact analysis and possibly some advantages, including a chance of better integration into the planning process. There are already regional catastrophes where cumulative impacts have exceeded environmental limits; ultimate environmental thresholds assessment might have helped avoid such damage. An example is the ruination of the Aral Sea and, to a lesser extent, the Aegean Sea. Its application might prevent similar fates for the Baltic, the Mediterranean and the Inland Sea of Japan (the latter has recently been attracting some impact assessment attention).

AESTHETIC IMPACT ASSESSMENT

Aesthetic impact assessment is used within land capability assessment and landscape architecture and planning, and may feed into environmental impact assessment. Although concern for visual impact has dominated aesthetic assessment, it also involves other parameters, often ones that are very difficult to evaluate, and may draw on psychology and the arts (Bagley, 1972; Pearson, N. Associates, 1984; Roebig, 1983; Hyman and Stiftel, 1988: 111–36; Institute of Environmental Assessment, 1995; Institute of Environmental Assessment and Landscape Institute, 1995; Canter, 1996: 467). A particular landscape or architecture may give a sense of surprise, a tactile or other sensory effect, not just a visual impact. For example, some people may rate a flowing stream more highly than a slow-flowing river; or a perfectly good Tudor-style house may not 'fit' a neighbourhood. Culture plays a crucial role: an eighteenth-century middle-class Englishman would probably have disliked wild scenic landscapes, preferring tamed and cultivated vistas; his late-twentieth-century equivalent may well profess the opposite view.

Visual impact can be assessed by using photo-montage or video simulation techniques, or by estimating degree of intrusion (in the zone of visual influence), surprise effects, etc. from construction plans and local topographic maps. By selecting the right site, construction style and colour, visual aesthetic impacts may be controlled. In effect, consumer tastes may be satisfied or the development camouflaged.

STATE-OF-THE-ENVIRONMENT ACCOUNTS; EVALUATION; AUDIT AND ASSESSMENT

There is some confusion over the use of expressions such as 'environmental auditing'. This particular term has been applied to mean a sort of stocktaking, eco-review, eco-survey, environmental assessment (itself a vague expression), state-of-the-environment assessment, the 'production of green charters' and the checking of impact assessments to determine their effectiveness (Edwards, 1992).

State-of-the-environment accounts, and *environmental quality evaluation* make use of knowledge of how the ecosystem is structured and functions in order to collect relevant data showing the state of an area. This need not be a terrestrial survey; seas like the Baltic or Aegean have been assessed. State-of-the-environment accounts attempt to assess whether economic development is consistent with sustainable development. Results, which may be presented in the form of an inventory or as a map, seek to establish the status of a whole ecosystem, whereas environmental impact assessment focuses on the effects of development (Nip and Udo deHaes, 1995). Ecological evaluation seeks to establish what is of value: for example, habitat, resources, services, etc. – often things to which it is difficult to assign monetary values. The US Fish and Wildlife Service made such evaluations as early as 1976.

There is some ambiguity in the terms *environmental audit, environmental assessment* and *ecological assessment*. The first, which can be applied at company or institution, state, national or global levels, may mean:

1. an auditing (i.e. inventory-focus) approach to the environment which seeks to review conditions and evaluate impacts of development. One form is 'new systems of national accounts' (these are national state-of-the-environment accounts);
2. studies aimed at avoiding or reducing environmental damage; or
3. a means by which a body systematically and holistically monitors the quality of the environment with which it interacts or for which it is responsible (something increasingly seen as vital in any quest for sustainable development).

The third of these processes is now more likely to be termed an *eco-audit* – an 'internal' review of the activities and plans of a company or body, as explained later in this chapter (Wisemann, 1994). State-of-the-environment accounts set out a region's or nation's environmental, social and economic assets. Norway, France, the Netherlands, Canada, Denmark, the UK and the World Bank have each developed national state-of-the-environment accounts systems, an example being the French Comptes du Patrimoine Natural (national heritage accounts) developed since 1978. National state-of-the-environment accounts make use of environmental assessment and have been promoted as improvements on the use of indicators such as gross domestic product to document 'development status' (Thompson and Wilson, 1994: 613). They may prove important in future trade agreements to ensure that environmental effects are counted, and could play a part in the quest for sustainable development (Ahmad *et al.*, 1989). Pearce *et al.* (1989) promoted environmental accounting procedures that treat the environment as natural capital and measure its depletion or enhancement.

In the UK, *environmental assessment*, is used by government bodies to mean environmental impact assessment. The term may also apply to a form of pre-development stock-taking; for example, site selection for a nuclear waste repository. In the USA, environmental assessment means a concise public document which should provide enough evidence for a decision to be made on whether or not to proceed to a full environmental impact assessment and publication of an environmental impact statement. Environmental assessment has been applied to surveillance or screening tasks such as checking drugs or industrial activities for impacts (Vincent, 1993). It is also used for study that seeks to establish the state of an environment with less focus on impacts than environmental impact assessment has (Institute of Environmental Assessment, 1995).

Environmental appraisal is a generic term used in the UK for the evaluation of the environmental implications of proposals. The process of evaluating how accurate and useful an impact assessment has been is termed *post-environmental impact assessment audit* or *environmental impact assessment review* (or sometimes, post-project analysis), and is dealt with in Chapter 4.

Some people are critical of accounting methods, arguing that they are just 'stock-taking' and stop short of encouraging a crucial change of attitude towards environment and development. For a review of the relationship between environmental audits, environmental impact assessment, state-of-the-environment reports and new systems of national accounts (Fig. 2.5), *see* Thompson and Wilson (1994).

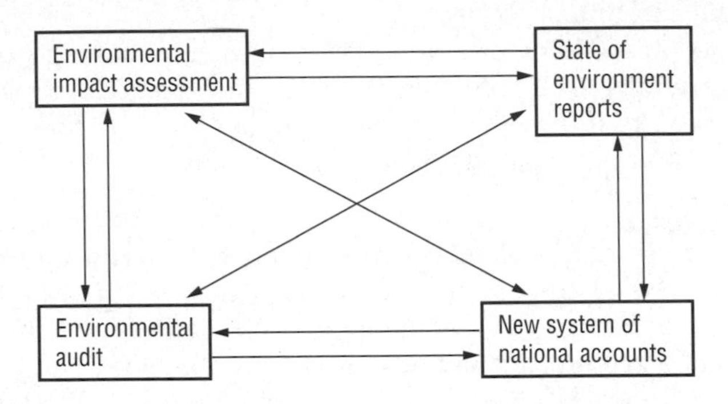

FIGURE 2.5 Relationships, possible exchanges of information and methodologies for environmental impact assessments, environmental audits, New System of National Accounts, state-of-the-environment reports
Source: Thompson and Wilson (1994: 612, Fig. 5)

ECO-AUDITING

Eco-auditing, sometimes termed corporate environmental auditing or environmental management systems auditing, may be defined as a multidisciplinary methodology used to assess periodically and objectively the environmental performance of an organization, an authority or, in some instances, a region. For a review of the management aspects of UK and European eco-auditing, *see* McKenna & Co., 1993. It is a management tool based on periodic, objective evaluation of organizations to see whether they meet regulatory requirements (Willig, 1994: 205–16; Buckley, 1995; Gilpin, 1995: 170).

Eco-auditing has so far been mainly a voluntary process – whereas in relation to finance and company matters, 'auditing' is usually involuntary – that seeks to increase public awareness and aid the quest for sustainable development. Environmental impact assessment and social impact assessments examine *potential* effects of proposed developments; eco-auditing focuses more on assessing *actual* effects of existing activities. Both impact assessment and eco-auditing are tools of environmental management. Terminological confusion might be reduced if 'assessment' were used only for *ex ante* studies of proposals and 'eco-auditing' for environmental review and periodic audit of *existing* developments or activities.

Eco-auditing evolved in the USA as commercial bodies were made more responsible for the damage they caused, a change prompted by the 'polluter-pays principle' (or 'precautionary principle'), support for which spread in the 1970s (EPA, 1988). Handbooks and guidebooks on the topic were appearing by the mid-1980s (e.g. Harrison, 1984; Blakeslee and Grabowski, 1985). In 1986 the US Environmental Protection Agency issued an Environmental Auditing Policy Statement, designed to encourage the use of eco-audits by US companies, which laid down guidelines. In the same year it defined an eco-audit as 'a systematic, documented, periodic and objective review by a regulated entity of facility operations and practices related to meeting environmental requirements'. The International Chamber of Commerce attempted a definition in 1988: 'a management tool comprising a systematic, documented, periodic and objective evaluation of the organization, management and equipment . . . with the aim of safeguarding the environment' (Spedding *et al.*, 1993: 15).

Eco-auditing is increasingly practised in the USA, Europe, Australia and other developed areas, some impetus being given by the publication of *Agenda 21* (Local Government Management Board, 1992; Keating, 1993) following the 1992 Rio 'Earth Summit' and by the passing of the European Commission's Fifth Environmental Action Programme in 1992; this seeks to promote 'shared responsibility' (by people, commerce and government) for the environment, popular environmental awareness, and a move toward sustainable development. The use of eco-auditing marks a growing shift from the requirement merely to comply with environmental regulations to the development of forward-looking environmental management strategies (Willig, 1994: 7). Within the past few years a growing number of guides to and reviews of the subject have been published in Europe and North America (Local Government Management Board, 1991, 1992; Thompson and Thérivel, 1991; Grayson, 1992; Spedding *et al.*, 1993; Woolston, 1993; Richards and Biddick, 1994; Sharratt, 1995: 59). There has been less progress in developing countries, although India has modified its Companies Act to include a requirement for eco-auditing (Spedding *et al.*, 1993) and Indonesia has since 1995 required companies to conduct eco-audits.

Eco-auditing provides a systematic approach to environmental management, embracing the 'precautionary principle', through a process of review and assessment of an organization's *existing* impacts and threats, and is becoming a valuable part of corporate and government environmental management. Eco-audits offer some or all of the following benefits:

- They generate valuable data for regional or national state-of-the-environment reports.
- They may form a valuable way of monitoring (Thompson and Wilson, 1994: 612).
- They allow the bodies using them to assess their environmental performance.
- They help establish an effective environmental protection scheme.
- They assist efforts for sustainable development.
- They inform the public about a body's environmental performance.

- They may involve the public in environmental management.
- They help identify cost recovery through recycling, sale of by-products, etc.
- They reduce the risk of a body being accused of negligence.

In the USA and some other developed countries the public tends not to trust its government to take sufficient action to minimize environmental risks and damage. Eco-auditing has evolved as a means of addressing those worries (Sharratt, 1995: 1). It is not only government and businesses that have adopted it. Some non-governmental organizations, including Friends of the Earth, have assisted companies and authorities to prepare eco-audits. For a review of motivation and legislation, *see* Turpin and Frears (1993). There are sound business reasons for undertaking eco-auditing. Insurance companies may offer favourable rates, there can be public image cachet and it can help an organization avoid court actions for environmental damage.

There are two broad categories of eco-audits: industrial or private-sector corporate eco-audits, and local authority or higher-level government ones. The latter are sometimes called 'green charters', and these are more standardized than private-sector corporate eco-audits. They are mainly commissioned by local authorities to show environmental quality (Glasson *et al.*, 1994: 294; Leu *et al.*, 1995). Box 2.1 provides a more detailed breakdown of the various categories of eco-audit.

Health and safety management and eco-auditing are related. However, whereas eco-auditing is (presently) normally voluntary, an employer is usu-

Box 2.1 Types of eco-audit

- *Site* or *facility audit* - the company or body carries out an audit to see how well it conforms to safety and other regulations and cares for the environment.
- *Compliance audit* – audit to assess whether regulations are being heeded.
- *Issues audit* - assessment of the impact of a body's activities on a specific environmental or social issue, e.g. rain forest loss (Grayson, 1992: 40).
- *Property transfer audit* (sometimes called acquisition or transactional audits) - a company or body audits prior to disvestiture, taking over another body, entering a joint-venture alliance, altering a lease, etc., to check whether there are any problems such as contaminated land.
- *Waste audit* - to see whether regulations are met, if costs can be reduced by sale of by-products, etc. (Ledgerwood *et al.*, 1992; Thompson and Wilson, 1994). The motivation to audit may be to comply with legislation or come from a genuine desire to preempt problems. Eco-auditing may extend to checking the environmental impacts of suppliers or subsidiaries, and the use and disposal of products and packaging.
- *Life cycle assessment/analysis* involves evaluation that extends beyond the time horizon of a single owner, company or even government - 'cradle-to-grave', e.g. impacts of products from manufacture, through use to disposal (as waste or by recycling) (Fava, 1994).

ally responsible under law to a government for employee health and safety. In areas like the safeguarding of workplace quality, there is considerable overlap between eco-auditing and health and safety management. Some countries now require the environmental quality of new buildings to be tested to ensure that they do not harm employees, specify the use of eco-friendly construction materials and legislate to reduce the wasting of energy. Energy savings and better employment conditions means savings on power bills and less absenteeism.

Increasingly the emphasis is shifting from simple reduction of environmental damage to pursuit of sustainable development. Barton and Bruder (1995: xv) see local eco-auditing as a *key* measure in the delivery of sustainable development; as 'a process for establishing what sustainable development means in practice – how to interpret it locally, how to test whether you are achieving it'. In the UK the first eco-audit by a local authority took place in 1989; it was carried out by Kirklees District Council, assisted by Friends of the Earth. Roughly 87 per cent of local government authorities in the UK had used eco-auditing, or planned to, in the early 1990s, encouraged by the UK's Environmental Protection Act 1990 (Grayson, 1992: 50). Unfortunately, some of the eco-audits produced are little more than publicity documents. Eco-audits alone are 'snapshot' views; they are more effective if part of a structured environmental management system. The same is true of environmental impact assessment and risk assessment.

Barton and Bruder (1995: 12) recognize two components in eco-auditing. The first is external – the collation of available data to produce a state-of-the-environment report). The second is internal – after the state-of-the-environment report has been produced it provides a foundation for efforts to assess policies and practices. First-time audits are usually more complex than follow-up audits. Some bodies and companies publish eco-auditing guidelines or manuals that can help other auditors, and there is also interest in the use of computers and expert systems or information technology, as for example retrieval systems like LEXIS (used by lawyers) or conceptual or hypertext searching.

Environmental auditing may be done 'in-house' or by consultants or government agencies. There is a trend towards better training and accreditation and registration or licensing of auditors (*see* Buckley, 1995: 292–3; Thompson *et al.*, 1995). Accreditation and independent auditors are vital if objective auditing is required and eco-auditing is to be respected. There is also a need for internationally recognized standards or even a single world standard for eco-auditing, so that when an organization or company develops an environmental quality management system it can register with something that is reliable and well-known. The following standards are steps along that path. For a recent review of standards and codes used in various countries, *see* Lister and Tinsley (1996).

BS 7750

In early 1992 the world's first eco-auditing standard was published – the British Standards Institution's BS 7750 Specification for Environmental

Management Systems (British Standards Institution, 1992, 1994a, 1994b), derived from an earlier Management Quality System, BS 5750. A number of countries have adopted it and it was revised in 1993 and 1994 to try to make it more compatible with the more recently introduced Eco-management and Audit Scheme (EMAS) which is dealt with in the next subsection (for details of various standards, *see* Bohoris and O'Mahoney, 1994; Sharratt, 1995: 41–53; Willig, 1994: 33–42; Buckley, 1995). BS 7750 is a means for an organization to establish an environmental management system, whereas EMAS is more of an environmental protection system. To obtain BS 7750 an organization has to establish and maintain environmental procedures and an environmental protection system that meets BS 7750 specifications, and demonstrate compliance with it. It must also be committed to cycles of self-improvement through internal eco-auditing. There are three elements to BS 7750: possession of an environmental policy; having a documented environmental management system; and keeping a register of effects on the environment (Johnston, 1993).

Critics of BS 7750 argue that it is possible to obtain the standard by *promising* to do better and then to release relatively little information to the public (it is not as open as, say, the US Toxic Releases Inventory). At the time of writing, BS 7750 did not provide for a publicity logo indicating that a company has gained the standard.

There have been signs in the UK that small companies may find the cost of BS 7750 more of a challenge than larger companies.

Thompson and Wilson (1994: 612) list forces that prompt bodies to embark on eco-auditing, such as pressure from banks, insurance companies, shareholders, etc. Some companies may hesitate to participate if they are afraid of having to disclose confidential information or trade secrets, or because they fear adverse publicity if their eco-audit reveals problems. Alternatively, they may be self-satisfied, content with the *status quo*, afraid of the costs of eco-auditing or reluctant to learn of problems and challenges.

EMAS

The European Union's Eco-management and Audit Scheme (EMAS; Regulation EEC 1836/93) was launched in 1993, although it was not until 10 April 1995 that it came into force in the UK (EEC, 1993; Brown, 1995). EMAS seeks to encourage industry in all European Union states to adopt a site-specific proactive approach to environmental management and improve its performance. EMAS is in some ways similar to and is broadly compatible with the already established British Standard Specification for Environmental Management Systems, BS 7750, but is much broader in scope. In addition, it requires greater public reporting of audits, is stronger than BS 7750 on environmental protection and is aimed more at industrial activities. EMAS places more emphasis on *ensuring* that a body regulates its environmental impacts.

EMAS registration is voluntary, but is established in the European Union by Regulation so that consistent rules and regulations are supposed to be set for all those participating. Participants write and adopt an environmental policy that includes commitments to:

- meeting all legislative requirements and continued improvement of performance;
- implementation of an environmental programme with objectives and targets derived from a comprehensive review process;
- establishing a management system (which includes future environmental audits) to deliver these objectives and targets;
- issuing public environmental statements (EMAS does not insist on *full* publication of audits).

Originally it had been planned to make full public disclosure compulsory but this idea was abandoned. An accredited third party verifies all these things; for details, *see Journal of the Institution of Environmental Sciences* 1995: **4(3)**, 4–7. If these terms are broken, the organization may be suspended from EMAS and so lose its right to a special logo indicative of 'green credentials', which means loss of considerable publicity advantage and, possibly, increased insurance premiums. It could even mean a boycott by suppliers, investors or sales outlets (*see* Fig. 2.6a).

There have been criticisms of EMAS, notably the charges that its auditing criteria are too vague (Karl, 1994), that it badly disrupts the activities of an organization, that it costs too much, especially for small companies (which EMAS assists), that it may reveal trade secrets and that it may generate hostility from the public or workforce.

FIGURE 2.6 (a) The European Union's EMAS (Eco-management and Audit Scheme) eco-audit award logo. This may be used on a company's environmental statements, brochures and reports, headed paper and advertisements. The latter must contain no reference to specific products or services, and the logo may not be used on products or product packages. (b) The European Union's eco-labelling scheme: the logo can be used on products and packaging

Europe is moving to improve on EMAS by gradually introducing strategic environmental assessment, to be applied to all plans, policies and programmes (Barton and Bruder, 1995: 11) (*see also* Chapter 3). If the European Union adopts an 'Environmental Charter', which is a possibility, eco-auditing will probably become more widespread, possibly even compulsory.

ISO 14000

The International Standards Organization (ISO) has been seeking to develop a standard broadly compatible with EMAS and BS 7750. At the time of writing the closest link was ISO/CD 14000 (updated to ISO[DIS] 14001 in 1996). This is roughly equivalent to BS 7750 and EMAS, but is more 'user-friendly' and easier to understand (for details, *see* Rothery, 1993; Baxter and Bacon, 1996; Sayre, 1996; Jackson, 1997). These ISO standards are related to the ISO 9000 series (roughly equivalent to BS 5750), which are widely used in business all over the world and deal with quality systems (total quality management – TQM) registration (Willig, 1994: 33–42).

International Environmental Rating System (IERS)

IERS is run by SGS Australia and Det Norske Veritas, offering 10 different levels of corporate environmental audit standard (S1–S5 and A1–A5). The level S5 is roughly equivalent to BS7750 or EMAS, according to Buckley (1995: 291).

ECO-LABELLING

Eco-labelling has been adopted in many countries, including Canada, the USA, Germany and Sweden (Grayson, 1992: 24; Spedding *et al.*, 1993: 71; Karl, 1994). The European Union introduced an eco-labelling scheme in the 1990s that provides a label for products judged against other similar products to have less impact on the environment (*see* Fig. 2.6b). Eco-labelling does not require an eco-auditing, only the assessment of the product by an independent judge. It is nevertheless a process that assesses environmental impact and communicates an assessment to the consumer and, where sales are not direct from a manufacturer, the merchant. However, the focus is on the product and often nothing is said about the process of production; for example, an eco-friendly product may come from an environmentally damaging factory or pose a disposal problem after use. So far there is little standardization or real 'policing' of eco-labelling, although Friends of the Earth, a non-governmental organization, monitors product claims and awards an annual and well-publicized 'Green Con Award' for 'eco-deception' (perhaps this could be extended to impact assessment!). West (1995) feels that eco-labelling needs back-up from adequate legally enforced standards, without which it could become just a marketing gimmick. He also expressed concern that world trade agreements such as GATT, the General Agreement on Tariffs and Trade (now the World Trade Organization) might make meaningful eco-labelling difficult.

Assessment of contaminated land

There have been increased efforts, mainly in developed countries, to recognize, document and assess contaminated land better than has been the practice in the past. Some impetus has been given by a number of disasters involving contaminated land like those at Love Canal or Lekkerkerk (Barrow, 1995: 270). (Love Canal is near Niagara City in New York State. In this notorious case, an abandoned canal was used for landfill disposal of industrial chemicals, mainly from 1942 to 1953, with inadequate monitoring and record-keeping. In 1953 the site was infilled and landscaped and a housing estate of several hundred homes, with a school, was built. In the 1970s people started to notice a high incidence of cancer, birth defects, etc. In 1977 the link with pollution became clear, but local people had to campaign before in 1978 there was evacuation and compensation for 240 families. By 1992 clean-up costs and legal costs had exceeded $250 million. Lekkerkerk is a town in the Netherlands where hazardous wastes were burned, again with inadequate monitoring and record-keeping. In the 1970s over 200 houses were built on the site. Households became ill, and after investigations the estate was evacuated at a cost, at 1981 prices, of £156 million.) Traditionally in Europe and North America, those buying land are subject to the rule of *caveat emptor* ('let the buyer beware') and should commission assessments. However, it is not easy to check the pollution history of a site. The risk of 'planning blight' (reduced land values if there is a hint of contamination) and fear of future legal damages awards has led to less than perfect record-keeping or official reluctance to release records (the latter is less of a problem in the USA). There are additional problems of fly-tipping, meaning the illegal and unrecorded disposal of material, and unexpected cumulative effects as a number of materials mix and react with each other. With developers being increasingly held more responsible for damages if they fail to spot contamination, there has been a trend towards vendors commissioning contamination assessments and then issuing 'green warranties'.

Decision analysis

Decision analysis is an approach sometimes used in impact assessment to estimate the amount of risk or impact people will accept. In effect it establishes preferences. There are various methods. For example, the 'standard gamble method' asks what chance of getting a preferred outcome A (compared with getting nothing) a person would be willing to exchange for getting outcome B, or C, or whatever, with certainty. By making pair-wise comparisons for all the considered alternatives it is possible to arrive at an ordinal scale ranking of preference for alternatives; for example, A might be preferred to B, which in turn is preferred to F, which is preferred to C, which is preferred to D, etc. (Hyman and Stiftel, 1988: 39; James, 1994: 25; Tamura *et al.*, 1994).

PROJECT AND PROGRAMME APPRAISAL AND PROJECT AND PROGRAMME EVALUATION

There is a large literature on project and programme appraisal and evaluation, which are widely used for consultancy, business and management studies. I do not consider them in depth, although they do have some overlap with impact assessment (*see*, for example, the journal *Project Appraisal*, and O'Riordan and Sewell, 1981; and for a review of the application of impact assessment to project appraisal in the UK, *see* Lichfield, 1988).

POLICY EVALUATION, POLICY ASSESSMENT AND POLICY ANALYSIS

Although interest has increased in the 1990s, policy evaluation(or assessment, or analysis) was developed before 1970 (DoE, 1991). It involves the evaluation of policies as they are planned and/or implemented, in terms of the objectives they are intended to reach. Boothroyd (1995: 83–95) argued for a synthesis of policy evaluation and impact assessment as a way of (perhaps more effectively) dealing with the issues addressed by strategic environmental assessment (*see* Chapter 3).

POST-PROJECT ANALYSIS

According to Gilpin (1995: 171), post-project analysis is an environmental study undertaken during the operational stage of a project (or programme) to assess compliance with terms required by an environmental impact statement (effectively a type of review; *see* Chapter 4), and to consider the quality and possible improvement of environmental management. It thus has broader aims than environmental auditing and overlaps a little with post-environmental impact assessment audit (the procedure for checking whether impact assessment has been effective; *see* Chapter 4).

EXPERT SYSTEMS APPROACH AND INITIAL DECISION ANALYSIS

The *expert systems approach* relies on computer techniques (*see also* Chapter 5). Used in environmental studies, eco-auditing, environmental impact assessment, social impact assessment, health care, etc., expert systems can be valuable once perfected but may take a great deal of research and time to develop (Loehle and Osteen, 1990; Fedra, 1991; Geraghty, 1992, 1993). They are particularly useful when there is a shortage of expertise to conduct assessment, and may have potential for improving public involvement (Schibuola and Byer, 1991). The approach involves developing a computer program that stores a body of knowledge and with it helps a user perform tasks that usually demand input from a human expert – for example, impact assessment or risk analysis.

Initial decision analysis is a computer-based decision-making technique organized so that a rational step-by-step process can be followed to use existing knowledge to develop policies. It is useful for complex problems where various groups are in conflict. The initial decision analysis process uses a 'biosocial systems model' to solve resource management problems. Part of the process involves workshops with panels of experts and others involved in the proposed development, who are briefed in all aspects of the development and on the biosocial systems model. Panelists participate in a 'white-box' (i.e. supposedly open to scrutiny) policy-making process (Bonnicksen, 1985).

REFERENCES

Abelson, P. 1979: *Cost–benefit analysis of environmental problems*. Teakfield: Saxon House (reprinted 1980: Farnborough: Gower).

Adams, J. 1996: Cost–benefit analysis: the problem, not the solution. *The Ecologist* **26(1)**, 2–4

Ahmad, A. 1993: Environmental impact assessment in the Himalayas: an ecosystems approach. *Ambio* **XXII(1)**, 4–9.

Ahmad, Y-J., El Serafy, M. and **Lutz, E.** (eds) 1989: *Environmental accounting for sustainable development*. Washington, DC: World Bank.

Ament, R.H. 1970: Comparison of Delphi forecasting studies in 1964 and 1969. *Futures* **2(1)**, 35–44.

Andrews, R.N.L. 1988: Environmental impact assessment and risk assessment: learning from each other. In Wathern, P. (ed.), *Environmental impact assessment: theory and practice*. London: Unwin Hyman, 85–97.

Anon. 1978: Appendix A. The Delphi inquiry. In Enzer, S., Drobnick, R. and Alter, S. (eds), *Neither feast nor famine: food conditions to the year 2000*. Lexington, MA: Lexington Books, 105–26.

Armstrong, J. and **Harmon, W.H.** 1980: *Strategies for conducting technology assessments*. Boulder: Westview.

ASCE 1977: *Environmental impacts of international civil engineering projects and practices*. Proceedings of a Seminar on Environmental Impact Assessment, ASCE Convention, San Francisco, 17–21 October, 1977. New York: American Society of Civil Engineers.

Auerbach, S.I. 1978: Current perceptions and applicability of ecosystem analysis to impact assessment. *Ohio Journal of Science* **78(4)**, 163–75.

Bagley, M.D. 1972: *Aesthetic assessment methodology for environmental impact analysis*. Technical Note TN-OED-004. Menlo Park, CA: Stanford Research Institute.

Barrow, C.J. 1995: *Developing the environment: problems and management*. Harlow: Longman.

Bartel, S., Gardner, R. and **O'Neill, R.** 1992: *Ecological risk assessment*. Ann Arbor, MI: Lewis Publishers.

Barton, H. and **Bruder, N.** 1995: *A guide to local environmental auditing*. London: Earthscan.

Baxter, M. and **Bacon, R.** 1996: Which EMS? – your questions answered. *Environmental Assessment* **4(1)**, 5–6.

Beek, K.J. 1978: *Land evaluation for agricultural development*. Wageningen: International Institute for Land Reclamation and Improvement.

Biswas, A.K. and **Qu Geping.** (eds) 1987: *Environmental impact assessment for developing countries*. London: Tycooly International.

Blaikie, P., Cannon, T., Davis, I. and **Wisner, B.** 1994: *At risk: natural hazards, people's vulnerability, and disasters*. London: Routledge.

Blakeslee, H.W. and **Grabowski, T.M.** 1985: *A practical guide to plant environmental audits*. New York: Van Nostrand Reinhold.

Bohm, P. and **Henry, C.** 1979: Cost–benefit analysis and environmental effects. *Ambio* **XIII(1)**: 18–24.

Bohoris, G.A. and **O'Mahoney, E.** 1994: BS 7750, BS 5750 and the EC's Eco-Management and Audit Scheme. *Industrial Management and Data Systems* **94(2)**, 3–6.

Bonnicksen, T.M. 1985: Initial decision analysis (IDA): a participatory approach for developing research policies. *Environmental Management* **9(5)**, 579–92.

Boothroyd, P. 1995: Policy assessment. In Vanclay, F. and Bronstein, D.A. (eds), *Environmental and social impact assessment*. Chichester: Wiley, 83–128.

British Standards Institution 1992: *BS 7750 British standard for environmental management*. London: British Standards Institution.

British Standards Institution 1994a: *Part I: Auditing of environmental management systems*. Guidelines for environmental auditing – audit procedures, ISO/CD 14011/1, Document 94/400414DC. London: British Standards Institution.

British Standards Institution 1994b: *Environmental management systems BS 7750*. London: British Standards Institution.

Brown, D.J.A. 1995: EU Eco-Management and Audit Scheme (EMAS). *Journal of the Institution of Environmental Sciences* **4(3)**, 4–7.

Bruckner, S.M., Hastings, S.E. and **Latham, W.R.** 1987: Regional input–output analysis: a comparison of five ready-made model systems. *Review of Regional Studies* **17(2)**, 1–16.

Brush, S.B. 1986: Farming systems research. *Human Organization* **45(3)**: 220–38.

Bryant, E. 1991: *Natural hazards*. Cambridge: Cambridge University Press.

Buckley, R.C. 1995: Environmental auditing. In Vanclay, F. and Bronstein, D.A. (eds), *Environmental and social impact assessment*. Chichester: Wiley, 283–301.

Burchell, R.W., Listokin, D. and **Dolphin, W.R.** 1985: *The new practitioner's guide to fiscal impact analysis*. Piscataway, NJ: Rutgers University, Center for Urban Planning and Research.

Cairns, J. Jr and **Dickson, K.L.** 1980: Risk analysis for aquatic ecosystems. In *Symposium proceedings on biological evaluation of environmental impacts*. FWS/OBS-80/26. Washington DC: US Department of the Interior, Council on Environmental Quality/Fish and Wildlife Service.

Canter, L.W. 1993: Pragmatic suggestions for incorporating risk assessment principles in EIA studies. *Environmental Professional* **15(1)**, 125–38.

Canter, L.W. 1996: *Environmental impact assessment*, 2nd edn (1st edn 1977). New York: McGraw-Hill.

Carpenter, R.A. 1995a: Risk assessment. *Impact Assessment* **13(2)**, 153–87.

Carpenter, R.A. 1995b: Risk assessment. In Vanclay, F. and Bronstein, D.A. (eds), *Environmental and social impact assessment*. Chichester: Wiley, 193–219.

Cavallin, A., Marchetti, M., Panizza, M. and **Soldati, M.** 1994: The role of geomorphology in environmental impact assessment. *Geomorphology* **9(2)**, 143–53.

Chambers, R. 1992: *Rural appraisal: rapid, relaxed and participatory*. Discussion Paper 311. Brighton: University of Sussex, Institute of Development Studies.

Chambers, R. 1994a: The origins and practice of participatory rural appraisal. *World Development* **22(7)**, 953–69.

Chambers, R. 1994b: Participatory rural appraisal (PRA): an analysis of experience. *World Development* **22(9)**, 1253–68.

Chambers, R. 1994c: Participatory rural appraisal (PRA): challenges, potentials and paradigm. *World Development* **22(10)**, 1437–54.

Clark, B.D., Gilad, A., Bisset, R. and **Tomlinson, P.** (eds) 1984: *Perspectives on environmental impact assessment*. Dordrecht: D. Reidel.

Clark, N. 1990: Development policy, technology assessment and the new technologies. *Futures* **22(3)**, 225–31.

Coates, J.F. 1971: Technology assessment: the benefits . . . the costs . . . the consequences. *The Futurist* **5(4)**, 225–31.

Coates, V.T. and **Coates, J.F.** 1989: Making technology assessment an effective tool to influence policy. In Bartlett, R.V. (ed.), *Policy through impact assessment: institutionalized analysis as a policy strategy*. New York: Greenwood Press, 17–25.

Conrad, J. (ed.) 1980: *Society, technology and risk assessment*. London: Academic Press.

Conway, G.R. and **Barbier, E.** 1990: *After the green revolution: sustainable agriculture for development*. London: Earthscan.

Conway, G.R. and **McCracken, J.A.** 1990: Rapid rural appraisal and agroecosystem analysis. In Altieri, M.A. and Hecht, S.B. (eds), *Agroecology and small farm development*. Boca Raton, FL: Westview/CRC, 221–36.

Cooper, C. 1976: Ecosystem models and environmental policy. *Simulation* **26(2)**, 133–8.

Cooper, C. 1981: *Economic evaluation and the environment: a methodological discussion with particular reference to developing countries*. London: Hodder & Stoughton.

Covello, V.T., Mumpower, J.L., Stallen, P.J.M. and **Uppuluri, V.R.R.** (eds) 1985: *Environmental impact assessment, technology assessment, and risk analysis: contributions from the psychological and decision sciences*. Berlin: Springer-Verlag.

Davidson, D.A. 1980: *Soils and land use planning*. London: Longman.

Degg, M. 1992: Natural disasters: recent trends and future prospects. *Geography* **77(3)**, 198–209.

Dierkes, M., Edwards, S. and **Coppock, R.** (eds) 1980: *Technological risk*. Cambridge, MA: Oelgeschlager, Gunn & Hain.

Dixon, J.A., Carpenter, R.A., Fallon, L.A., Sherman, P.B. and **Manipomoke, S.** 1986: *Economic analysis of the environmental impacts of development projects*. London: Earthscan (with the Asian Development Bank).

DoE 1991: *Policy appraisal and the environment* (UK Department of the Environment). London: HMSO.

Doremus, C., McNaught, P.C., Fuist, E. and **Stanley, E.** 1978: An ecological approach to environmental impact management. *Environmental Management* **2(3)**, 245–8.

Douglas, M. and **Wildavsky, A.** 1982: *Risk and culture*. Berkeley, CA: University of California Press.

Duinker, P.N. 1989: Ecological monitoring in environmental impact assessment: what can it accomplish? *Environmental Management* **13(6)**, 797–805.

Eberhardt, L.L. 1976: Quantitative ecology and impact assessment. *Journal of Environmental Management* **4(1)**, 27–70.

Edwards, F.N. (ed.) 1992: *Environmental auditing: the challenge of the 1990s*. Calgary: University of Calgary Press.

EEC 1993: *EMAS No.1836/93.* EEC Council Regulation. Luxembourg: Council of the European Communities.

EPA 1988: *Annotated bibliography on environmental auditing*. Washington DC: US Environmental Protection Agency, Office of Policy and Education.

FAO 1978: *Report on the agroecosystems zones project,* vol. 1. *Methodology and results for Africa*. World Soil Research Paper 48. Rome: Food and Agriculture Organization.

Farvar, M.T. and **Milton, J.P.** (eds) 1972: *The careless technology: ecology and international development*. New York: Garden City Press. (First published 1969.)

Fava, J.A. 1994: Life-cycle assessment: a new way of thinking. *Environmental Toxicology and Chemistry* **13(6)**, 853–4.

Fedra, K. 1991: *Expert systems for environmental screening*. Laxenburg: International Institute for Applied Systems Analysis.

Fischhoff, B., Lichtenstein, S., Slovic, P., Derby, S.L. and **Keeney, R.L.** 1981: *Acceptable risk*. Cambridge: Cambridge University Press.

Forrester, J. 1971: *World dynamics*. Cambridge, MA: Wright-Allen Press.

Furlong, J. 1995: EC approach to environmental risk assessment of new substances. *Science of the Total Environment* **171(1–3)**, 275–9.

Geraghty, P.J. 1992: Environmental assessment and the application of an expert systems approach. *Town Planning Review* **63(2)**, 123–42.

Geraghty, P.J. 1993: Environmental assessment and the application of expert systems – an overview. *Journal of Environmental Management* **39(1)**, 27–38.

Gilbert, E.H., *et al.* 1980: *Farming systems research: a critical appraisal*. MSU Rural Development Paper 6. East Lansing: Michigan State University, Department of Agricultural Economics.

Gilpin, A. 1995: *Environmental impact assessment (EIA): cutting edge for the twenty-first century*. Cambridge: Cambridge University Press.

Ginzberg, L.R. (ed.) 1991: *Assessing ecological risks of biotechnology*. Stoneham, MA: Butterworth-Heinemann.

Glasson, J., Thérivel, R. and **Chadwick, A.** 1994: *Introduction to environmental impact assessment: principles and procedures, process, practice and prospects*. London: university College London Press.

Gonzalez, A., Diaz de Teran, J.R., Frances, E. and **Cendrero, A.** 1995: The incorporation of geomorphological factors into environmental impact assessment for master plans: a methodological proposal. In McGregor, D.F.M. and Thompson, D.A. (eds), *Geomorphology and land management in a changing environment*. Chichester: Wiley, 179–93.

Gordon, T.J. and **Helmer, O.** 1964: *Report on a long range forecasting study*. RAND Corporation Paper P-2982. Santa Monica, CA: RAND Corporation.

Graham Smith, L. 1993: *Impact assessment and sustainable resource management*. Harlow: Longman.

Grayson, L. (ed). 1992: *Environmental auditing: a guide to best practice in the UK and Europe*. Letchworth: Technical Communications and British Library Science and Information Service.

Green, H.C., Hunter, C. and **Moore, B.** 1990: Assessing the environmental impact of tourism development: the use of the Delphi technique. *International Journal of Environmental Studies* **35**, 51–62.

Harrison, L.L. (ed.) 1984: *The McGraw-Hill environmental auditing handbook: a guide to corporate environmental risk management*. New York: McGraw-Hill.

Hillborn, R. 1979: Some failures and successes in applying systems analysis to ecological systems. *Journal of Applied Systems Analysis* **6(1)**, 25–31.

Hirsch, A. 1993: Improving conservation of biodiversity in NEPA assessments. *Environmental Professional* **15(1)**, 103–15.

HMSO 1971: *Report of the Roskill Commission on the third London airport*. London: HMSO.

Holdridge, L.R. 1947: Determination of world plant formations from simple climate simulations. *Science* **105**, 367–8.

Holdridge, L.R. 1964: *Life zone ecology*. San Jose, CA: Tropical Science Center.

Holling, C.S. 1973: Resilience and stability of ecological systems. *Annual Review of Ecology and Systematics* **4**, 1–23.

Holling, C.S. (ed.) 1978: *Adaptive environmental planning and management*. Chichester: Wiley.

Holling, C.S. 1986: The resilience of terrestrial ecosystems: local surprise and global change. In Clark, W.C. and Munn, R.E. (eds), *Sustainable development of the biosphere*. Cambridge: Cambridge University Press, 292–320.

Horlick-Jones, T., Amendola, A. and **Casale, R.** (eds) 1995: *Natural risk and civil protection*. London: Chapman & Hall.

Hudson, N. 1981: *Soil conservation*. London: Batsford.

Hufschmidt, M.M. and **Hyman, E.L.** (eds) 1982. *Economic approaches to natural resource and environmental analysis*. Dublin: Tycooly.

Hundloe, T., McDonald, G.T., Ware, J. and **Wilkes, L.** 1990: Cost–benefit analysis and environmental impact assessment. *Environmental Impact Assessment Review* **10(1–2)**, 55–68.

Hyman, E.L. and **Stiftel, B.** 1988: *Combining facts and values in environmental impact assessment: theories and techniques*. Boulder, CO: Westview.

Institute of Environmental Assessment 1995. *Guidelines for baseline ecological assessment*. London: Spon (Chapman & Hall).

Institute of Environmental Assessment and **Landscape Institute** (eds) 1995: *Guidelines for landscape and visual impact assessment*. London: Spon (Chapman & Hall).

Jackson, S.L. 1997: *ISO 14000 implementation guide: creating an integrated management system*. Chichester: Wiley.

Jain, R.K., Urban, L.V. and **Stacey, G.S.** 1981: *Environmental impact assessment: a new dimension in decision making*. New York: Van Nostrand Reinhold.

James, D. (ed.) 1994: *The application of economic techniques in environmental impact assessment*. Dordrecht: Kluwer.

James, D. and **Morris, J.** 1995: The application of economic techniques in environmental impact assessment. *Project Appraisal* **10(2)**, 137.

Jiggins, J. 1995: Development impact assessment: impact assessment of aid projects in nonwestern countries. *Impact Assessment* **13(1)**, 47–70.

Johnston, B. 1993: Flying industry's green standard. *New Scientist* **138(1868)**, 21–2.

Kapp, W. 1950: *The social costs of private enterprise*. Cambridge, MA: Harvard University Press.

Karl, H. 1994: Better environmental future in Europe through environmental auditing. *Environmental Management* **18(4)**, 617–21.

Kates, R.W. 1978: *Risk assessment of environmental hazards*. Chichester: Wiley.

Kates, R.W. and **Hohenemser, C.** (eds) 1982: *Technological hazard assessment*. Cambridge, MA: Oelgeschlager, Gunn & Hain.

Keating, M. 1993: *The Earth Summit's agenda for change: a plain language version of Agenda 21*. Geneva: Centre for Our Common Future (Palais Wilson, 52 rue des Pâquis, CH-1201 Geneva).

Keyes, D.L. 1976: *Land development and the natural environment: estimating impacts*. Washington DC: Urban Institute.

Kozlowski, J.M. 1986: *Threshold approach in urban, regional and environmental planning*. St Lucia, Queensland: University of Queensland Press.

Krutilla, J. and **Eckstein, O.** 1958: *Multipurpose river development*. Baltimore: Johns Hopkins University Press.

Landon, J.R. (ed.) 1991: *Booker tropical soil manual*. Harlow: Longman. (1st edn 1984)

Ledgerwood, G., Street, E. and **Thérivel, R.** 1992: *Environmental audit and business strategy*. London: Pitman.

Leistritz, F.L. 1994: Economic and fiscal impact assessment. *Impact Assessment* **12(3)**, 305–18.

Leistritz, F.L. 1995: Economic and fiscal impact assessment. In Vanclay, F. and Bronstein, D.A. (eds), *Environmental and social impact assessment*. Chichester: Wiley, 129–39.

Leitmann, J. 1993: Rapid urban environmental assessment: toward environmental management in cities of the developing world. *Impact Assessment Review* **11(3)**, 225–60.

Leontief, W. 1986: *Input–output economics*, 2nd edn. New York: Oxford University Press.

Leu, W.S., Williams, W.P. and **Bark, A.W.** 1995: An environmental evaluation of the implementation of environmental assessment by UK local authorities. *Project Appraisal* **10(2)**, 91–102.

Lichfield, N. 1988: Environmental impact assessment in project appraisal in Britain. *Project Appraisal* **3(3)**, 133–42.

Linstone, H. 1985: The Delphi technique. In Covello, V.T., Mumpower, J.L., Stollen, P.J.M. and Uppuluri, V.R.R. (eds), *Environmental impact assessment, technology assessment, and risk analysis: contributions from the psychological and decision sciences*. Berlin: Springer-Verlag, 621–49.

Linstone, H.A. and **Simmons, W.H.** (eds) 1977: *Futures research: new directions*. London: Addison-Wesley.

Linstone, H.A. and **Turoff, M.** (eds) 1975: *The Delphi method: techniques and applications*. London and Reading, MA: Addison-Wesley.

Lister, N. and **Tinsley, H.** 1996: EMSs: contrasting international procedures. *Environmental Assessment* **4(2)**, 62–4.

Little, I.M.D. and **Mirrlees, J.A.** 1974: *Project appraisal and planning for developing countries*. London: Heinemann.

Local Government Management Board 1991: *Environmental auditing in local government: a guide and discussion paper*. Luton: Local Government Management Board.

Local Government Management Board 1992: *Local Agenda 21 – Agenda 21: a guide for local authorities in the UK*. Luton: Local Government Management Board.

Loehle, C. and **Osteen, R.** 1990: IMPACT – an expert system for environmental impact assessment. *AI Applications in Natural Resource Management* **4(1)**, 35–43.

Lowrance, W.W. 1976: *Of acceptable risk*. Los Altos, CA: Kaufmann.

McCracken, J.A., Pretty, J.N. and **Conway, G.R.** 1988: *An introduction to rapid rural appraisal for agricultural development*. London: Earthscan.

McHarg, I.L. 1969: *Design with nature*. Garden City, NY: Natural History Press.

McKenna & Co. 1993: *Environmental auditing: a management guide* (2 vols). London: Intelex Press.

Marsh, W.M. 1978: *Environmental analysis for land use and site planning*. New York: McGraw-Hill.

Maxwell, S. 1986: Farming systems research: hitting a moving target. *World Development* **14(1)**, 65–77.

Meadows, D.H., Meadows, D.L. and **Randers, J.** 1992: *Beyond the limits: global collapse or sustainable future?* London: Earthscan.

Meadows, D.H., Meadows, D.L., Randers, J. and **Behrens, W.W. III.** 1974: *The limits to growth*. London: Pan. (First published in 1972.)

Medford, D. 1973: *Environmental harassment or technology assessment?* Amsterdam: Elsevier.

Miller, A. and **Cuff, W.** 1986: The Delphi approach to the mediation of environmental disputes. *Environmental Management* **10(3)**, 321–30.

Munn, R.E. (ed.) 1979: *Environmental impact assessment: principles and procedures*, 2nd edn. SCOPE Report 5. Chichester: Wiley (1st edn 1975).

Nijkamp, P. 1986: Multiple criteria analysis and integrated impact analysis. *Impact Assessment Bulletin* **4(3–4)**, 226–61.

Nip, M.J. and **Udo deHaes, H.A.** 1995: Ecosystem approaches to environmental quality assessment. *Environmental Management* **19(1)**, 135–45.

O'Brien, D.M. and **Marchand, D.A.** (eds) 1982: *The politics of technology assessment: institutions, processes, and policy disputes*. Lexington, MA: Lexington Books.

ODA 1989: *Manual of environmental appraisal* (revised 1992). London: Overseas Development Administration.

O'Neill, J. 1996: Cost–benefit analysis, rationality and the plurality of values. *The Ecologist* **26(3)**, 98–103.

Ono, R. and **Wedemeyer, D.J.** 1994: Assessing the validity of the Delphi technique. *Futures* **26(3)**, 289–304.

O'Riordan, T. and **Sewell, W.R.D.** (eds) 1981: *Project appraisal and policy review*. Chichester: Wiley.

O'Riordan, T. and **Turner, R.K.** 1983: Traditional cost–benefit analysis and its critique. In O'Riordan, T. and Turner, R.K. (eds), *An annotated reader in environmental planning and management*. Oxford: Pergamon, 87–134.

Patricos, N.N. 1986: *International handbook on land use planning*. Westport, CT: Greenwood Press.

Pearce, D.W. 1971: *Cost–benefit analysis*. London: Macmillan.

Pearce, D.W., Markandya, A. and **Barbier, E.B.** 1989: *Blueprint for a green economy*. London: Earthscan.

Pearson, N. Associates 1984: *Wytch Farm oilfield development: Furzey Island visual impact analysis*. Sunbury: BP Development Ltd.

Pielou, E.C. 1981: The usefulness of ecological models: a stocktaking. *Quarterly Review of Biology* **56(1)**, 17–31.

Pigou, A. 1920: *The economics of welfare*. New York: Macmillan.

Pill, J. 1971: The Delphi method: substance, context, a critique and an annotated bibliography. *Socio-economic Planning* **5(1)**, 57–71.

Porter, A.L. 1995a: Technology assessment. *Impact Assessment* **13(2)**, 135–52.

Porter, A.L. 1995b: Technology assessment. In Vanclay, F. and Bronstein, D.A. (eds), *Environmental and social impact assessment*. Chichester, Wiley, 67–81.

Porter, A.L. and **Rossini, F.A.** 1980: Technology assessment/environmental impact assessment: towards integrated impact assessment. *IEEE Transactions on Systems, Man and Cybernetics* **10(8)**, 417–24.

Porter, A.L., Rossini, F.A., Carpenter, S.R. and **Roper, A.T.** 1980: *A guidebook for technology assessment and impact analysis*. New York: North-Holland.

Raferty, J. 1993: *Risk analysis in project management*. London: Chapman & Hall.

Rakos, C. 1988: Recent developments of technology assessment and environmental impact assessment in Europe. *Project Appraisal* **3(4)**, 205–9.

Ratanachai, C. 1991: Environmental impact assessment as a tool for risk management. *Toxicology and Industrial Health* **7(5–6)**, 379–91.

Renwick, W.H. 1988: The eclipse of NEPA as environmental policy. *Environmental Management* **12(3)**, 267–72.

Ricci, R.F. (ed.) 1981: *Technological risk assessment*. Boston: Martinus Nijhoff.

Richards, L. and **Biddick, I.** 1994: Sustainable economic development and environmental auditing: a local authority perspective. *Journal of Environmental Planning and Management* **37(4)**, 487–94.

Richey, J.S., Horner, R.R. and **Mar, B.W.** 1985a: The Delphi technique in environmental assessment 1: Implementation and effectiveness. *Journal of Environmental Management* **21(2)**, 135–46.

Richey, J.S., Horner, R.R. and **Mar, B.W.** 1985b: The Delphi technique in environmental assessment 2: Consensus of critical issues in environmental monitoring program design. *Journal of Environmental Management* **21(2)**, 147–60.

Rivas, V., Gonzales, A., Fischer, D.W. and **Cendrero, A.** 1994: An approach to environmental assessment within the land-use planning process: northern Spanish experiences. *Journal of Environmental Planning and Management* **37(3)**, 305–22.

Roebig, J.H. 1983: An aesthetic impact assessment technique. *Impact Assessment Bulletin* **2(3)**, 29–40.

Rothery, B. 1993: *BS 7750: implementing the environment management standard and the EC eco-management scheme*. Aldershot: Gower.

Save the Children 1995: *Toolkits: a practical guide to assessment, monitoring, review and evaluation*. Save the Children Development Manual 5. London: Save the Children.

Sayre, D. 1996: *Inside ISO 14000: the competitive advantage of environmental management*. London: Earthscan.

Schibuola, S. and **Byer, P.H.** 1991: Use of knowledge-based systems for the review of environmental impact assessments. *Environmental Impact Assessment Review* **11(1)**, 11–27.

Schmid, E., Schaad, W. and **Cochrane, S.** 1995: Earthquake risk assessment for reinsurance portfolios: a database for worldwide seismicity quantification. In Duma, D. (ed.), *Proceedings of the 10th European Conference on Earthquake Engineering 28 August–2 September, 1994* (4 vols). Vienna: Technical University of Vienna, 23–8.

Shaner, W.W., Philipp, P.F. and **Schmehl, W.R.** 1982: *Farming systems research and development: guidelines for developing countries*. Boulder, CO: Westview.

Sharratt, P. (ed.) 1995: *Environmental management systems*. Rugby: Institution of Chemical Engineers.

Soderstrom, E.J. 1981: *Social impact assessment: experimental methods and approaches*. New York: Praeger.

Solomon, B.D. 1985: Regional econometric models for environmental impact assessment. *Progress in Human Geography* **9(3)**, 378–99.

Spedding, L.S., Jones, D.M. and **Dering, C.J.** (eds) 1993: *Eco-management and eco-auditing: environmental issues in business*. London: Chancery Law Publishers (Chichester: Wiley).

Stocking, M. 1984: The geomorphologist's role in the environmental impact assessment of agricultural development in Zambia. *Zeitschrift für Geomorphologie* **28(1)**, 41–51.

Stouth, R., Sowman, M. and **Grindley, S.** 1993: The panel evaluation method: an approach to evaluating controversial resource allocation proposals. *Environmental Impact Assessment Review* **13(1)**, 13–35.

Stromquist, L. 1992: Environmental impact assessment of natural disasters: a case-study of the recent Lake Balati floods in northern Tanzania. *Geografisker Annaler* ser. A: *Physical Geography* **74(2–3)**: 81–91.

Suter, G.W. II (ed.) 1993: *Ecological risk assessment*. Boca Raton, FL: Lewis.

Tamura, H., Fujita, S.I. and **Koi, H.** 1994: Decision analysis for environmental impact assessment and concensus formation among conflicting multiple agents – including case studies for road traffic. *Science of the Total Environment* **153(3)**, 203–10.

Taylor, J.R. 1993: *Risk analysis for process plant, pipelines and transport*. London: Chapman & Hall.

Thakur, M.S., Kennedy, M.J. and **Koranth, N.G.** 1991: An environmental evaluation of biotechnological processes. *Advances in Applied Microbiology* **36**, 67–86.

Thompson, D. and **Wilson, M.J.** 1994: Environmental auditing: theory and applications. *Environmental Management* **18(4)**, 605–15.

Thompson, D.R., Bacon, R.A., Tarling, J.P. and **Baverstock, S.J.** (eds) 1995: *The EARA register of environmental auditors*, 1st edn. London: Earthscan.

Thompson, S and **Thérivel, R.** (eds) 1991: *Environmental auditing*. Oxford Polytechnic, School of Planning, Working Paper 130. Oxford: Oxford Polytechnic.

Treweek, J. 1995a: Ecological impact assessment. *Impact Assessment* **13(3)**, 289–316.

Treweek, J. 1995b: Ecological impact assessment. In Vanklay, F. and Bronstein, D.A. (eds), *Environmental and social impact assessment*. Chichester: Wiley, 171–91.

Turpin, T. and **Frears, F.** 1993: EC and UK legislation on environmental assessment and its effect on the UK water industry. *Journal of the Institution of Water and Environmental Management* **7(3)**, 275–82.

UN Branch for Science and Technology Development 1991: *UN workshop on technology assessment for developing countries*. Washington DC: Office of Technology Assessment.

Vincent, P.G. 1993: Environmental impact assessment: US requirements in new drug applications. *Journal of Hazardous Materials* **35(2)**, 211–6.

Vizayakumar, K. and **Mohapatra, P.K.J.** 1989: An approach to environmental impact assessment by using cross impact simulation. *Environment and Planning A* **21(6)**, 831–7.

Ward, D.V. 1978: *Biological environmental impact studies: theory and methods*. New York: Academic Press.

West, K. 1995: Eco-labels: the industrialization of environmental standards. *The Ecologist* **25(1)**, 16–20.

Westman, W.E. 1978: Measuring the inertia and resilience of ecosystems. *Bioscience* **28**, 705–10.

Westman, W.E. 1985: *Ecology, impact assessment and environmental planning*. Chichester: Wiley.

White G.F. and **Haas, J.E.** 1975: *Assessment of research on natural hazards*. Cambridge, MA: MIT Press.

Whitehead, P. 1992: Examples of recent models in environmental impact assessment. *Journal of the Institution of Water and Environmental Management* **6(4)**, 475–84.

Whitney, J.B. 1985: Integrated economic–environmental models in environmental impact assessment. In Maclaren, V.W. and Whitney, J.B. (eds), *New directions in environmental impact assessment in Canada*. Toronto: Methuen, 53–86.

Whyte, A.V. and **Burton, I.** (eds) 1980: *Environmental risk assessment*. SCOPE 15. New York: Wiley.

Wiesbecker, L.W. and **Porter, A.L.** 1993: *Issues in performing technology assessment for developing countries*. Proceedings of a UN Expert Group Meeting on Technology Assessment, Monitoring and Forecasting. Paris: United Nations.

Wiggins, S. and **Shields, D.** 1995: Clarifying the 'logical framework' as a tool for planning and managing development projects. *Project Appraisal* **10(1)**, 2–12.

Willig, J.T. (ed.) 1994: *Environmental TQM*, 2nd edn. (1st edn 1992). New York: McGraw-Hill.

Winpenny, J. 1991: *Values for the environment: a guide to economic appraisal* (ODI). London: HMSO.

Wisemann, G. 1994: Europe braces for eco-audit: standardizing the acronyms. *Chemical Week* **154(13)**, 54–6.

Woolston, H. (ed.) 1993: *Environmental auditing: an introduction and practical guide*. London: British Library, Science Reference and Information Service.

World Bank 1985: *Manual of industrial hazard assessment techniques*. Washington DC: World Bank.

World Bank 1991a: *Environmental assessment sourcebook*, vol. 1: *Policies, procedures, and cross-sectoral issues*. World Bank Technical Paper 139. Washington DC: World Bank (Environment Department).

World Bank 1991b: *Environmental assessment sourcebook*, vol. 2: *Sectoral guidelines*. World Bank Technical Paper 140. Washington DC: World Bank (Environment Department).

Yap, N.T. 1990: Round the peg or square the hole? Populists, technocrats and environmental assessment in third world countries. *Impact Assessment Bulletin* **8(1–2)**, 69–84.

3

IMPACT ASSESSMENT: ROLE AND RELATIONSHIP WITH PLANNING, POLICY, POLITICS AND MANAGEMENT

'To decide means to take actions
And actions rock the boat
And if your actions don't succeed
You may not stay afloat'

(Anon. – the decision-maker's plight)

How impact assessment fits into planning and management

There are two extreme views of impact assessment. It can be seen, first, as a 'technocratic' planning tool or, second, as a politicized process that improves decision-making (Formby, 1990). The reality is perhaps best summed up by Hall (1980: 86), who noted: 'EIA [environmental impact assessment] is unlikely to become the long-sought-after scientific tool which, plugged in, will solve all planning problems . . . neither is it a medieval philosopher's stone . . . which promises everything and yields nothing. Rather it is a useful additional battery of techniques that can aid the planner – and also the political decision maker and the affected public – to conduct a more rational, more structured debate about the effects of proposals for development.'

Anything is better than *laissez-faire* consideration of environmental and social issues, which was the usual situation before 1970 when environmental impact assessment appeared.

When environmental impact assessment was first introduced there was widespread scepticism about its value. Many saw it as delaying progress and increasing the costs of development, of benefit mainly to assessors and lawyers involved in related litigation. Since the 1970s there has developed a much wider acceptance of the need for environmental and social concern, and impact assessment is today seen as a very valuable input to planning and decision-making. However, there is still opposition, and a good deal of half-hearted support or attempts to use environmental impact assessment to manipulate development.

The degree to which environmental awareness is integrated into planning and policy-making is still unsatisfactory, and impact assessment could help correct this lack of integration (Kozlowski, 1989; Lichfield, 1992a; 1992b). Closer integration of impact assessment into planning and environmental management should help counter many of its problems. There should be more use of impact assessment as a 'planning tool' and as a 'decision-making tool' (Armour, 1990: 4; *Impact Assessment Bulletin* 1990: **8(1–2)**, special issue entitled: 'Integrating impact assessment into the planning process: international perspectives and experience'; Graham Smith, 1993: 12). Note that 'integration' is an imprecise term, and can be interpreted in different ways. For example, it is possible for impact assessors to adopt either an 'anthropocentric' or an 'ecocentric' stance, that is, to put concern for humans before or after their concern for the environment respectively. The latter had almost always been the case before 1970.

Impact assessment can warn of impacts that threaten environmental quality, life-support systems, human welfare and social stability, and can help unite the science of environmental analysis with the politics of resource management (Graham Smith, 1993: 95). Impact and risk assessment should ideally be a pre-development, predictive evaluation of alternatives, but in practice is often the selection and justification of an already chosen development option (Suter, 1993: 6).

Impact assessment can help ensure that planners and decision-makers are more accountable for their actions, and thus encourage them to think carefully about proposals. As has already been stressed in Chapter 1, attempts to achieve sustainable development will be difficult, if not impossible, without effective impact assessment.

Retrospective impact assessment (that initiated after a development is under way) provides opportunities for documenting the effects of development, which in the past have seldom been well recorded. Such hindsight is valuable for understanding scenarios and supports better planning and decision-making. There are frequently situations where it is difficult to understand a problem, and retrospective assessment can help, ensuring that responses are directed at causes rather than symptoms. Retrospective 'damage assessments' are widely used by the insurance industry for loss adjustment and by those seeking to establish compensation payments. These 'postdictive' (as opposed to predictive) studies can be simple checks, such as counts of dead animals or lists of damage, or (less often) sophisticated modelling to establish how and why impacts occurred. So far, especially in developing countries, impact assessment has mostly been applied to projects already under way.

An expanding field is *issue definition assessment* – the evaluation of a perceived environmental or socio-economic problem that has yet to be demonstrated to have actually happened. A good example of such a problem is global warming. Issue definition assessment determines (usually depending a good deal on expert judgement) the state of knowledge and shows what research is needed or whether the risk can be safely dismissed (Suter, 1993: 8).

There are, as mentioned earlier, sceptics. Some feel that many environmental impact assessments are of little real value in practice, that they represent a procedural burden and that there is little consensus on how to conduct impact assessment (Hyman and Stiftel, 1988). Slocombe (1993: 291) felt that rather than the integration of biogeophysical concerns into development planning and urban and regional planning, a separate discipline of environmental planning had emerged, in part at least because of impact assessment.

After more than two decades of evolution, impact assessment is still not a perfect means of building bridges between environmental concern and development planning (including planning authorities, private developers and administrative bodies), but it could be, and little alternative seems to be available (Hollick, 1986: 157; Armour, 1991: 30; Graham Smith, 1993: 12, 328). Adams (1990: 149) observed that it is 'one thing to have an EIA [environmental impact assessment] commissioned and carried out, but quite another to integrate it into decision making'.

Impact assessment has offered some countries, especially less developed countries, a way of introducing a degree of environmental planning that would otherwise probably have been unlikely. Since the 1970s there has been adoption in a wide range of states, including the former communist bloc with its centralized planning systems. Impact assessment is evolving – and in a complex and dynamic world should be more proactive, dynamic and adaptive, emphasize evaluation rather than inventory, be better integrated with environmental management, and adopt a broader outlook than one restricted to the project level. Impact assessment faces increasing challenges, which include:

- serious transboundary impacts (effects involving more than one state), some of them major global problems;
- impacts that are often complex and potentially serious;
- the need to support the goal of sustainable development;
- problems of jurisdiction (can assessment extend to cover all ministries, another state, internationally owned resources, etc?);
- the need to identify and consider cumulative impacts.

How impact assessment fits into planning in the sense of its relationship to land use planning and pollution control can be easily illustrated (*see* Fig. 3.1); how it interrelates with the planning and management process is less easy to pin down. Planning has been defined in many ways: for example, as

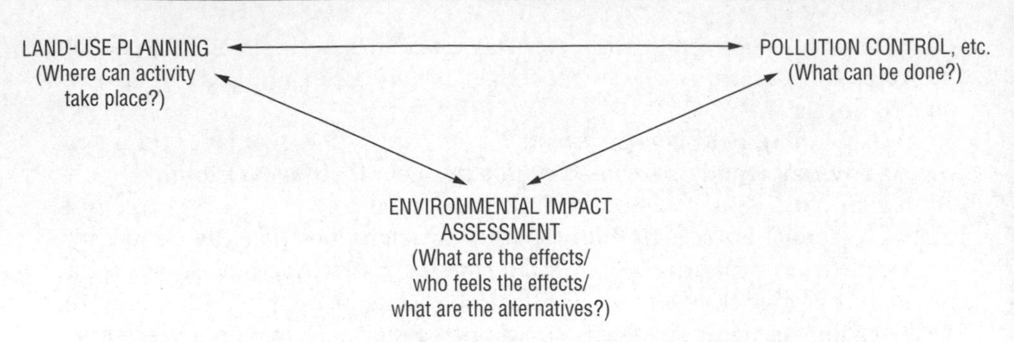

Figure 3.1 How environmental impact assessment 'fits' into planning

Box 3.1 Types of planning approach

- 'Classic' prescriptive approach (comprehensive and rational)
- Incremental (disjointed incremental) approach
- Adaptive approach
- Contingency approach
- Advocacy approach
- Participatory and consensual approach

Based in part on Briassoulis (1989)

systematic problem-solving and decision-making, or as a process that seeks solutions to problems and needs or one that develops actions that will satisfy goals and objectives. There are many models of planning (*see* Box 3.1). The dominant 'classical' prescriptive planning process, an ideal seldom fully achieved in practice, involves clarification of needs, specification of goals and objectives; development of alternative means to attain goals or objectives, assessment of costs for each alternative, and selection of the most promising means (Faludi, 1973) (*see* Fig. 3.2). Planning can be crudely divided into 'blueprint planning', which expects little change and has clumsy, inflexible implementation, and 'process planning', which accepts that the plan is likely to be modified during implementation. The latter approach makes more sense where there is uncertainty of data, management, etc., and is suited to environmental planning (Hollick, 1981: 80). Impact assessment tends to function as if it is dealing with blueprint-type planning, giving a 'snapshot' assessment. There have therefore been arguments for more adaptive approaches (e.g. Holling, 1978); *see* the discussion of adaptive environmental assessment and management later in this chapter.

There are many reasons why impact assessment is often marginalized from planning and management (Adams, 1990: 149–60):

- It is unusual to find an ecologist or sociologist in charge of the planning or management process; an economist or lawyer is more likely.
- Impact assessment is not seen as central, so is not fully integrated into planning and decision-making.

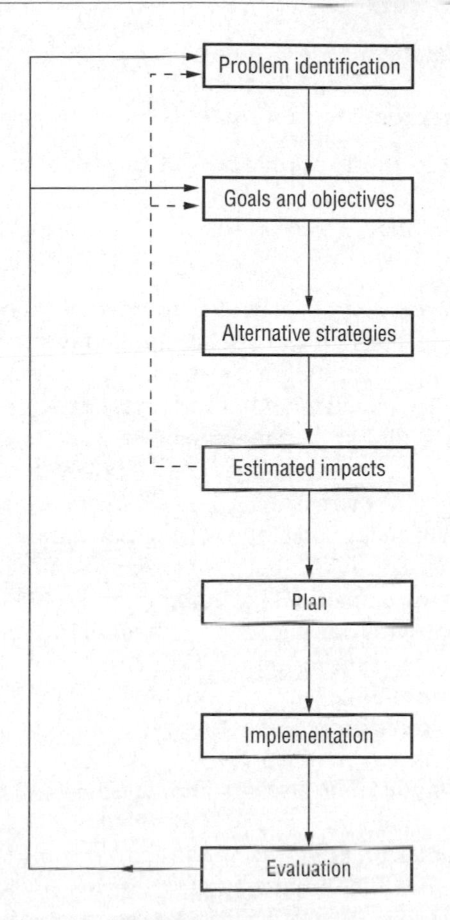

Figure 3.2 The rational comprehensive model of decision-making – one of the best known prescriptive models
Source: Based on Graham Smith (1993: 58, Fig. 4.2)

- Development contracts are bid for, which means keeping costs down, and this may mean that impact assessment is seen as peripheral and therefore underfunded or cut back.
- Managerial classes in developing countries seek to control development and select approaches that profit them, but which may also fail.

Graham Smith (1993: 77–94) argued that impact assessment could be integrated into 'classical' planning by a combination of three approaches:

1. use of synoptic planning with impact assessment as the basis for a sequential approach. Synoptic planning is a rational, comprehensive approach whereby proposals for the future are devised and assessed via a series of steps:

   (a) identify needs;
   (b) specify grades and objectives;

   (c) develop alternative means to attain each goal;
   (d) estimate the costs of each alternative;
   (e) select the most promising alternative.
2. develop a manual to guide application of impact assessment in planning;
3. integrate planning and impact assessment by adopting a scientific approach.

Impact assessment is not only a set of environmental management 'tools' that can be well used or misused; it can also be applied to plans and the planning process – that is, it can be an integral, 'shaping' part of planning and management (Herington, 1979; Armour, 1990). Lindblom (1959) argued that much planning followed the disjointed incrementalism model – a sort of scientific 'muddling through', in which sudden changes in policy punctuate ongoing slight sequential shifting, and there is a high level of consensus. Hollick (1981: 83) and Fuggle (1989: 37) have argued that disjointed incrementalism may work with policies, programmes and plans that allow reversal of mistakes; however, environmental problems are sometimes irreversible, and so it is not a good idea to link impact assessment with this approach to planning. Projects that escape impact assessment or that *individually* appear to offer no adverse impacts might have an incremental (cumulative) impact, such as the gradual unrestricted spread of housing to a scenic area. Impact assessment therefore needs to be used at levels in addition to project level (*see* the discussion of the tiered approach later in this chapter) and perhaps should be linked to fields such as land use planning, pollution control, etc. (Wood, 1980).

There are many variables that determine the character and effectiveness of impact assessment (*see* Box 3.2). Impact assessment involves value assumptions and can be influenced by politics and powerful interest groups. It should therefore be practised in a way that as far as possible is 'transparent' (Clark, 1988; Beattie, 1995: 113). If the public opposes environmental control, or if politicians are out of step with public wishes for impact assessment, clearly its execution will be less than enthusiastic. A holistic approach, optimization of development, participation by the public, etc. are goals, not reality, as most environmental planning is too fragmentary to address these issues sufficiently (Westman, 1985: 91).

Some developers see impact assessment as a hurdle to be jumped in the planning process, rather than an aid; others (possibly the majority at present) regard it as a 'necessary evil', a technique or tool that might have some benefit but that also causes delay and increased expenditure; yet others accept it as a vital process, more than a technique or tool, that warns of risks, feeds into contingency planning and aids the quest for sustainable development (Ortolano and Shepherd, 1995; Glasson *et al.*, 1994: 12).

The integration of impact assessment and planning must be done with care. The fear has been voiced that there is a risk that existing, obstructive planning attitudes might dominate and that, if kept separate, assessment may overcome inertia and have more power. There have been criticisms that

Box 3.2 Variables that help determine the character and effectiveness of environmental impact assessment (EIA)

- Is EIA independent of decision making or integrated within it?
- Is EIA an administrative or legislative procedure?
- Is EIA mandatory (with enforced rules and procedures) or encouraged (by recommended guidelines)?
- Is EIA an addition to pre-existing procedures, an alternative or a replacement?
- Is EIA respected by decision-makers, the public, etc.?
- Is EIA applied at project, programme, plan or policy level?
- Is EIA applied before development decisions are made, during implementation or after completion?
- Is EIA applied at local, regional, federal, national or international level?
- Is EIA confined mainly to the assessment of physical impacts or does it adopt a total, comprehensive, holistic approach (including social and economic factors and, ideally, a consideration of all facets of a development)?
- Does the EIA provide adequate information; is it accurate?
- How has EIA been promoted. Was it embraced by planners or engineers as a way of ensuring professional standards or as part of planning? Insisted on by government? Demanded by an aid agency or funding body as a condition for making a loan or grant? Promoted by a non-governmental organization? Ordered by a court? Requested by local people? Mutually requested by more than one of the above?
- Is EIA driven from the 'top down' or the 'bottom up'?
- Do different bodies involved in EIA co-operate with and trust each other?

Based in part on UNECE (1981: 3).

impact assessment is self reinforcing prophecy. Integration is difficult; it is worth noting that:

- disciplines approach problems in different ways, probably vary in terminology and may have incompatible data;
- there may be 'disciplinary chauvinism' (Burdge and Opryszik, 1983);
- development planning is dominated by economists and urban and regional planners, who may find it difficult to relate to environmental planners;
- there may be no suitable integrative framework;
- many of those active in environmental impact assessment, social impact assessments and related fields are not trained as planners;
- many of those involved in impact assessment and related fields do not belong to professional planning bodies.

One or more of these problems may hinder real integration. It is probably easier to integrate impact assessment into project planning than at the programme or policy level. So far, it has mostly been applied to projects (Wood, 1995: 8), although there has been some application at programme and policy level, for example in foreign aid programmes and policies, transport programmes and policies, energy policy and trade policy (*see* the section on strategic environmental assessment later in this chapter). There has been some interest in programme and policy evaluation in business and management studies. Beckler and Porter (1986) called for more standardization of programme evaluation.

Figure 3.3 illustrates an ideal project planning helix, showing how impact assessment relates to policy and longer-term planning (*see* Fig. 3.3 inset). In an ideal world, as skills develop the helix should taper towards the right. In practice, planners have often been slow to learn from hindsight, so it is left untapered.

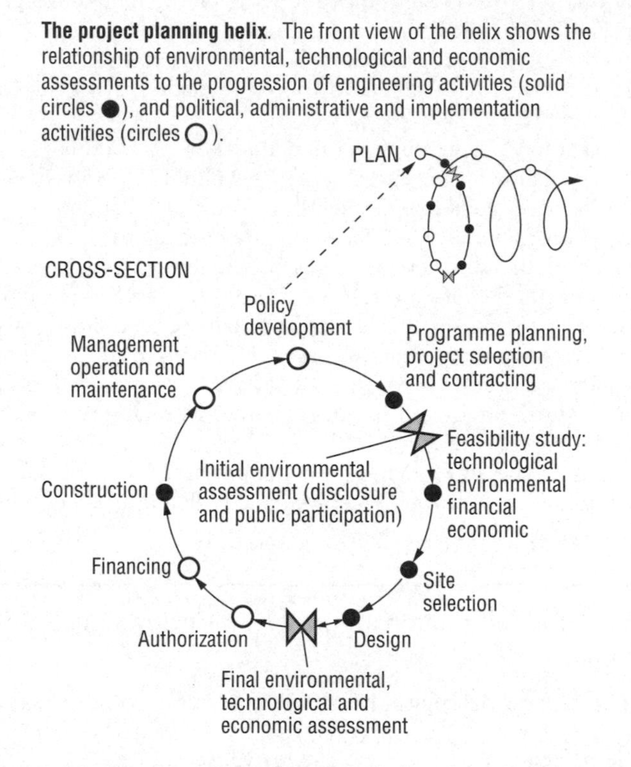

FIGURE 3.3 An idealized project planning helix. Successive cycles, at least for similar types of project, policy or programme, should be 'tighter' as hindsight and data collection mean fewer difficulties and errors
Source: Based on Barrow (1987: 66, Fig. 2.7)

What does impact assessment need from a planning system? Clearly, a supportive policy and planning system and a supportive implementation and management system – although it is really impact assessment that is, or should be 'supporting' those systems. Three things seem to stand out as vital if impact assessment is to be effective:

- rational decisions on what deserves impact assessment and what escapes;
- a competent, incorruptible review body to oversee and assess the impact assessment process and environmental impact statements;

- enforcement of the recommendations of environmental impact statements: the agency must be strong and skilled enough to prevent the statements being ignored or modified.

WHEN AND AT WHAT LEVEL SHOULD IMPACT ASSESSMENT BE DONE?

Impact assessment should be at the heart of the environmental management process. It should be integrated into the planning and design process, and initiated before development decisions are made so that alternatives can be explored, thereby making available its full potential as a preventive measure. Often that is not the case, and impact assessment is applied at a later stage and is poorly integrated into planning and management. There are, however, some developments that encourage optimism. For example, some engineers have called for environmental impact assessment to be incorporated into the design process (Brown, 1992).

As discussed earlier, the bulk of assessment is at the project level, at the local scale and carried out during the implementation stage. Clark and Herington (1988: 3) are among those who have argued that impact assessment should be adopted at all levels of the planning process (assessment at different levels should be integrated; *see* the discussion of tiered environmental impact assessment later in this chapter). It would also make sense to apply impact assessment to the development of research programmes, and some aid donor agencies have begun to do this (Kennett and Perl, 1995).

Wood (1988: 107–14) examined the difficulties of applying impact assessment above project level (to programmes, policies or plans), highlighting the following (modified by the present author):

1. Plans often relate to a number of possible developments in various places, which makes matters much more complex than when there is a single project focus.
2. In view of 1 above, it is difficult to be precise in predictions, which makes impact assessment much more of a challenge.
3. There is less hindsight experience of plan, policy and programme impact assessment than there is at the project level.
4. Scoping at levels above project impact assessment is difficult; what should be focused upon? (For a discussion of scoping, *see* Chapter 4)

Given these difficulties, Wood felt it better to integrate impact assessment into planning than to modify environmental impact assessment to deal with plans.

Figure 3.4 illustrates the chronological sequence of actions within a comprehensive impact assessment system. Although project-level impact assessment initiated after development decisions have been made is far less satisfactory, it still has a useful role in stock-taking and the identification of

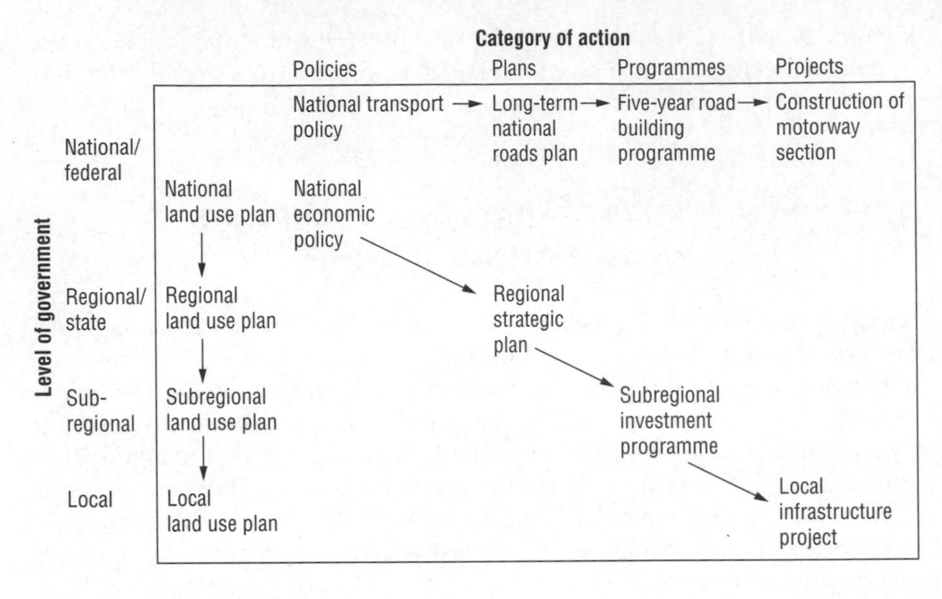

FIGURE 3.4 Chronological sequence of actions within a comprehensive environmental impact assessment system.
Source: Based on Wood (1979: 15a, Fig. 2)

remedial actions (Armour, 1990: 5). A useful categorization of impact assessment, according to timing of application, motivation and role is the following:

- Genuinely proactive, carried out *ex ante* (i.e. before a decision is made to develop). Gives the full advantages of impact assessment.
- Perfunctory endorsement – the decisions having been made, the impact assessment is undertaken to aid licensing, quell local opposition or opposition from a non-governmental organization, achieve political legitimation, or satisfy funding agency demands. Identifies problems, but is probably too late to allow a change of development approach, especially if large-scale engineering is involved.
- 'Cosmetic' – again, the decision to develop has been made. The timing is similar to that of assessments in the previous category but the motive is more dubious. May deliberately hide problems or threats (often presenting them in an environmental impact statement in an obscure way), or even set up 'straw men' – diversionary issues to attract attention away. Probably the most serious abuse of impact assessment.
- Retrospective – undertaken too late to allow consideration of alternatives or impact avoidance. Provides an inventory of what has happened and what might yet come, and clarifies problems, so assisting remedial action (UNECE, 1990). Documents the situation so that the planners and decision-makers have a better chance to learn from hindsight.

Impact assessment should not be a single event, nor a 'bolt-on' extra, which in developing countries is often entirely in the hands of a team of consultants who are not familiar with local conditions. It must not be divorced from the planning and decision-making process; to function fully, it must be carefully integrated. Armour (1991: 30) felt that impact assessment had focused too much on procedures and should be better integrated with planning and management if it was to give better results. Clark and Herington (1988: 3) described impact assessments as 'reformist tools' that can improve planning and decision-making as well as make environmental predictions.

Since 1970 a diversity of approaches to impact assessment have been adopted by various countries and bodies. Details may differ, but the core aims and difficulties faced are broadly the same. Impact assessment has had to adapt to a wide range of legal systems (Sheate, 1994), customs, expectations, degrees of public freedom and involvement, levels of media attention and freedom of information access. Impact assessment may be commissioned by a variety of bodies, including national or local government, funding bodies, the developer, public 'watchdog' organizations and non-governmental organizations. (Heuber [1992] examines input from the last of these.) Assessors have access to different levels of expertise and funding, and face varying time constraints and levels of supervision. The assessment may make use of one or more groups of assessors, such as consultants, specialist academics, government assessors and aid agency assessors. There is thus a diversity of approach, competence, independence and resourcing.

An assessment team may consist of employees of a developer, consultants hired by the developer or another body (e.g. a government or aid agency); it may be independent or it may be a mixture of two or more of these groups. Teams may adopt a multidisciplinary approach, which implies the need for a team of individual specialists who must somehow communicate on as near as possible equal terms, or an interdisciplinary one, which implies more interchange, perhaps overlap of duties and stronger co-ordination).

Many impact assessments, and associated approaches, are conducted by consultants. Consultants may, in relation to affected people and development decision-makers, either be 'insiders' or 'outsiders', usually the latter. There are both problems and benefits associated with using consultants. For example, a consultant who is not an 'insider' may be able to stand back and ask what others would not dare; a consultant may be the best in their field; a consultant is able to give intensive, full-time attention to a task. The negative qualities of consultants may include unfamiliarity with the local environment and people's needs; perhaps a tendency to present findings that please those commissioning the assessment; and unavailability soon after completion of an assessment (*see* Save the Children, 1995: 217–24 for a fuller review of these strengths and weaknesses). There may well be advantages in mixing local assessors and consultants so that they interact and learn from each other.

It is not unknown for agencies or companies to try to persuade a 'host' government to modify or nullify assessment regulations (renegotiate regulations – 'regneg') to make assessment easier for them or even avoidable.

The co-ordination of assessment can at least ensure that the team(s) involved:

- have the same information on the assessment (especially if there are phases – those doing early studies probably have a different conception and face different problems from those that they or others face later);
- use common, recognized units in data collection and assessment;
- share a common conception of the task and ethos.

The impact assessment process itself can cause impacts: research into local attitudes and needs may generate local concern, land speculation, opportunistic squatter settlement, and so on. Mitigation efforts recommended by an environmental impact statement have been known to generate more problems than the predicted impacts would probably have caused (McCoy, 1975; Clark, 1980).

PUBLIC INVOLVEMENT WITH IMPACT ASSESSMENT

The 'public' may be the people affected by a development, or it may be pressure groups, members of non-governmental organizations or quasi-statutory bodies ('public watchdog committees'), the media, etc. who are most affected. Sometimes the public are locals, but they could live in distant places: the protest against the French nuclear tests on a Pacific island in 1995 is an example of the latter, and protesters were sophisticated enough to make use of the Internet to pass on information and rally support. The 'public' can be a wide spectrum of different groups, with varying resources and skills and different approaches to involvement. It is possible for the same people to be encountered in an impact assessment in more than one way, perhaps as members of a group of fishermen, as residents of a village, as members of a particular age or sex group, as belonging to a particular economic group, as part of a cultural group. Women usually make up a considerable proportion of the public affected by development but are often not consulted.

The degree and nature of public involvement varies a lot, even between democracies. For example, in the UK the impact assessments for the Channel Tunnel were mainly presented to Select Committees of Parliament, so there was comparatively little public involvement (Bartlett, 1989: 34–6). When expert opinion is used, as is often the case in environmental impact assessment, social impact assessment, eco-auditing, etc., it can be divorced from the real-life experience. There has been little co-ordination of public involvement, consequently it tends to be *ad hoc* (Roberts, 1995; FEARO, 1988).

Public involvement can pose problems:

- It can be manipulated by developers to legitimize their decisions.
- Public support for a development does not mean that it will have no negative environmental, social or economic impacts; assessment may need to convince people that they should not support a proposal (Burdge and Vanclay, 1996: 73).
- A public may lack education, current awareness, or some other skills, necessitating careful involvement, perhaps preparation.

- An informed and educated public might resort to litigation or effective protest.
- Administrators may have to overrule short-term interest-group wishes and local demands to satisfy longer-term, national- or global-scale welfare or environmental quality, maximize equity and achieve sustainable development.
- It may be necessary to manage public involvement according to the stage of development, for example at early, middle and late stages. At each stage public awareness and interest are likely to differ.
- There may be risks of breach of confidentiality or disclosure of trade or strategic secrets.
- Professionals and decision-makers fear that public involvement may mean they lose control of decision-making.
- Effective involvement can demand skilled briefings from the assessors, and the ability to brief effectively is often a rare skill.
- Speculation can occur; for example, if people become aware of an urban improvement or resettlement scheme they may buy land in the hope that it rises in value, or try to join the ranks of those to be compensated, or exploit the relocatees.
- There may be a fear that some groups will protest if they are told too much.
- People often adopt a 'not in my backyard' (NIMBY) attitude – narrow self-interest that prompts them to oppose or refuse to co-operate with development in their locality but to accept it elsewhere. Conflict between groups may result from this.
- The public may lose interest during a long assessment process.
- Special-interest groups may manipulate public opinion.

It is worth noting that some of the difficulties just listed are impacts caused by the process of impact assessment.

In some countries, special interest groups, even some with relatively few members, can be vociferous, and sometimes have had considerable success in disrupting and delaying development of roads, airports, military installations, etc. For example, construction work at Narita Airport, Tokyo, was delayed by about 5 years by bands of protesters in the late 1960s and early 1970s. Hyman and Stiftel (1988: 43) observed, 'special interest groups find it economically worthwhile to muster political clout, while the rest of society lacks the incentive and resources to effectively counteract those groups'. Some assessors apply the expression 'collective action problem' to this sort of disproportionate special-interest influence. The assessor faces two challenges: getting representative views from the public, and communicating impact assessment results to the public.

People participate for different reasons. These include a sense of duty; a chance to promote themselves; a desire to advance wider causes; hope for material gain; and religious or political motivation. It is possible to split public opinion into two likely categories: community public values and consumer values (i.e. individual preferences). The assessor has to decide what cross-section of the public will be manipulated, informed or consulted: all or elected representatives; selected representatives; or special-interest groups.

The public may be involved in assessment, planning and management without necessarily participating or even being consulted; that is, data may be gathered from them, information disseminated or there may be public relations activity. Canter (1996: 587–622), in a review of public involvement in impact assessment, notes a spectrum ranging from developers controlling the citizenry to, at the other extreme, partnership and empowerment. Between the extremes there may be tokenism, manipulation, one-way communication, two-way communication but without the public having influence and full participation (Sewell and Coppock, 1977; Hughes and Dalal–Clayton, 1996). The involvement may be direct or indirect.

Assessors and those commissioning assessment may forget that the public can often tell if they are 'rubber-stamping' decision-making and react accordingly. When public involvement is supported, the assessors must find effective means for achieving it (Connor, 1977). It is possible to involve selected groups or individuals without the public in general knowing about or influencing how things are done (Fagence, 1977; Hollick, 1986: 172). Degrees of involvement may vary at different stages of a development: there may be little public involvement in planning but some during implementation and running, or there can be involvement at all stages. Participation or consultation might – it is now widely argued that it should – take place at one or all of the stages of screening, scoping, data collection, assessment, preparation of an environmental impact statement or final release of the statement. Often it is just at the last of these stages that consultation or participation ocurs, though the USA does involve people at the scoping stage. Consultation and participation are likely to be more difficult in the case of comprehensive, integrated and strategic assessment because of the complexity of impacts, alternatives, etc. (Blahna and Yonts-Shepard, 1989).

Unfortunately the view that 'EIA [environmental impact assessment] is not EIA without consultation and public participation' (Wood, 1995: 225) is still held by far too few planners and decision-makers. The USA has freedom of information laws, notably the Freedom of Information Act, and, from the start, the National Environmental Policy Act encouraged citizen participation. This is probably why so much litigation related to environmental impact assessment has occurred in the USA as compared with other nations. The Canadian government has also been at the forefront of involving the public in assessment (Roberts, 1988; Gariepy, 1991). France, on the other hand, in spite of a history of citizen involvement in government, has been relatively slow to develop public participation in environmental impact assessment (Ortolano and Shepherd, 1995: 11). The UK has a history of informing and consulting on many of its major projects via public inquiries. Unfortunately, these tend to involve people once the assessment is well under way or completed, and can be slow: in the case of the Roskill Commission Inquiry in 1971 into the site of the third London airport there were

258 days of public meetings. Also, administrators can usually control public meetings to get their own way.

Non-democratic governments, in contrast to the above cases, are unlikely to worry too much about public involvement.

What are the advantages of public involvement?

- The quest for sustainable development is less likely to break down where the public support efforts and are kept informed and, ideally, involved. Impact assessment and community development approaches have much to offer to those seeking sustainable development (Ghai and Vivian, 1992).
- People may provide valuable data and improve the monitoring process, especially if they are involved early enough (Yap, 1990). There have been many cases where the public could have warned developers of risks and needs, but were not consulted.
- Consultation may be needed to determine alternative development options.
- Consultation may give assessors more chance of modifying developments so that they work.
- Public involvement may make planners and decision-makers more accountable and more careful.
- Involvement may 'empower' people and they may gain confidence and skills from participation (Yap, 1990; Graham Smith, 1993: 66).
- An involved public might be more likely to support a development; if uninformed it may neglect or oppose it.
- Involvement can be vital in order to resolve conflicts.
- An advantage for the developer is that participation can help 'legitimize' decisions.

Public involvement may not mean communication with the public as a whole. It can take place via representatives; special-interest groups, such as women's groups; non-governmental organizations; neighbourhood groups, such as residents' associations; youth groups, labour unions, etc.; or a mediator or mediators. The media too may act as an intermediate 'filter'. The media are adept at buffering or motivating a public, although they vary in political, moral or social stance, resources (financial and access to modern networks) and degree of freedom.

Indirect public involvement through a trained mediator is common practice in the USA (Lake, 1971, Gilpin, 1995: 70) and also in Canada (Sadler, 1993), where provisions for mediation were made in the Environmental Assessment Act 1992. Mediators can sometimes be effective at solving conflict and deadlocks by assisting parties to reach agreement through negotiation. Often the negotiations take place in an intensive group meeting in which all parties, on an equal footing and faced by a time constraint, seek to find answers to problems. Such a meeting is termed a *charette* (Lake, 1980; Amy, 1983a, 1983b; Curtis, 1983). There has been considerable development of environmental mediation in North America, where there are journals devoted to the field, such as the *Canadian Environmental Mediation Newsletter* and *Disputes Resolution Forum* (Susskind *et al.*, 1987).

Wood (1995: 228) listed means for achieving public participation, including:

- questionnaires/surveys;
- advertisements;
- interviews with the public (which may allow the interviewer to see non-verbal clues);
- leafleting;
- media statements/presentations;
- exhibitions;
- telephone phone-in or 'hot-line' for advice;
- community liaison staff;
- public meetings/inquiries;
- community advisory committees;
- circulation of documents such as environmental impact statements, (this may be limited by cost – which is often over £100 a copy) or, to cut costs, précis documents (Hollick, 1986: 170);
- simulation exercises;
- the Delphi technique (*see* Chapter 2);
- public referendum;
- having public representatives on planning and decision-making bodies.

The USA circulates impact assessment documents, making environmental impact statements available to the public for comments for a period of 45 days, and Australia does so for 28 days (Westman, 1985: 36). The World Bank (1991) asks (but does not compel) its borrowers to circulate environmental impact statements and suggests release at mid-point in the review. Often, when lenders (usually in less developed countries) leave public involvement to the recipient government, the task is neglected. To try to get round this problem, at least partially, the World Bank's Operational Directive 4.01 requires the Bank to consult local non-governmental organizations. Recently, some researchers have been exploring the possibilities offered by interactive multimedia for public participation (Hughes and Schirmer, 1994). At present it is probably fair to say that most impact assessment neglects public participation and relies on expert opinion (Hyman and Stiftel, 1988: 52). And, when public opinion is considered, it is probably best to show the range of views, not aggregate them all.

Impact assessment can be used to increase government control over development or to try to ensure that developers assume more responsibility (Clark and Herington, 1988: 4). What is needed is an approach to impact assessment that ensures that public participation is not marginalized – indeed, one that uses local people as a data source and means of control of the environmental impact assessment process. Rapid rural appraisal and participatory rural appraisal may offer opportunities for involving people, and may also help ensure adequate multidisciplinary study. Participatory impact assessment and management have already been developed from such roots (Yap, 1990).

Save the Children (1995; 207–16) provides a concise evaluation of the effectiveness of public participation.

IMPACT ASSESSMENT COSTS AND THE QUESTION OF 'WHO PAYS?'

Estimates of costs can be subjective because impact assessment is to varying degree integrated into planning and management, so that full costs can be hidden, and also because each assessment is, to some extent, unique. A 'successful' assessment is one that costs less than the problems it prevented, but it is likely to be difficult to establish and place a value on what was prevented (Wood, 1995: 253). There have been enough costly mistakes that might have been avoided by applying impact assessment to make impact assessment attractive even if it is only partly successful. A good example of avoidable failure is the Tanganyika Groundnut Scheme. In this scheme, the UK tried during the late 1940s and early 1950s to grow groundnuts on a large scale in Tanganyika (which now, together with Zanzibar, forms Tanzania) in East Africa, with modern techniques making use of the savannah soils. Huge investments were made in the hope that groundnut oil would be produced to help meet the UK demand for fats and margarine. Unfortunately, little effort was made to check the scanty rainfall records and soils data. The land was prepared but rainfall 'failed' and soils were 'unsuitable', so the scheme was not a success.

There are aspects of impact assessment that are difficult to cost: the amount of time added to a development (if any); improved public awareness and perhaps support for authorities; a 'green' image for the developer; and employment for assessment workers. Some costs would probably occur anyway even without impact assessment, for example planning delays.

It is possible to add some or all of the following in order to obtain an idea of costs:

- data collection/studies;
- report writing and circulation;
- delay to development costs;
- legal costs;
- mitigation costs;
- cost of difficulties caused by assessment (particularly changes to development plans).

Factors that affect costs include (Kakonge and Imevbore, 1993: 302–3):

- whether it is application of assessment to a new field;
- remoteness and difficulty of the site (remote and difficult sites push up costs);
- size of the project or programme (there may be economies of scale);
- availability of data;
- quality of assessment required (careful versus 'quick-and-dirty').

In practice it is seldom possible, but the ideal would be to conduct an impact assessment for at least 12 months so that seasonal conditions and cycles become apparent. It is scoping, discussed in Chapter 4, that shapes an impact assessment and largely determines costs. Boundaries between administrative regions may affect scope of assessment rather than 'best practice'.

A study of 50 construction projects in the USA by the United Nations Economic Commission for Europe (UNECE 1981: 59) showed that on average an environmental impact statement costs less than 0.19 per cent of total project cost, and concluded that impact assessment led to considerable savings by averting problems. Hollick (1986: 171) looked at two lengthy and 'expensive' assessments – the Berger Inquiry, Canada, and the Ranger Uranium Inquiry, Australia – and concluded that they were within the range 0.1 to 1.0 per cent of total project costs, and that in general, impact assessment and related costs seldom exceed 3.0 per cent of total project or programme costs. Wathern (1988: 283) estimated that the environmental assessment for the Mahawelli Development Project in Sri Lanka involved 95 man-months of assessment between 1979 and 1981, and cost 0.08 per cent of the total project budget – probably a relatively expensive exercise. The cost of an environmental impact assessment for a large hydro project in Thailand has been estimated to have been 0.3 per cent of total costs, that for a Thai industrial estate 0.48 per cent, and for a Thai pulp-mill 0.034 per cent (Nay Htun, 1988: 233). Clark (1984) suggested that a typical figure is 0.5 per cent to 0.2 per cent of total project costs. Lohani and Halim (1987) presented costs for selected impact assessments in Thailand, the Philippines, China and Vietnam; the values range from 0.01 to 0.48 per cent of total project costs. Coles *et al.* (1992) suggested environmental impact assessment costs in the UK ranged from 0.000 025 to 5.0 per cent of total project costs. Wood (1995: 254, 265) felt that environmental impact assessment costs, as a proportion of total project costs, generally appeared to range from about 0.1 to 1.0 per cent with an average of about 0.5 per cent and was convinced that environmental impact assessment benefits outweigh costs. Gilpin (1995: 25–6) cites direct costs of environmental impact assessment in Australia in 1990 that are generally less than 1.0 per cent of total project costs; as the Australian government has, since 1991, allowed a tax deduction to cover capital expenditure on environmental impact assessment, this is not at all discouraging. Virtually all the available data are for costs of environmental impact assessment applied to projects, rather than programmes or policies.

A significant component of impact assessment costs is data acquisition. This means that once a database is established and kept updated, successive impact assessments spend less on data; the background information has already been collected and it is mainly just project- or programme-specific information that is needed. Developing countries are less likely to have adequate data, which increases impact assessment costs. Faunal and floral studies can be demanding of time and expertise. Delay can be a major cost (which might be countered by submitting draft environmental impact statements) but tends to be less of a risk as databases are built up (Hollick, 1986).

Impact assessment provides employment: in California alone there are several thousand professionals involved in environmental impact assessment (Wood, 1995: 257), something that should be weighed against costs.

The cost of impact assessment having been discussed, the question arises, 'who pays?' Increasingly the 'polluter pays' principle is applied, according to which the developer bears the direct cost. There are problems with this: an investor faced with having to pay for assessment may be tempted to locate the development where impact assessment is lax or not required. Sometimes an authority is prepared to accept a poor impact assessment and adverse impacts in return for job creation or bribes, or for strategic advantages. A country may be tempted by the lure of tied aid (*see* the Glossary) and insist on less than stringent impact assessment. In the end, one is forced to ask, 'Can the polluter (i.e. developer) be trusted to pay for a good job?' Table 3.1 lists ways in which assessment may be paid for.

TABLE 3.1 Who pays for assessment?

1. 'The polluter pays' – the developer bears the *direct* cost and may be allowed to commission assessors.
2. The developer pays *via taxes* or a *levy* for an independent assessment body.
3. The government pays via taxation – local or national; for example, the USA's National Environmental Policy Act made provision for the federal authorities to pay.
4. A funding body, such as a lending bank or aid agency, requires and funds impact assessment.
5. A non-governmental organization carries out assessment and bears the cost.
6. Local people commission and pay for the assessment.

APPROACHES TO IMPACT ASSESSMENT

Impact assessment has to be technically effective and socially and politically appropriate. Sometimes assessment is unsatisfactory because it is technically weak, and sometimes because it has been applied in a politically inept way. Ideally, a correct technical, political and social balance can be achieved, but there will need to be considerable adaptation to each specific situation.

Adaptive environmental assessment and adaptive environmental assessment and management

An important point made by Hyman and Stiftel (1988: 21) is that impact assessment often adopts a static, 'snapshot' approach; it tends to assume that causal relationships are constant over a period of time, whereas often they are not, and so assessment is ineffective. For example, monetary units may be devalued, the environment may alter, decision-making objectives may change, and so on. The environment, societies and their economies are not static, are often not even stable; hence, there is a need to be adaptive in impact assessment and ensure assessment is continuous or, at least, regularly repeated (Holling, 1978, 1986; IIASA, 1979; Regier, 1985; Gilmour and Walkerden, 1994). Hollick (1981: 83) warned that impact assessment could discourage planners from carrying out adequate monitoring if they were satisfied with a static 'snapshot' view.

Adaptive environmental assessment and adaptive environmental assessment and management are broader than 'mainstream' environmental impact assessment and have a bias toward coping with uncertainty, rather than just impact prediction; in short, they seek to 'expect the unexpected' and assist policy-makers as well as project developers (Environmental and Social Systems Analysis Ltd, 1982; Everitt, 1983; Walters, 1984; Jones and Greig, 1985; Bisset, 1987: 48–58; Graham Smith, 1993: 25; *see also* Chapter 5). There is an attempt to integrate environmental, social and economic understanding and, with adaptive environmental assessment and management, to link assessment and management.

Adaptive environmental assessment and management was pioneered in the early 1970s by Holling and colleagues (Holling, 1978), Environment Canada, the University of British Columbia in Vancouver and the Austrian-based International Institute for Applied Systems Analysis (IIASA, 1979). It uses a series of carefully designed research periods followed by multidisciplinary modelling workshops that are attended by science and social science experts, planners, managers, resource users and locals. The workshops develop alternative scenarios and management strategies; these are then compared to arrive at the best problem-solving approach. The workshops seek to ensure that the assessment team and participants continually review efforts to predict and model policy options for decision-makers, and also provide a 'bridge' for different disciplines and competing perceptions. The end result is a computer-based systems model that can be tested and 'tuned' until it supports adaptive management, and can help identify indirect impacts. Jones and Greig (1985) list over 30 applications of adaptive environmental assessment and management with references; *see also* Gilmour and Walkerden (1994).

Adaptive environmental assessment and management can be viable where baseline data are poor; it is, however, quite demanding in terms of research expertise and time for completion. It does encourage and facilitate multidisciplinary assessment. Some see adaptive impact assessment as particularly supportive of sustainable development (Grayson *et al.*, 1994). It has been successfully applied to the Obergurgl Valley (Austria), starting in 1974 when a joint UNESCO (Man and Biosphere Program)/IIASA/University of British Columbia assessment team conducted a 3-month study using an approach developed by Holling (1978).

Comprehensive, tiered, integrated, regional and holistic impact assessment

The approaches listed in the subheading all broadly share the goal of seeking to cover more than a restricted range of impacts over more than a 'snapshot' of time, including those that are indirect and cumulative (Nijkamp, 1986) (*see*

Chapter 4). Some of these approaches also seek to assess local, regional and global scales; some seek to assess project, programme and policy activities; and some try to adopt a short-term through to a long-term view.

It is not easy to find an effective, comprehensive, integrated approach that can be easily adapted to particular situations. The most promising is probably, *tiered assessment*, (Lee, 1978, 1982: 73–5; Wood, 1988). This is a sequential process starting with broad assessment at policy level, tier 3 (the highest level) (for example, impact assessment of national road policy), followed by more specific assessment at the programme level, tier 2 (for example, regional road programmes) and finally, even more specific assessment of individual (road) projects, which constitutes tier 1. Efforts are made to cross-reference broad and specific assessment. Events in tier 1 are conditioned by prior events or parallel events in higher tiers; it is unsatisfactory to look at a lower tier without considering the higher ones. Tiered impact assessment can also adopt a multi-sectoral approach in which the tiers are horizontal; if sectors were considered in isolation, cumulative impacts might be missed, or a particular sector might be missed. This requires a holistic approach to avoid missing interactive effects, and Harvey *et al.* (1995) have explored such an approach. Tiered impact assessment should acquire data that make subsequent or related impact assessments easier, faster and cheaper to conduct. Tiered impact assessments should complement each other so as to avoid duplication. It may be possible with some types of development to do broad impact assessments and dispense with a plethora of individual assessments; for example, instead of factory-by factory impact assessment it might be possible to carry out a single assessment for an industrial estate as an entity. The USA tries to encourage a tiered approach, and in other countries (e.g. the Netherlands and more recently the European Union as a whole) the trend is towards such an approach.

Integrated impact assessment is a generic term for the study of the full range of ecological and socio-economic consequences of an action (Davos, 1977; Nijkamp, 1980: 11; Rossini and Porter, 1983; Westman, 1985: 6; Lang, 1986; McDonald, 1990). It is also difficult to predict development impacts if no account is taken of other current and planned developments (hence the tiered approach just discussed). There is also the possibility of closer integration of impact assessment into planning, policy-making and management if a tiered approach is adopted (McDonald and Brown, 1990; Parson, 1995).

To consider cumulative impacts a *regional impact assessment* approach can be an advantage, as for example in cases where successive tourism developments lead to regional problems or numerous irrigation developments cause difficulties. Cocklin *et al.* (1992); and Burde *et al.* (1994) applied an expert systems approach to environmental impact assessment for regional planning. It is also a good approach for establishing planning objectives, for example, the impacts of a new shopping centre or mall are considered by Norris (1990). Within a regional framework it is likely that individual project impact assessments will be better conducted (Merrill, 1981; Briassoulis, 1986; Gilpin, 1995: 29). Clearly it makes sense to assess development projects in their spatial or regional setting than in isolation; it allows the interfacing of planning and environmental management at the regional level and offers possibilities for assessing exogenous impacts on the region (Cooper and Zedler, 1979; Rauschelbach, 1991). Economists have used econometrics and input–output analysis to explore economics and environmental linkages at regional level: for example, the impacts of an irrigation development on a region, like Malaysia's Muda Scheme (Bell and Hazel, 1980) (*see also* Isard, 1972; Bell and Slade, 1982; Bell *et al.*, 1982; Solomon, 1985). James (1994) presented case-studies of environmental impact assessment with a regional focus.

Integrated regional environmental assessment, has the following objectives (*see* Ballard *et al.*, 1982: 6):

- to provide a broad, integrated perspective of a region about to undergo or undergoing developments;
- to identify cumulative impacts from multiple developments in the region;
- to help establish priorities for environmental protection;
- to assess policy options;
- to identify information gaps and research needs.

Integrated regional environmental assessment has no single methodology and can be more difficult than normal 'mainstream' environmental impact assessment. A solution might be to subdivide regions into smaller units for assessment (perhaps ecosystems or riverbasins, although there may be situations where administrative regions offer better possibilities).

Regional risk assessment, seeks to establish the risks likely to affect a region (Suter, 1993: 365–75). *Regional economic modelling* is already quite well established and can feed information and some techniques into integrated regional environmental assessment (James *et al.*, 1983; Gilpin, 1995). Davies and Webster (1981) have applied integrated regional environmental assessment to socio-economic assessment.

Integrated environmental management seeks to reconcile conflicting interests and concerns, minimize negative impacts and enhance positive results. Impact assessment plays an important part in this process (Fuggle, 1989). Crudely, it is an approach that seeks to put impact assessment and evaluation into the planning and decision-making process. For an example of an integrated environmental management procedure (proposed for South Africa), *see* Sowman *et al.* (1995).

Assessment at programme, policy and planning levels: strategic environmental assessment

Assessment is increasingly applied above project level, for example to study the impacts of:

- overseas aid provision (e.g. Goodland and Tillman, 1995);
- structural adjustment (mainly to review what has happened, rather than predict the effects; *see* Kone, 1993; Russel, 1993);

- free-trade developments;
- grant provision for farmers;
- changes in public transport;
- conservation management (Mangun, 1989);
- flood control efforts (Beaufort, 1992).

Programme impact assessment covers a wider field than project assessment; it tries to include cumulative impacts triggered by a set of related projects or proposals. Experience of programme-level, policy-level and planning-level impact assessment, especially for proactive studies, is more limited than for projects but it is growing (Wathern *et al.*, 1987; Dietz and Pfund, 1988; Wandesforde-Smith, 1989; FEARO, 1992). Bartlett (1989) has examined the way in which impact assessment could support policy strategy.

Assessment must allow for the fact that other policies and cultural and other forces affect a given policy. Although by no means carried out in a vacuum, projects are often more isolated than programmes and far more isolated than policies. Alburo and Koppel (1984) discussed the problem of assessing policy impacts in the Philippines, and Wathern *et al.* (1988: 104) warned that it is often difficult to isolate causes of environmental or social change and link them to policy.

Strategic environmental assessment is a form of policy- or programme-focussed tiered, nested or sequential environmental impact assessment that seeks to provide a framework within which project, programme and policy impact assessment can take place. Environmental impact assessment can be used at the project level, tiered with strategic environmental assessment to link it to programme and policy levels. Since the 1980s there has been an increasing demand for project-focussed environmental impact assessment to be extended 'upward' to cover programme and policy levels. Most of the tasks involved in strategic environmental assessment are the same as those undertaken for project level environmental impact assessment. Therefore, many of the methods and techniques are transferable, although they may need some modification. *See* Lee and Walsh, 1992; UNECE, 1992; Wood, 1992, 1995: 266–88; Wood and Djeddour, 1992; Buckley, 1994; Connor and Ruddy, 1994; Dalal-Clayton and Sadler, 1995; Gilpin, 1995: 28; Partidário, 1996; Thérivel and Partidário, 1996; *Project Appraisal* 1992: **7(3)** – special issue on strategic environmental assessment. Partidário (1996: 31) defined strategic environmental assessment as a process that 'attempts to assess systematically the environmental impacts of decisions made at, what is conveniently called, [the] level of strategic decisions'. Interest in the process has been expanding considerably.

Strategic environmental assessment can be applied:

- with a sectoral focus (e.g. to waste disposal, drainage programmes, transport programmes – Sheate, 1992);
- with a regional focus (e.g. to regional plans, rural plans, national plans);
- with an indirect focus (e.g. to technology, fiscal policies, justice and enforcement, sustainable development).

It can be applied to a higher, earlier, more 'strategic' tier of decision-making than project environmental impact assessment (Thérivel *et al.*, 1992; Thérivel, 1993; Connor and Ruddy, 1994; Glasson *et al.*, 1994: 6); *see* Figs 3.5a and 3.5b.

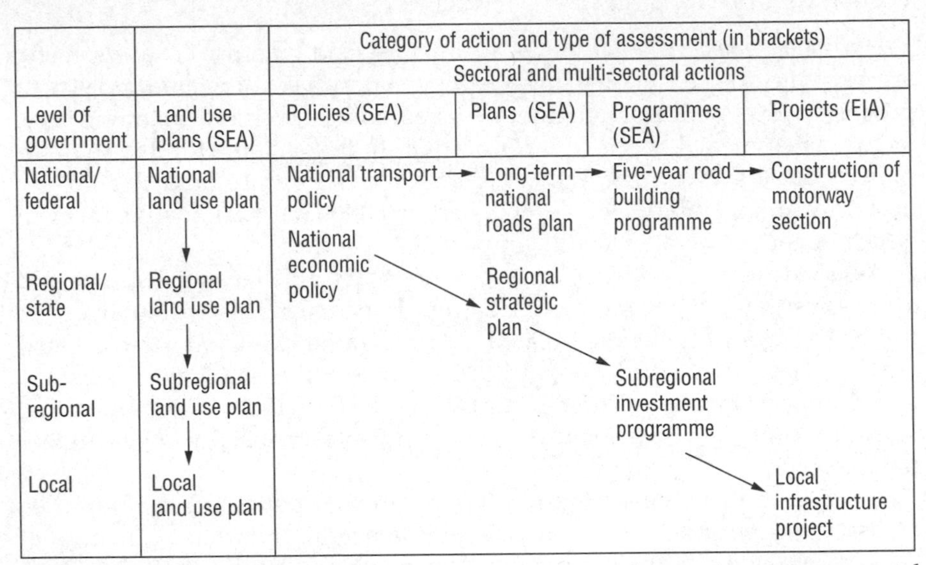

Category of action and type of assessment (in brackets)
Sectoral and multi-sectoral actions
Level of government
Land use plans (SEA)
Policies (SEA)
Plans (SEA)
Programmes (SEA)
Projects (EIA)
National/ federal
National land use plan
National transport policy
Long-term national roads plan
Five-year road building programme
Construction of motorway section
National economic policy
Regional/ state
Regional land use plan
Regional strategic plan
Sub- regional
Subregional land use plan
Subregional investment programme
Local
Local land use plan
Local infrastructure project

FIGURE 3.5 (a) Tiered planning and assessment (strategic environmental assessment, SEA). NB. This is a simplified representation of what, in reality, could be a more complex set of relationships. In general, those actions at the highest tier or level (e.g. national policies) are likely to require the broadest and least detailed form of SEA. EIA = environmental impact assessment.

Strategic environmental assessment can, for example, be applied to a national energy policy, a whole industrial development zone or to an area of scenic value. Thompson *et al.* (1995) applied it to the regulation of salmon (*Salmo salar*) fish farming in Scottish lochs – an activity that was having serious impacts but which was inadequately controlled by legislation. Reed (1996: 357) called for countries undergoing macroeconomic reform to be subject to strategic environmental assessment, listing several reasons why.

Provision for strategic environmental assessment was made by the USA's National Environmental Policy Act in 1970 and in California's Environmental Quality Act of 1985, and it is now in use in various countries, including Canada, the Netherlands, the USA (notably California), Germany and New Zealand. The Netherlands has had a statutory strategic environmental assessment system in force since 1987 for waste management, drinking water supply, energy and electricity, and some land use plans, and formal requirements were strengthened in 1991 under the National Environmental Policy Act (Verheem, 1992; Thérivel, 1993). New Zealand has had laws on strategic environmental assessment since 1991, under Part V of the Resource Management Act. That same year, the European Community (now the European

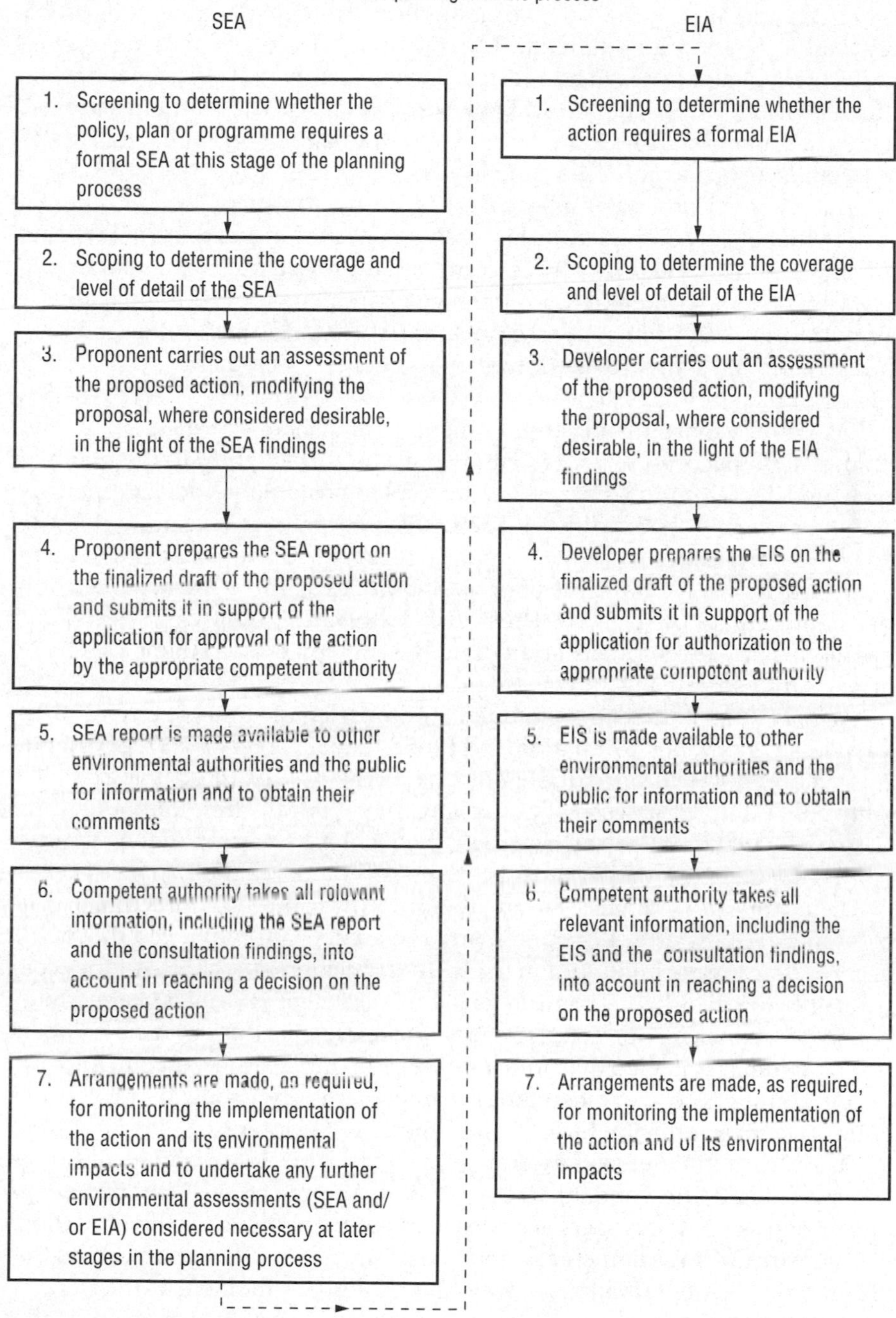

Principal stages in the process
SEA
EIA
1. Screening to determine whether the policy, plan or programme requires a formal SEA at this stage of the planning process
2. Scoping to determine the coverage and level of detail of the SEA
3. Proponent carries out an assessment of the proposed action, modifying the proposal, where considered desirable, in the light of the SEA findings
4. Proponent prepares the SEA report on the finalized draft of the proposed action and submits it in support of the application for approval of the action by the appropriate competent authority
5. SEA report is made available to other environmental authorities and the public for information and to obtain their comments
6. Competent authority takes all relevant information, including the SEA report and the consultation findings, into account in reaching a decision on the proposed action
7. Arrangements are made, as required, for monitoring the implementation of the action and its environmental impacts and to undertake any further environmental assessments (SEA and/or EIA) considered necessary at later stages in the planning process
1. Screening to determine whether the action requires a formal EIA
2. Scoping to determine the coverage and level of detail of the EIA
3. Developer carries out an assessment of the proposed action, modifying the proposal, where considered desirable, in the light of the EIA findings
4. Developer prepares the EIS on the finalized draft of the proposed action and submits it in support of the application for authorization to the appropriate competent authority
5. EIS is made available to other environmental authorities and the public for information and to obtain their comments
6. Competent authority takes all relevant information, including the EIS and the consultation findings, into account in reaching a decision on the proposed action
7. Arrangements are made, as required, for monitoring the implementation of the action and of its environmental impacts

FIGURE 3.5 (b) A comparison of strategic environmental assessment (SEA) and environmental impact assessment (EIA). EIS = environmental impact statement
Source: Based on Lee and Walsh (1992: 131, 132, Figs 2 and 3)

Union) and the UK published proposals for its uses, although Thérivel *et al.* (1992: 32) note that in the UK poor long-term strategic planning will probably make its adoption difficult. The European Union is moving toward requiring member states to adopt strategic environmental assessment procedures (Wilson, 1993). The World Bank supports the process and has applied it on a sectoral basis, for example to a large drainage programme in Pakistan and to a locust control programme in Africa.

Strategic environmental assessment is useful for site selection, and the conducting of such a 'higher-order' assessment may mean there is less need for, and less depth required from, component project environmental impact assessments. The strategic environmental assessment approach can cope better with cumulative impacts, assessment of alternatives and mitigation measures than can standard environmental impact assessment, according to Glasson *et al.* (1994: 22, 300).

It is claimed that strategic environmental assessment can ensure that environmental impact assessment is initiated at the correct point in the planning cycle and therefore makes it easier to pursue sustainable development by helping prevent possibly difficult-to-reverse problems. Increasingly, it is seen as a key approach for implementing the concept of sustainable development, because it allows the principle of sustainability to be carried down from policies to individual projects (*see* Thérivel *et al.*, 1992: 22, 126). The Netherlands has linked its approach to strategic environmental assessment closely to principles of sustainable development.

There is some parallel development and overlap of strategic environmental assessment, policy evaluation and policy impact assessment (*see* Chapter 2). Boothroyd (1995) provided a critique of the first of these, comparing it with policy impact assessment, and stressing that all these methods must also ensure that they are not 'captured' by development proponents (in other words, the implementation must be sufficiently independent).

The European Union is planning to adopt strategic environmental assessment more widely after 1995, and many researchers are convinced that it will become much more important in the future. (Incidentally, choice of the word 'strategic' for this sort of comprehensive assessment is ironical; assessment has so far often been neglected for strategic reasons – issues of national security or trade). Strategic environmental assessment, at least in principle, can enable countries to work together on transboundary problems (*see* 'Transboundary environmental impact assessment' in Chapter 6).

There is growing interest in strategic environmental assessment, although it is more demanding of data and expertise than project-level environmental impact assessment. It is seen as a way of overcoming many of the limitations of 'mainstream' environmental impact assessment. A difficulty it faces is that programmes evolve in a subtle way and at a given moment it may not be easy to see what actually constitutes a programme. Another problem is that policy-makers may not want to give potential opponents or competitors a perspective of their strategy, so public involvement is a problem. Strategic environmental assessment methodology is in need of development; one group of workers felt that it suffered from the problems of environmental impact assessment and policy analysis (Thérivel *et al.*, 1992: 41). A core difficulty is that strategic environmental assessment must make accurate assessments in spite of the fact that proposals and policies are often vague (compare the project-level situation); system boundaries are uncertain; information on existing and future developments is limited; there are a large number of possible alternatives to consider; a number of possibly uncooperative bodies are involved; and there are probably more political pressures than are felt at project level (Thérivel *et al.* 1992: 41–2).

Global-scale environmental impact assessment

Interest in global impact assessment has increased over the past ten years. Much is associated with concern for the effects of global warming, pollution, nuclear accident or warfare, vulnerability to drought and famine, and the effects of conflicts such as the Gulf War. An important step toward establishing transboundary environmental impact assessment has been made through the Espoo Convention (*see* Chapter 6), and there has been co-operation between countries to conduct environmental impact assessments, in Antarctica and the Arctic (*see* Chapter 6).

References

Adams, W.M. 1990: *Green development: environment and sustainability in the third world*. London: Routledge.

Alburo, J. and **Koppel, B.** 1984: Impact assessment and project development. *Journal of Philippine Development* **11(1)**, 1–25.

Amy, D.J. 1983a: Environmental mediation: an alternative approach to policy statements. *Policy Sciences* **15**, 343–65.

Amy, D.J. 1983b: *The politics of environmental mediation*. New York: Columbia University Press.

Armour, A. 1990: Integrating impact assessment in the planning process: from rhetoric to reality? *Impact Assessment Bulletin* **8(1–2)**, 3–15.

Armour, A. 1991: Impact assessment and the planning process: a status report. *Impact Assessment Bulletin* **9(4)**, 27–33.

Ballard, S.C., Devine, M.D., James, T.E. Jnr and **Charlock, M.A.** 1982: Integrated regional environmental assessments: purposes, scope, and products. *Impact Assessment Bulletin* **2(1)**, 5–13.

Barrow, C.J. 1987: *Water resources and agricultural development in the tropics*. Harlow: Longman.

Bartlett, R.V. (ed.) 1989: *Policy through impact assessment: institutionalized analysis as a policy strategy*. New York: Greenwood Press.

Beattie, R.B. 1995: Everything you already know about EIA (but don't want to admit). *Environmental Impact Assessment Review* **15(3)**, 109–14.

Beaufort, G. 1992: Environmental impact assessment (EIA) and policy analysis as tools for combating flooding in the Netherlands. In Saul, A.J. (ed.) *Floods and flood* *management*. Proceedings of the 3rd International Conference on Floods and Flood Management, Florence, Italy, 24–26 November, 1992. Dordrecht: Kluwer, 91–103.

Beckler, H.A. and **Porter, A.H.** (eds) 1986: *Program evaluation today*. Utrecht: Van Arkel.

Bell, C.L.G and **Hazel, P.** 1980: Measuring the indirect effects of an agricultural investment project on its surrounding region. *Australian Journal of Agricultural Economics* **62(1)**, 75–86.

Bell, C.L.G., **Hazel, P.** and **Slade, C.** 1982: *Project evaluation in regional perspective: a study of an irrigation project in Northwestern Malaysia*. Baltimore: Johns Hopkins University Press.

Bell, C.L.G. and **Slade, R.** 1982: *Project evaluation in regional perspective*. Cambridge, MA: Harvard University Press.

Bisset, R. 1987: Methods for environmental impact assessment: a selective survey with case studies. In Biswas, A.K. and Qu Geping (eds), *Environmental impact assessment for developing countries*. London: Tycooly International, 3–64.

Blahna, D.J. and **Yonts-Shepard, S.** 1989: Public involvement in resource planning: towards bridging the gap between policy and implementation. *Society and Natural Resources* **2(3)**, 209–27.

Boothroyd, P. 1995: Policy assessment. In Vanclay, F. and Bronstein, D.A. (eds), *Environmental and social impact assessment*. Chichester: Wiley, 83–126.

Breakell, M. and **Glasson, J.** (eds) 1981: *Environmental impact assessment: from theory to practice*. Oxford: Oxford Polytechnic, Department of Town Planning.

Briassoulis, H. 1986: Integrated economic–environmental–policy modelling at the regional and multiregional level: methodological characteristics and issues. *Growth and Change* **17**, 22–34.

Briassoulis, H. 1989: Theoretical orientations in environmental planning: an inquiry into alternative approaches. *Environmental Management* **13(4)**, 381–92.

Brown, A.L. 1992: *Beyond EIA: incorporating environment into the engineering design process*. Proceedings of the National Conference on Environmental Engineering – the Global Environment, 17–19 June 1992 (part 5), St Leonards, Australia. St Leonards: E.A. Books for the Australian Institute of Engineers, 229–33.

Buckley, R.C. 1994: Strategic environmental assessment. *Environmental Planning and Law Journal* **11(2)**, 166–8.

Burde, M., **Jackel, T.**, **Dieckmann, R.** and **Hemker, H.** 1994: Environmental impact assessment for regional planning with SAFRAN. In Guariso, G. and Page, B. (eds), *Computer support for environmental impact assessment*. IFIP Transactions B: Applications in Technology. Proceedings of the Conference on Computer Support for EIA–CSEIA, Como, Italy, 6–8 October, 1993. Amsterdam: North-Holland, 245–56.

Burdge, R. and **Opryszik, P.** 1983: On mixing apples and oranges: the sociologist does impact assessment with biologists and economists. In Rossini, F.A. and Porter, A. (eds), *Integrated impact assessment*. Boulder, CO: Westview, 107–17.

Burdge, R.J. and **Vanclay, F.** 1996: Social impact assessment: a contribution to the state-of-the-art series. *Impact Assessment* **14(1)**, 59–86.

Canter, L.W. 1996: *Environmental impact assessment*, 2nd edn. New York: McGraw-Hill. (1st edn 1977)

Carpenter, R.A. 1980: Using ecological knowledge for development planning. *Environmental Management* **4(1)**, 13–20.

Clark, B.D. 1980: The impact of environmental impact assessment. In Elkington, J. (ed.), *Environmental impact assessment*. London: Oyez Publishing, 3–16.

Clark, B.D. 1984: Environmental impact assessment (EIA): scope and objectives. In Clark, B.D., Bisset, R., Gilad, A. and Tomlinson, P. (eds), *Perspectives on environmental impact assessment*. Dordrecht: D. Reidel, 3–14.

Clark, L. 1988: Politics and bias in risk assessment. *Social Science Journal* **25**, 155–65.

Clark, M. and **Herington, J.M.** (eds) 1988: *The role of environmental impact assessment in the planning process*. London: Mansell.

Cocklin, C., Parker, S. and **Hay, J.** 1992: Notes on cumulative environmental change. I: Concepts and issues. *Journal of Environmental Management* **35(1)**: 31–49.

Coles, T.K., Fuller, K. and **Slater, M.** 1992: *Practical experience of environmental assessment in the UK*. East Kirkby, Lincolnshire: Institute of Environmental Assessment.

Connor, D.M. 1977: Constructive citizen participation: the development and setting of a concept. *Social Impact Assessment* **13(8)**, 8–16.

Connor, J. and **Ruddy, G.** 1994: SEA: evaluating the policies EIA cannot reach. *Town and Country Planning* **63(2)**, 45–8.

Cooper, F. and **Zedler, H.** 1979: Ecological assessment for regional development. *Journal of Environmental Management* **10(4)**, 285–96.

Curtis, F.A. 1983: Integrating environmental mediation into EIA. *Impact Assessment Bulletin* **2(3)**, 17–25.

Dalal-Clayton, B. and **Sadler, B.** 1995: *Strategic environmental assessment: a briefing paper*. London: International Institute for Environment and Development.

Davies, H.C. and **Webster, D.R.** 1981: A compositional approach to regional socioeconomic EIA. *Socio-economic Planning Science* **15(4)**, 159–63.

Davos, C.A. 1977: Towards an integrated environmental impact assessment within a social context. *Journal of Environmental Management* **5(4)**, 297–306.

Dietz, T. and **Pfund, A.** 1988: An impact identification method for development programme evaluation. *Impact Assessment Bulletin* **6(3–4)**, 137–48.

Environmental and Social Systems Analysis Ltd (ed.) 1982: *Review and evaluation of adaptive environmental assessment and management*. Vancouver: Environment Canada.

Everitt, R.R. 1983: Adaptive environmental assessment and management: some current applications. In University of Aberdeen, Project and Development Control Unit (ed.), *Environmental impact assessment*. The Hague: Martinus Nijhoff, 293–306.

Fagence, M. 1977: *Citizen participation in planning*. Oxford: Pergamon.

Faludi, A. 1973: *Planning theory*. Oxford: Pergamon.

FEARO 1988: *Manual on public involvement in environmental assessment*. Hull, Quebec: Federal Environmental Assessment Review Office.

FEARO 1992: *Environmental assessment in policy and programme planning: a sourcebook*. Ottawa: Federal Environmental Assessment Review Office.

Formby, J. 1990: The politics of environmental impact assessment. *Impact Assessment Bulletin* **8(1–2)**, 191–6.

Fuggle, R.F. 1989: Integrated environmental management: an appropriate approach to environmental concern in developing countries. *Impact Assessment Bulletin* **8(1–2)**, 31–45.

Gariepy, M. 1991: Toward a dual-influence system: assessing the effects of public participation in environmental impact assessment for Hydro-Quebec projects. *Environmental Impact Assessment Review* **11(4)**, 353–74.

Ghai, D. and **Vivian, J.M.** 1992: *Grassroots environmental action: people's participation in sustainable development*. London: Routledge.

Gilmour, A.J. and **Walkerden, G.** 1994: A structured approach to conflict resolution in EIA: the use of adaptive environmental assessment and management (AEAM). In Gilmour, A.J. and Page, B. (eds), *Computer support for environmental assessment,* vol. 16. Proceedings of the IFIP Working Conference on Computer Support for Environmental Assessment, 6–8 October 1993, Como, Italy. TC5/WG5.11 transactions: B, Applications in Technology, 199–210.

Gilpin, A. 1995: *Environmental impact assessment* (EIA)*: cutting-edge for the twenty-first century*. Cambridge: Cambridge University Press.

Glasson, J., Thérivel, R. and **Chadwick A.** 1994: *Introduction to environmental impact assessment: principles and procedures, process, practice and prospects*. London: University College London Press.

Goodland, R.G. and **Tillman, G.** 1995: *Strategic environmental assessment: strengthening the environmental assessment process.* Discussion draft. Washington DC: World Bank.

Graham Smith, L. 1993: *Impact assessment and sustainable resource management*. Harlow: Longman.

Grayson, R.B., Doolan, J.M. and **Blake, T.** 1994: Applications of adaptive environmental assessment and management (AEAM) to water quality in the Latrobe River catchment. *Journal of Environmental Management* **41(3)**, 245–58.

Hall, P. 1980: Environmental impact analysis: scientific tool or philosopher's stone? *Built Environment* **4(2)**, 85–93.

Harvey, T., Mahaffey, K.R., Velazquez, S. and **Dourson, M.** 1995: Holistic risk assessment: an emerging process for environmental decisions. *Regulatory Toxicology and Pharmacology* **22(2)**: 110–17.

Herington, J.M. (ed.) 1979: *The role of environmental impact assessment (EIA) in the planning process.* Proceedings of an Institute of British Geographers Conference, Loughborough, 17–20 April 1979. Loughborough: University of Loughborough, Department of Geography.

Heuber, R. 1992: The World Bank and environmental assessment: the role of non-governmental organizations. *Environmental Impact Assessment Review* **12(4)**, 331–48.

Hollick, M. 1981: Environmental impact assessment as a planning tool. *Journal of Environmental Management* **12(1)**, 79–90.

Hollick, M. 1986: Environmental impact assessment: an international evaluation. *Environmental Management* **10(2)**, 157–78.

Holling, C.S. 1978: *Adaptive environmental assessment and management*. Chichester: Wiley.

Holling C.S. 1986: The resilience of terrestrial ecosystems: local surprise and global change. In Clark, W.C. and Munn, R.E. (eds), *Sustainable development of the biosphere*. Cambridge: Cambridge University Press, 292–320.

Hughes, R. and **Dalal-Clayton, B.** (1996) *Participation and environmental assessment: a review of issues*. EPG Issues Planning series II. London: International Institution for Environment and Development.

Hughes, G. and **Schirmer, D.** 1994: Interactive multimedia, public participation and environmental assessment. *Town Planning Review* **65(4)**, 399–414.

Hyman, E.L. and **Stiftel, B.** 1988: *Combining facts and values in environmental impact assessment: theories and techniques*. Boulder, CO: Westview.

IIASA 1979: *Expect the unexpected: an adaptive approach to environmental management.* Executive Report 1. Laxenburg, Austria: International Institute for Applied Systems Analysis.

Isard, W. 1972: *Ecologic–economic analysis for regional development*. New York: Free Press.

James, D. (ed.) 1994: *The application of economic techniques in environmental impact assessment*. Dordrecht: Kluwer.

James, T.E Jr, Ballard, S.C. and Devine, M.D. 1983: Regional environmental assessments for policy making and research and development. *Environmental Impact Assessment Review* **4(1)**, 9–24.

Jones, M.L. and **Greig, L.A.** 1985: Adaptive environmental assessment and management: a new approach to environmental impact assessment. In Maclaren, V.W. and Whitney, J.B. (eds), *New directions in environmental impact assessment in Canada*. Toronto: Methuen, 21–42.

Kakonge, J.O. and **Imevbore, A.M.** 1993: Constraints on implementing environmental impact assessments in Africa. *Environmental Impact Assessment Review* **13(4)**, 299–308.

Kennett, S.A. and **Perl, A.** 1995: Environmental impact assessment of development oriented research. *Environmental Impact Assessment Review* **15(4)**, 341–60.

Kone, T. 1993: Ajustement structural et politique en Côte d'Ivoire: l'impact environnemental. *Labour, Capital and Society* **26(1–SI)**, 86–101.

Kozlowski, J. 1989: *Integrating ecological thinking into the planning process: a comparison of the EIA and UET concepts*. Paper FS-II-89–404. Berlin: Wissenschaftszentrum Berlin für Sozialforschung.

Lake, L.M. 1971: Mediating environmental disputes. *Ekistics* **44(267)**, 164–70.

Lake, L.M. 1980: *Environmental mediation: the search for consensus*. Boulder, CO: Westview.

Lang, R. (ed.) 1986: *Integrated approaches to resource planning and management*. Calgary: University of Calgary Press.

Lee, N. 1978: Environmental impact assessment of projects in EEC countries. *Journal of Environmental Management* **6(1)**, 57–71.

Lee, N. 1982: The future development of environmental impact assessment. *Journal of Environmental Management* **14(1)**, 71–90.

Lee, N. and **Walsh, F.** 1992: Strategic environmental assessment: an overview. *Project Appraisal* **7(3)**, 126–37.

Lee, N and **Wood, C.** 1978: EIA: a European perspective. *Built Environment* **4(2)**, 101–10.

Lichfield, N. 1992a: The integration of environmental assessment into development planning – Part 1, Some principles. *Project Appraisal* **7(2)**, 58–66.

Lichfield, N. 1992b: The integration of environmental assessment into development planning – Part 2, A case study. *Project Appraisal* **7(3)**, 175–83.

Lindblom, C.E. 1959: The science of muddling through. *Public Administration* **19**, 17–88.

Lohani, B.N. and Halim, N. 1987: Recommended methodologies for rapid environmental impact assessment in developing countries: experiences derived from case studies in Thailand. In Biswas, A.K. and Qu Geping (eds), *Environmental impact assessment in developing countries*. London: Tycooly International, 65–111.

McCoy, C.B. 1975: The impact of an impact study. *Environment and Behaviour* **7**, 358–72.

McDonald, G.T. 1990: Regional economic and social impact assessment. *Environmental Impact Assessment Review* **10(1–2)**, 23–36.

McDonald, G.T. and **Brown, A.L.** 1990: Planning and management processes and environmental assessment. *Impact Assessment Bulletin* **8(1–2)**: 261–74.

Mangun, W.R. 1989: Environmental impact assessment as a tool for wildlife policy management. In Bartlett, R.V. (ed.), *Policy through impact assessment: institutionalized analysis as a policy strategy*. New York: Greenwood Press, 51–62.

Merrill, F. 1981: Area-wide environmental impact assessment guidebook. *Environmental Impact Assessment Review* **2(3)**, 204–7.

Nay Htun 1988: The EIA process in Asia and the Pacific region. In Wathern, P. (ed.), *Environmental impact assessment: theory and practice*. London: Unwin Hyman, 225–38.

Nijkamp, P. 1980: *Environmental policy analysis: operational methods and models*. Chichester: Wiley.

Nijkamp, P. 1986: Multiple criteria analysis and integrated impact analysis. *Impact Assessment Bulletin* **4(3–4)**, 226–61.

Norris, S. 1990: The return of impact assessment: assessing the impact of regional shopping centre proposals in the United Kingdom. *Papers in Regional Science* **69**, 101–19.

OECD 1979: *Environmental impact assessment*. Paris: Organisation for Economic Co-operation and Development.

Ortolano, L. and **Shepherd, A.** 1995: Environmental impact assessment: challenges and opportunities. *Impact Assessment* **13(1)**, 3–30.

Parson, E.A. 1995: Integrated assessment and environmental policy making: in pursuit of usefulness. *Energy Policy* **23(4–5)**, 463–76.

Partidário, M. do R. 1996: Strategic environmental assessment: key issues emerging from recent practice. *Environmental Impact Assessment Review* **16(1)**, 31–55.

Rauschelbach, B. 1991: Die Umweltvertraglichkeitsprüfung als raumbezogene Planungsaufgabe: Fragestellung, Inhalt und Methodik (EIA as a spatial planning task: problems, objectives, methodology). *Geographische Rundschau*, **43(1)**: 363.

Reed, D. (ed.) 1996: *Structural adjustment, the environment, and sustainable development*. London: Earthscan.

Regier, H.A. 1985: Concepts and methods of AEAM and Holling's science of surprise. In Maclaren, V.W. and Whitney, J.B. (eds) *New directions in environmental impact assessment in Canada*. Toronto: Methuen, 21–42.

Roberts, R. 1988: Public involvement: a Canadian Government manual for planning and implementing public involvement programs. *Environmental Impact Assessment Planning Review* 8(1), 3–7.

Roberts, R.C. 1995: Public involvement: from consultation to participation. In Vanclay, F. and Bronstein, D.A. (eds), *Environmental and social impact assessment*. Chichester: Wiley, 221–46.

Rossini, F.A. and **Porter, A.** (eds) 1983: *Integrated impact assessment*. Boulder, CO: Westview.

Russel, S.C. 1993: Applications of environmental impact assessment principles to policies, plans and programmes. *Project Appraisal* **8(4)**, 257.

Sadler, B. 1993: Mediation provisions and options in Canadian environmental assessment. *Impact Assessment Review* 13(6), 375–90.

Sanchez, L. 1993: Environmental impact assessment in France. *Environmental Impact Assessment Review* **13(4)**, 255–65.

Save the Children 1995: *Toolkits: a practical guide to assessment, monitoring, review and evaluation*. Save the Children Development Manual 5. London: Save the Children.

Sewell, W.R.D. and **Coppock, J.J.** (eds) 1977: *Public participation in planning*. London: Wiley.

Sheate, W. 1992: Strategic environmental assessment in the transport sector. *Project Appraisal* **7(3)**, 170–3.

Sheate, W. 1994: *Making an impact: a guide to EIA law and policy*. London: Cameron May.

Slocombe, D.S. 1993: Environmental planning, ecosystem science, and ecosystem approaches for integrating environment and development. *Environmental Management* **17(3)**, 289–303.

Solomon, B.D. 1985: Regional econometric models for environmental impact assessment. *Progress in Human Geography* **9(3)**, 378–99.

Sowman, M., Fuggle, R. and **Preston, G.** 1995: A review of the evolution of environmental evaluation procedures in South Africa. *Environmental Impact Assessment Review* **15(1)**, 45–68.

Susskind, L., McMahon, E. and **Rolley, S.** 1987: Mediating development disputes: some barriers and bridges to successful negotiation.*Impact Assessment Review* **7(2)**, 127–38.

Suter, G.W. II (ed.) 1993: *Ecological risk assessment*. Boca Raton, FL: Lewis.

Thérivel, R. 1993: Systems of strategic environmental assessment. *Environmental Impact Assessment Review* **13(3)**, 145–68.

Thérivel, R. and **Partidário, M.R.** 1996: *The practice of strategic environmental assessment*. London: Earthscan.

Thérivel, R., Wilson, E., Thompson, S., Heaney, D. and **Pritchard, D.** 1992: *Strategic environmental assessment*. London: Earthscan.

Thompson, S., Treweek, J.R. and **Thurling, D.J.** 1995: The potential application of strategic environmental assessment (SEA) to the farming of Atlantic salmon (*Salmo salar* L.) in mainland Scotland. *Journal of Environmental Management* **45(3)**, 219–29.

UNECE 1981: *Environmental impact assessment*. Proceedings of a Seminar of the UN Economic Commission for Europe, Villach, Austria, September 1979 – ENV/SEM 10/2). Oxford: Pergamon.

UNECE 1990: *Post-project analysis in environmental impact assessment*. New York: United Nations Economic Commission for Europe.

UNECE 1992: *Application of environmental impact assessment principles to policies, plans and programmes*. New York: United Nations Economic Commission for Europe.

Verheem, R. 1992: Environmental assessment at the strategic level in the Netherlands. *Project Appraisal* 7(3), 150–7.

Walters, C. 1984: *Adaptive management of renewable resources*. New York: Macmillan.

Wandesforde-Smith, G. 1989: Environmental impact assessment, entrepreneurship and policy change. In Bartlett, R.V. (ed.), *Policy through impact assessment: institutionalized analysis as a policy strategy*. New York: Greenwood Press, 155–66.

Wathern, P. (ed.) 1988. *Environmental impact assessment: theory and practice*. London: Unwin Hyman.

Wathern, P., Brown, I., Roberts, D. and **Young, S.** 1988: Assessing the environmental impacts of policy. In Clark, M. and Herington, J.M. (eds), *The role of environmental impact assessment in the planning process*. London: Mansell, 103–23.

Wathern, P., Young, S.N., Brown, I.W. and **Roberts, D.A.** 1987: Assessing the impacts of policy: a framework and an application. *Landscape and Urban Planning* 13, 321–30.

Westman, W.E. 1985: *Ecology, impact assessment and environmental planning*. Chichester, Wiley.

Wilson, E. 1993: Strategic–environmental assessment of policies, plans and programmes. *European Environment* **3(2)**, 2–6.

Wood, C.M. 1979: The European perspective. In Herington, J.M. (ed.), *The role of environmental impact assessment (EIA) in the planning process*. Proceedings of an Institute of British Geographers Conference, Loughborough, 17–20 April 1979. Loughborough: University of Loughborough, Department of Geography, 9–21.

Wood, C.M. 1980: Environmental impact assessment, pollution control and planning. *Clean Air* **10(3)**, 71–80.

Wood, C.M. 1988: EIA in plan making. In Wathern, P. (ed.), *Environmental impact assessment: theory and practice*. London: Unwin Hyman, 98–114.

Wood, C.M. 1992: Strategic environmental assessment in Australia and New Zealand. *Project Appraisal* **7(3)**, 137–43.

Wood, C.M. 1995: *Environmental impact assessment: a comparative review*. Harlow: Longman.

Wood, C.M. and **Djeddour, M.** 1992: Strategic environmental assessment: EA of policies, plans and programs. *Impact Assessment Bulletin* **10(1)**, 3–23.

World Bank 1991: *Environmental assessment sourcebook*, vol. 2. *Sectoral guidelines*. World Bank Technical Paper 140. Washington DC: World Bank (Environment Department).

Yap, N.T. 1990: Round the peg or square the hole? Populists, technocrats and environmental assessment in Third World countries. *Impact Assessment Bulletin* **8(1–2)**, 69–84.

4

THE IMPACT ASSESSMENT PROCESS

> Many politicians have been quick to grasp that the quickest way to silence critical 'ecofreaks' is to allocate a small proportion of funds for any engineering project for ecological studies. Someone is inevitably available to receive those funds, conduct the studies regardless of how quickly results are demanded, write large reports containing diffuse reams of uninterpreted and incomplete descriptive data, and in some cases, construct 'predictive' models, irrespective of the quality of the data base. These reports have formed a 'gray literature' so diffuse, so limited in distribution that its conclusions and recommendations are never scrutinized by the scientific community at large. (Schindler, 1976)

INTRODUCTION

Assessment is a process (*see* Box 4.1a for a definition of this term) whose 'end-product', is the environmental impact statement – although arguably the environmental impact assessment is a stage, not the finish, with mitigation and avoidance following (Hollick, 1981a). The 'by-products' of impact assessment include: feedback to planning, public relations, monitoring, data gathering and processing. As the environmental impact assessment process has diffused from the USA since 1970 it has diversified into a variety of systems; even within the USA there is considerable variation from one state to another. They nevertheless share 'core' similarities, in that each adopts a multidisciplinary approach and uses iterative steps – a pattern broadly similar to that shown in Fig. 4.1. Shared similarities have suggested a mnemonic that describes the most commonly adopted approach stages, namely ISIC: identify development requiring assessment; scope; interpret; communicate. Differences in the process are most marked concerning the degree of public involvement; whether there are

Box 4.1A Terms relating to the assessment process

- *Process* – system of administration or series of steps.
- *Phase* – not a precise term; generally indicates whether the stage under discussion is pre-development, during development or post-development. Alternatively it might indicate study design, data-gathering, evaluation, communication, post-communication.
- *Methodology* – a structure for organizing a process.
- *Procedure* – steps suggested or enforced by law or an overseeing body.
- *Methods* – a technique or assemblage of techniques to do one or more of the following: provide information, assess information, present information.
- *Actors* – the government; the public; specifically affected parties; the developer(s); non-governmental organizations; consultants engaged by one or more of the previous groups; regulatory bodies; the media (*see* Glasson *et al.*, 1994: 21).

Box 4.1B Types of data

- 'Hard data' – reliable, *quantitative*, verifiable. Can be examined with mathematical techniques.
- 'Soft data' – *qualitative* information; may be difficult to get exactly the same observations if the study is repeated. Cannot always be verified.

- *Continuous variable* – whatever two values are mentioned, it is always possible to imagine more possible values in between them (e.g. length measurements).
- *Discrete variable* - possible values are clearly separated from one another (e.g. number of children – one cannot measure a half-child or one-third child).

adequate feedbacks in the process; whether the process is cyclic; the degree of 'openness'; and the extent of review and audit.

In practice, the assessment process takes place in the wider context of local, regional and national 'development' and is likely to be less ordered and systematic than the literature suggests. Often it is rushed, under pressure from many different interests; it can be inadequate or even only partly finished, and tends to rely on outmoded concepts from 1970s ecologists. (Box 4.2 suggests ways in which objectivity might be maintained.) Fairweather (1989: 141) had grave misgivings about the state of environmental impact assessment in Australia (and elsewhere), like the author of the passage quoted at the start of this chapter, fearing that scientific research was not solving policy-oriented problems. Rather it was being used to give impact assessment 'a cloak of scientific respectability'. Yet Fairweather acknowledged that science embraces observation, hypothesis and testing – all of which *should* be part of impact assessment. Morris and Thérivel (1995: 6) feel that too much stress is put on the fact that assessment process takes place step by step, and that it is an iterative process (that is, there are feedback loops at various points that may prompt a repetition or a change in planned steps as need demands; Munn, 1979: 15).

Box 4.2 Checklist of suggestions for improving the objectivity of the environmental impact assessment process:

- Seek to balance development with environment.
- Review legal, political and media situation with which environmental impact assessment has to 'fit'.
- Note public expectations and reservations at scoping meeting(s).
- Note performance of similar developments elsewhere and also effectiveness of their assessment.
- Check assessors and others involved in assessment to establish their proponents' track record.
- Seek views of those with best proven professional judgement.
- When considering the distributional effects of development, bear in mind the principles of natural justice.
- Note relevant religious, cultural, archaeological, historical, social and economic factors.
- Resist inertia, intimidation, threats, inducements, 'advice'.
- Establish liaison groups with local community, relevant non-governmental organizations and other proponents, to act as sounding boards.

Source: Based on Gilpin (1995: 23 – checklist 2.5)

Box 4.3 Broad phases in the assessment process

Key:
{?} indicates possibility of public involvement (but not a universal practice)
{P} indicates likelihood of public involvement

- *Pre-development monitoring* – ongoing part of planning and management to warn of problems and provide data for assessment (makes screening and scoping easier) and planning {?}.
- *Screening* – preliminary review to decide whether to carry out a full assessment (may lead to no assessment, a partial assessment report or a full assessment). A number of countries require initial environmental evaluation to determine whether a full environmental impact assessment is required. May involve public meeting(s) {?}.
- *Scoping* – definition of parameters, setting goals, drawing boundaries of study (depth, subjects and time available), deciding of approach and team, costs, etc. {?}.
- *Gathering of data* (baseline monitoring) – scoping having established what was needed, measure likely impacts, establish risk, etc. Include information on development alternatives (if not already committed to a course of action) {P}.
- *Identifying of impacts* – using observation, forecasting, desk-research, etc., determine whether there are impacts {?}.
- *Evaluating/assessing* – interpret/determine significance (scale and importance), ideally with indication of reliability and probabilities of the interpretation. This

phase may involve weightings and transformations of data to permit comparisons and/or communication. Consider impacts of alternative developments, including non-development, and impacts of impacts {?}.

- *Reviewing of progress and standard of assessment* – check how assessment is proceeding and whether there are problems {?}.
- *Communicate* – release of a draft or final environmental impact statement and/or public meeting(s) {P}. If the public are consulted, there will be feedback and a return to evaluation/assessment in order to prepare a final revised communication {?}.
- *Post-environmental impact analysis audit* – assess the accuracy and value of the assessment techniques, processes, procedures, administrations and management. There is likely to be feedback to improve future assessment {?}.
- *Monitoring* – ongoing watch for new impacts {?}.

Figures 4.1 and 4.2 and Box 4.3 present information on the environmental impact assessment approach. Figure 4.1 is a simplification, omitting feedbacks, monitoring and post-environmental impact assessment audit.

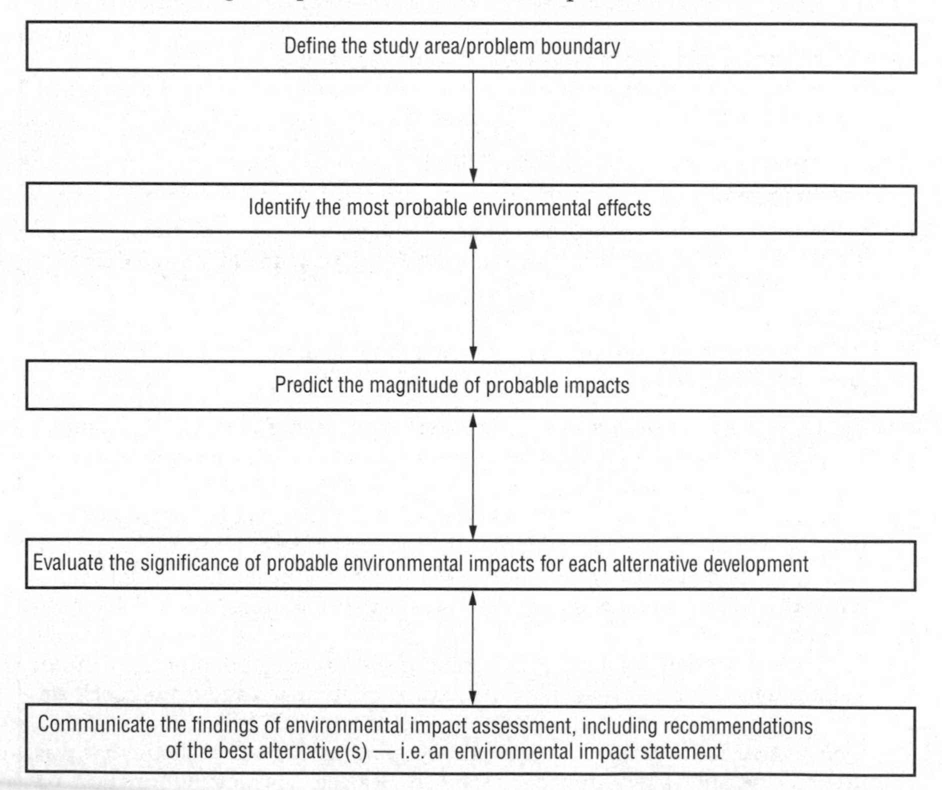

FIGURE 4.1 Typical pattern of phases, steps (or stages) in the assessment process. NB. Post-environmental impact assessment audit and review/monitoring and feedbacks are not shown

Source: Drawn by the author using information from Erickson (1979: 90); Wood and Gazidellis (1985: 14); Graham Smith (1993: 16)

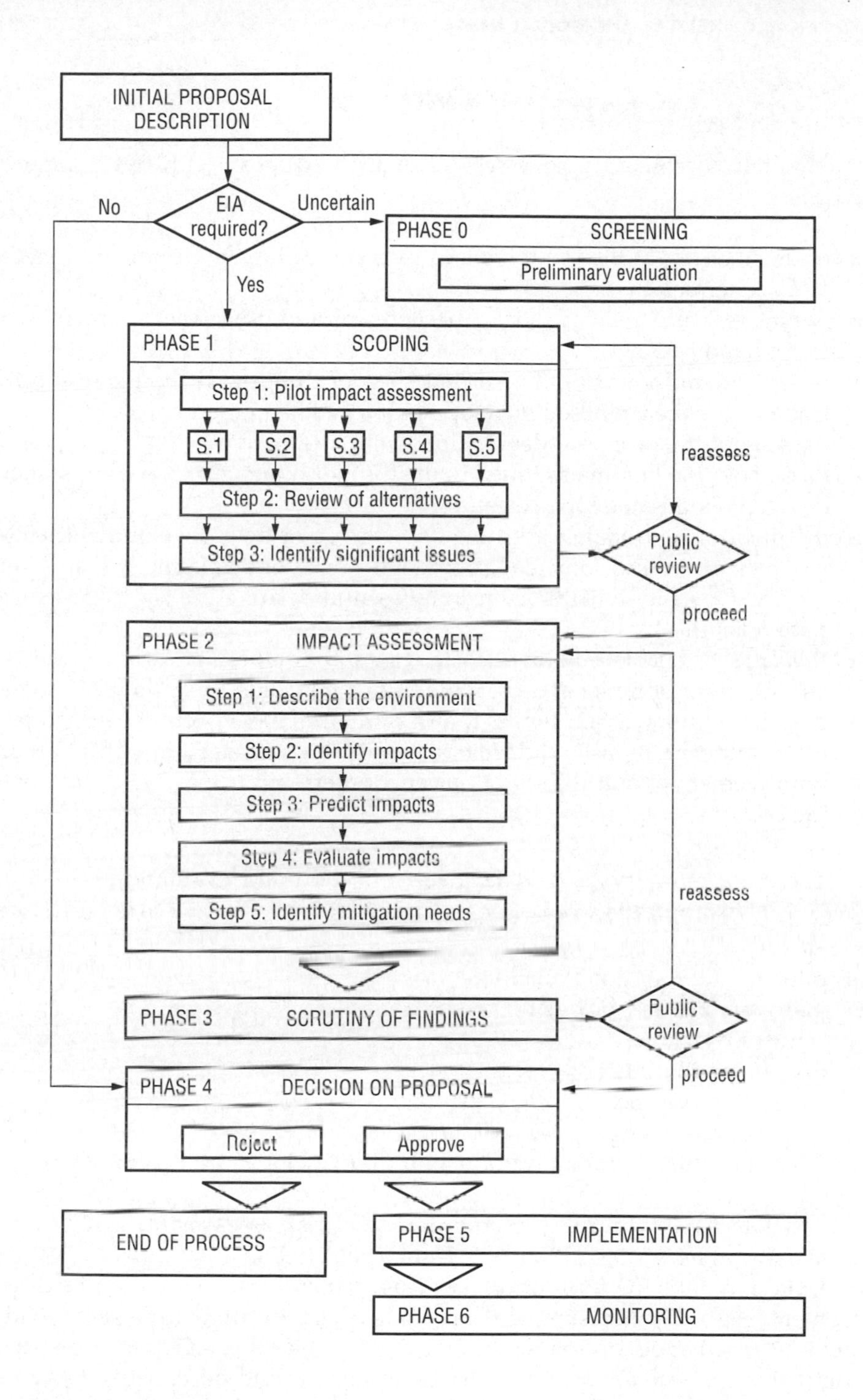

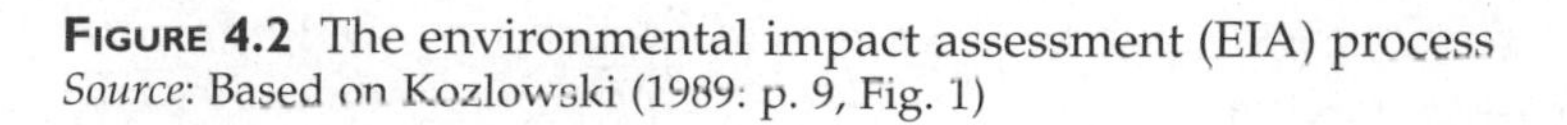

FIGURE 4.2 The environmental impact assessment (EIA) process
Source: Based on Kozlowski (1989: p. 9, Fig. 1)

DATA

Adequately to forecast and assess changes associated with development requires:

- understanding of the structure and function of the environment, society and economics;
- willingness and ability to study mechanisms and processes, not merely to survey and catalogue;
- baseline data indicating the current state of things and the likely future trends were the proposed development not to be implemented;
- data about the proposed development and alternatives;
- data about likely impacts (often in the form of hindsight data from earlier developments somewhere else);
- the monitoring of data indicating any change or new development, usually by reference to some environmental, social or economic *indicator* (by which is meant something meaningful and relatively easy to measure with reliability);
- *standards* to enable objective exchanges and comparison of data and to allow setting of limits and recognition of thresholds. A standard is some criterion within certain limits. Many standards have been developed for temperate conditions and developed countries, and need modification or improvement to suit the situations encountered in tropical and developing countries.

There are many types of data, data collection and evaluation (Holling, 1978: 73; Nijkamp, 1986: 243). For example, data may be:

- poor, medium or good quality;
- 'hard' or 'soft' (*see* Box 4.1b);
- static or dynamic;
- discrete or continuous;
- single-step or process;
- evaluated by a single person or by a team;
- normal or 'fuzzy' (for an introduction to fuzzy logic, *see* Kosko, 1994).

Fuzzy data have been used in environmental risk assessment; *see Environmental Management* 1994: **18(2)**: 303; Kung *et al.* (1993).

Data may take the form of direct measurements, or may be indirect (*see* Canter, 1996: 122). Measurement may take place in qualitative steps (bad, neutral, good) or be continuous (. . . −1, 0, +1 . . .)(*see* Box 4.4) or fuzzy, meaning that there is no abrupt boundary between true and false, between 0 and 1; everything is a matter of degree (Kung *et al.*, 1993). Data-gathering can be poorly directed and encyclopaedic ('shotgun approach') or well-informed and carefully chosen. It is also desirable that data collection methods be standardized. Often the assessor must rely on an indicator; although indicators can be very accurate and valuable, one indicator seldom provides all that the assessor needs (OECD, 1991). Some indicators are based on single measurements, such as the use of gross national product as an indicator of a nation's economic performance; others, like the Human Development Index (UNDP, 1991), are composite values based on a variety of measurements.

Box 4.4 Types of scale

Scale	Nature	Measure of location	Example
Nominal	Classifies	Mode	Classify species, number soil types, etc.
Ordinal	Ranks	Median	Rank in order
Interval	Rates in units of equal difference	Arithmetic mean	Time (hours, minutes), temperature (°C)
Ratio	Rates in units of equal difference and ratio	Geometric/harmonic or arithmetic mean	Height, weight

Note:
Interval scales are calibrated so that each unit represents an equal size difference but the origin point is arbitrary. Interval data can be added or subtracted, but not multiplied or divided.

Nominal scales assign a name, label or description (qualitative categories). One can count the number of items in each category, but avoid mathematical manipulations of those data.

Ordinal scales are based on discrete orderings; that is, they indicate that a rank is higher or lower than another rank, but not how much higher or lower. Ordinal numbers cannot be added, subtracted, multiplied or divided. It is difficult to compare points on an ordinal scale (the Leopold matrix uses an ordinal scale, so one should not add its scores up).

Ratio scales have equal-sized differences between units as well as a meaningful origin point. Items on the scale can be added, subtracted, multiplied, divided (e.g. monetary units).

Interval and ratio scales can be dealt with by parametric statistics; ordinal data require non-parametric statistics. Parametric (or 'classical') tests dominated statistics until the 1930s. These can only be applied to data on an interval scale. Non-parametric tests can be applied to data on nominal and ordinal as well as interval scales. Crudely, non-parametric statistics is more flexible in the data it can use.

Source: Based partly on Westman (1985, table 4.3, p. 139)

It was stated above that data can either be measured directly, or be indirect. To identify an impact it is often necessary to monitor selected *indicators*, an indirect form of data. They may be physical, biological, social, cultural or economic. Some organisms act as valuable indicators, being sensitive to things the impact assessor wishes to monitor. Such organisms can be used as 'detectors'; for example, there are organisms sensitive to temperature change or pollutants. Other organisms may act as 'accumulator' indicators, including plants or animals that concentrate pollutants or radioactivity to give advance warning of slight changes in the environment at large. Another group of organisms are useful as indicators by virtue of some characteristic or quality. For example, the presence of a certain number of the gut bacterium *Escherichia coli* per litre of water is a widely used sign of pollution by sewage.

Because the causes and effects of environmental and socio-economic changes are complex and not fully understood, and because it deals with future uncertainties, environmental impact assessment cannot give wholly accurate results. Also, no matter how good the data and the understanding of situations, environmental impact assessment involves subjective judgement (Matthews, 1975). It could be said that impact assessment seeks to help decision-makers in the face of uncertainty, and precise answers are not possible or needed (De Jongh, 1988). The gathering and handling of data for environmental impact assessment may be complicated by demands for confidentiality (a development may be strategic or be commercially sensitive), or because groups of people or other things involved may each be affected in various ways, be easily missed or double-counted (e.g. people may be counted as local residents, as members of a special group – say fishermen, as members of an economic group or as members of a religious group). Double-counting may be a problem if an environmental impact assessment is carried out by a team that fails to liaise closely enough (for guidelines on environmental impact assessment data-gathering, *see* Fortlage, 1990). Recent developments in electronic data-handling, for example geographical information systems, are making it easier to gather, hold, update and use data.

Sometimes impact assessments give no indication as to the reliability of data and may repeat dubious sources. (This is a widespread problem in applied and academic studies and is not restricted to impact assessment.) Efforts should be made to check data, and if possible indicate confidence limits, probabilities, etc.

One may recognize a division between 'hard' (quantitative, objective, reliable and verifiable/testable) and 'soft' (qualitative,'subjective', difficult to test) data. This is basically a division between measurement and description. Examples of hard data include geological surveys, soil tests, etc.; soft data include opinion survey results, aesthetic judgements, social values, etc. Between the two extremes may be recognized 'intermediate data', which are reliable but not absolute or consistent: water quality tests, tests on the condition of vegetation etc. Quantitative data are often regarded as reliable and scientific, whereas qualitative data tend to be seen as less reliable and 'social'.

Within environmental impact assessment, there has been a trend towards 'hardening'; that is, towards the quantifying of qualitative measurements. An example is the attaching of numerical values by the use of a technique such as the Environmental Evaluation System (*see* Chapter 5) (Bisset, 1978; Hollick, 1981b; Hipel, 1988; Wathern, 1988: 21). Whatever technique is used, care needs to be taken to ensure that any transformation of qualitative information to quantitative is apparent.

There is nothing wrong with careful subjective judgement, provided those making the judgement and those using it are aware of the value judgement framework adopted. This is just as well, because impact assessment relies heavily on value judgements. Hyman and Stiftel (1988: 25) stress that assessments should keep facts separate from values. Hollick (1981a) argues that because rational evaluations must consider a wide range of issues and interlink them, quantitative data can be only partly adequate.

Ironically, science is increasingly being required to advise before it has acquired full understanding or adequate 'hard' data, and so has tended to face the same problems as those using 'soft' data (a point made by Sir Claus Moser in an address to the British Academy's annual meeting in 1990). Provided that data are collected and handled with care, qualitative and quantitative measurements should both have value and should complement each other. Lawrence (1993) argued the need for a blending of both.

Many fields are difficult to quantify, but could be assessed on a qualitative scale. Examples are:

- whether the chances of sustainable development have improved;
- the visual effects of development – issues such as aesthetics or visual intrusion;
- changes in job opportunities;
- social cohesion.

PHASES, STEPS AND STAGES OF THE IMPACT ASSESSMENT PROCESS

It is possible to recognize a number of (ideal) phases of the impact assessment process (*see* Fig. 4.1). However, they may not all be followed in a given assessment.

SCREENING

When interest in a possible development becomes strong enough, perhaps with the authorities issuing a statement or notice of intent, screening should be initiated. Screening is concerned with selecting developments that require assessment; it is, or should be, the 'trigger' for environmental impact assessment (Clark *et al.*, 1984; Wathern, 1988: 9). It ensures that unnecessary assessments are not carried out, but that developments that do need assessment are not missed. Put crudely, screening asks, 'How far and for how long is the proposed development likely to affect environment and people?'

Screening caused huge problems in the USA in the first few years after the passing of the National Environmental Policy Act. According to that act, environmental impact assessment was to be applied to 'major actions significantly affecting the quality of the human environment' (abbreviated in Section 102 of the Act to MASAQHE), but significantly, was not legally defined. Establishment of criteria to screen is thus important. Not all countries have mandatory screening, and some of those that do, use weak criteria and procedures.

In the USA nowadays, screening is mandatory and takes place by means of an *environmental assessment* (in the UK, 'environmental assessment' is used by the government to mean an environmental impact assessment) to establish, as required by the Act, whether a proposed development is a major action significantly affecting the quality of the human environment. If it is, a full environmental impact assessment is indicated. This screening process is known as an *initial environmental evaluation* and is conducted according to criteria set by federal agencies. Often it involves reviewing preliminary planning documents. If the initial environmental evaluation suggests that it is not necessary to proceed with an environmental impact assessment, a 'finding of no significant impact' (FONSI – *see* the Glossary) statement is issued and the environmental assessment is either released to the public 30 days before a final decision is made whether to proceed with the development, or listed in a Federal Register – which the public can see, but seldom trouble to do so.

A US-type initial environmental evaluation (termed in some countries as initial impact evaluation, initial impact investigation, initial impact assessment or preliminary assessment) can result in improvements in development without the full use of environmental impact assessment. However, caution is needed, as it is possible with this type of approach to release an environmental assessment or initial environmental evaluation rather than a full environmental impact assessment to reduce delay, and deflect public scrutiny (Wood, 1995: 118). Some countries, like the UK, have no 'half-way solution' like the initial environmental evaluation, and decide either to prepare or not to prepare a full environmental impact assessment. More and more countries are adopting national regulations that require screening, but it has often been far from well or widely practised. Screening needs to be carefully and objectively administered if it is to be effective. In Canada, initial environmental evaluations are quite demanding exercises, involving public involvement, and the country has an effective government screening panel. Screening should also be relatively quick and easy, and clearly identify what needs full environmental impact assessment.

Screening may not give a clear indication that impact assessment is required. Where screening gives an uncertain result it is probably wise to be safe rather than sorry and to favour assessment (Clark *et al.*, 1984: 165; Wood, 1995: 115–29). Sometimes authorities fail to screen adequately. It is not unusual simply to apply assessment in a more or less arbitrary way to developments of a certain cost or size – which may make little sense. For example, a reservoir of 99 km^2 may escape environmental impact assessment if a nation has decided to assess only those of 100 km^2 and over, and yet cause serious problems. Developing countries in particular may be lax about screening unless a funding agency insists, which is understandable when money and expertise are in short supply.

When screening is used, the need for an impact assessment may be indicated by criteria that might include one or more of the following:

- some recognized threshold value is reached;
- the site of development is sensitive;
- the development involves known or suspected dangers or costs;
- there is a risk that it will contribute to cumulative impacts;
- there are unattractive input–output considerations, for example labour migration to the development, pollution arising from it, heavy traffic generated by it.

Initial environmental evaluations are usually made by much the same methods as impact assessment proper: overlays, matrix methods, etc. For example, Canada's Federal Environmental Assessment Review Office has developed and published screening procedures, and the United Nations Environment Programme has published Initial Environmental Evaluation Procedures. In practice, initial environmental evaluation and impact assessment are likely to overlap. Screening should help in avoiding cumulative impacts: a small project might present little threat and thus escape full assessment, but more than one such development, separated in time or space, may have considerable effects and screening could spot this.

Scoping

There is sometimes confusion between screening and scoping, and there is some overlap. Scoping is an important stage that follows close on screening (or is the first activity if screening is neglected), and it *should* help determine the terms of reference for impact assessment. It is concerned with establishing the issues that impact assessment should consider and the parameters and approach – what the assessment is for, the limits of study, the timetable, the tactics, who should conduct the assessment (for a discussion of how to select personnel for assessment, *see* Save the Children, 1995: 15), the level of precision, etc. In short, it determines the focus of assessment – which should be on priority issues. A rapid way of scoping is to gather a group of experts and those associated with a proposal in order for them to 'brainstorm'; that is, hold a discussion aimed at reaching a mutual agreement.

Save the Children (1995) put forward a concept similar to scoping, in relation to assessment of proposals, in the form of two guiding mnemonics: SWOC – meaning, establish Strengths, Weaknesses, Opportunities and

Constraints – and SMART – a proposal should be Specific, Measurable, Achievable, Relevant and Time-bound.

Sometimes assessors devote excessive time and effort to obtaining baseline information. Assessment should not concentrate too much on data collection, nor try to consider all possible impacts. Such encyclopaedic efforts are costly and can be of relatively little use. What scoping must do is focus on identifying significant impacts – something of a 'catch-22' situation, in that the scoping has to decide what the impacts may be even though they may not be identifiable without a full impact assessment. In practice, assessors often resort to the use of checklists to help.

Rapid rural appraisal, participatory rural appraisal (*see* Chapter 2) and social evaluation face similar challenges, and those involved with these approaches have made calls for something similar to impact assessment screening, which would allow 'appropriate imprecision' or 'optimal ignorance' (that is, measure what is worth measuring and know what is worth knowing) rather than academic, in-depth study (Marsden *et al.*, 1994: 26).

Scoping can be valuable if continued throughout assessment, and not just carried out as a pre-assessment phase. Indeed, Kennedy and Ross (1992) reviewed the development of scoping and proposed the adoption of *focused environmental assessment*, which includes scoping during impact identification, assessment and management. There is a good reason for ongoing scoping. As development proceeds and impact studies progress there are sometimes unexpected problems, and assessment will therefore probably have to alter its 'boundaries'. If scoping is continued it can help in these circumstances to prevent expansion too far beyond the original terms of reference.

If conducted well, scoping prevents assessment from yielding unwieldy, badly assessed, encyclopaedic environmental impact statements; improves focus; helps prevent duplication of efforts; keeps costs and delays down; and improves co-ordination among the impact assessment team. In short, scoping can have a beneficial effect on virtually the *whole* assessment process.

Scoping was pioneered in the USA when the National Environmental Policy Act was revised in 1979, and has since spread. Litigation in the USA had encouraged impact assessors to adopt an 'encyclopaedic' approach so that in any defence 'thoroughness' could be argued even if the assessment was ineffective; scoping was one way of encouraging a return to a more considered approach (Wathern, 1988: 26). Scoping may be conducted by the developer, funding agency, government or some corporate body, but it is probably best done by an independent third party. It is possible to distinguish physical and ecological scoping, and socio-economic scoping. Ideally, scoping should include consultation with various affected groups.

As with screening, scoping may not be mandatory and is often poorly organized. If the latter is the case, impact assessment is likely to be somewhat *ad hoc* in approach. A common practice is to have scoping meetings between key actors. Some authorities involve the public (the National Environmental Policy Act required scoping to involve the public). Deciding what to omit is often very subjective and controversial, and specialists frequently want to add more than they are prepared to drop (Beanlands, 1988; Gilpin, 1995: 19). Biogeophysical scoping is probably easier than social scoping because the former can conclude that it is reasonable to study those things that current knowledge supports, whereas the latter depends more on public opinion and value judgements that are less fixed.

Canada has formal scoping legislation. In the USA, if screening suggests that an assessment is necessary, a notice of intent to carry out an environmental impact assessment is released, followed by a scoping meeting. The UK does not insist on scoping, although the Department of the Environment does strongly advise developers to scope and consult with local planning authorities. The World Bank and some other funding bodies require scoping before full release of money, but in developing countries scoping is often omitted or weak. Wood (1995: 130–42) reviewed the approach to scoping adopted by various countries, and refinements of the process were considered by Kennedy and Ross (1992).

CONSIDERATION OF ALTERNATIVES

Many assessments fail to consider all possible development alternatives, often because they are initiated too late, after planning decisions have been made and when implementation is about to begin or has started. (As discussed in Chapter 3, there is still value in such post-development assessments; for example, to check whether permits or agreements are complied with; to make a stocktaking of problems.) Failure to consider alternatives may happen if funds for assessment are limited, encouraging authorities to wait and see whether there is a need (UNECE, 1990).

IDENTIFICATION AND MEASUREMENT OF IMPACTS

Techniques and methods of impact identification, description, measurement, communication, review and audit are discussed in Chapter 5 (for a definition of 'impact', *see* the Glossary). There have been some moves to apply the term *effect* to natural results and *impact* to results caused by human activities, but the distinction has not had wide adoption.

Impact identification and measurement may be broken down into:

- identification of possible direct, indirect and (as far as possible) cumulative impacts;
- assessment of each impact significance (i.e. its extent and importance);
- evaluation of the likelihood that an impact or impacts will occur;
- a forecast of when or how often identified impacts might manifest themselves (*see* Box 4.5).

One of the characteristics of environmental impact assessment and related assessment is that impact identification and measurement are conducted in a

Box 4.5 Identification and assessment of impacts: the task

- *Identify impacts* – checklists, information on development proposals, social and physical baseline data.
- *Measure and predict* – what are the magnitude, the time frame (when will impact occur?), the extent? May require modelling.
- *Determine significance* – is it a danger, a nuisance, a benefit? Can it be reversed?
- *Establish reliability of predictions* – how likely is the impact?
- *Can identification be repeated objectively?*
- *Can impact be communicated?*

Source: Warner and Preston (1973); Skutsch and Flowerdew (1976); Lee (1982: 76)

structured manner so that there is less chance that something is overlooked or has wrong values attributed to it (Duinker and Baskerville, 1986); checklists, matrices and interactive computer programs are widely used as aids. Scoping should have indicated critical issues and helps prevent double-counting. An assessor should be objective, should try to establish 'confidence limits' on information in order to judge the reliability of predictions (Matthews, 1975; De Jongh, 1988), should be aware of positive or negative feedbacks, indirect impacts and cumulative impacts, and should exercise caution – the environment, society and economics seldom function according to the textbook (Munn, 1979: 33).

For most types of development there are checklists of possible impacts, or use can be made of computer techniques to help identify impacts (Julien *et al.*, 1992), and other approaches such as risk or hazard assessment may feed into impact identification. However, some developments break new ground or involve 'intangibles' that are difficult to identify and measure: quality of life, morals, sense of community, aesthetic conditions, etc. Useful ecological criteria like ecological resilience and function deserve more attention, according to Cairns and Niederlehner (1993), but, as already mentioned, there is a need to check the concepts and indicators being used to see that they are appropriate and reliable (Spellerberg and Minshul, 1992). Standards help to counter subjectivity, ease comparison and negotiations between bodies or states, and facilitate long-term monitoring, although there are problems because the real value of even some of the widely used standards, such as LD_{50} (the dose of a drug or other test substance at which 50 per cent of a group of organisms to which it is fed or administered die) is doubted by some experts.

There has been considerable effort to improve impact identification, description and measurement. The techniques available include field surveys; social surveys (Maclaren, 1987); literature searches; expert workshops; agency guidelines or checklists; interviews with specialists; and public consultations (*see* Chapter 5). Assessors tend to stress negative impacts and overlook positive and potentially positive impacts.

Impacts may be divided into: *direct impacts* (primary or first-order), *indirect impacts* (secondary or second-order, third-order, and so on) and *cumulative impacts*. Direct impacts are relatively easy to predict as they take place directly in space and time, but indirect and cumulative lie along or at the ends of chains of causation and so can be very difficult to predict; the effect is hidden some steps away in space and/or time from the cause. Cumulative impacts may be subdivided into: *additive* or *aggregate impacts* (basically, the sum effect of a number of impacts coming together) and *synergistic impacts* (resulting when an interaction of a number of impacts is 'greater than the sum of the parts'). Impacts may sometimes interact to cancel one another out; or perhaps solving one impact problem may trigger another. Ideally, assessors should consider more than just the impacts of 'their' development, policy, plan or programme, because external factors, such as other developments or policies, plans or programmes, may have an effect. Strategic environmental assessment and other tiered assessment approaches seek to do this.

Impact identification methods tend to operate either like a car's wide-reach dipped headlights, identifying a broad swathe of direct impacts, or like a 'pencil beam', probing well ahead to indirect impacts but on a limited front. Finding direct and indirect impacts over a wide front is very difficult in practice, and tracing possible cumulative impacts is even more so. Predicting physical impacts is not easy; predicting social impacts may be even more difficult. This is because of the complexity and fickleness of human behaviour, added to which are the facts that there may be interactions with physical and economic factors, that there are often different groups in a society that are affected in different ways, and that there may even be differences in impact between the two sexes. Crisis situations may attract more attention than slow, insidious trends, but these latter may reach a threshold to give a sudden 'time bomb' effect. Impacts often develop in a non-linear pattern with delays between cause and effect or sudden, easily overlooked shifts. Impact assessors must do the best they can and avoid giving clients a false sense of security.

Impact assessment often cannot proceed without an adequate set of baseline data. Baseline studies follow scoping and seek to establish the state of an environment, society or economy. Once that has been done, it can be established what would probably happen if no development took place and what would arise if it proceeded (Beanlands, 1988). It is often impossible to obtain a comprehensive collection of baseline data, so it is necessary to focus on selected ecosystem components or socio-economic factors – 'key' species or processes or established environmental indices (Ott, 1979). The assessor may have been provided with a set of pre-agreed 'limits of acceptable change' which the development must not exceed. Baseline data should give a picture of the status of the environment (and society) prior to development, which may not be a static situation: there may be a trend of natural degeneration or improvement, or a cyclic or periodic pattern of change. An impact assessment should present as one of the development options the situation that would prevail if no development took place. However, Graham Smith (1993: 17) warned that too much attention is often directed at obtaining baseline data and not enough on conditions during and following implementation or on assessment of information. Sometimes the wrong information is collected and is of little use to the assessor. As discussed earlier, excessive concentration on baseline studies wastes time and funds, and may lead to encyclopaedic and descriptive, rather than useful and analytical, environmental impact statements. Impacts can be caused by factors unrelated to developments.

Identified impacts should be logged with information on their magnitude (local, regional, global), significance (threat, nuisance or benefit), character (duration, reversibility, etc.) and reliability of prediction. It may be difficult to do all this, given that the assessor has to forecast future situations (Culhane *et al.*, 1987).

Following impact identification there is usually an attempt to measure or characterize, often quantify, then to organize and present findings (Skutsch and Flowerdew, 1976). Assessors may be tempted to characterize an impact by using a mean value, but an average may be meaningless where environment, society or economy is subject to considerable change. For example, mean monthly rainfall would not be a good indicator to use for agricultural development planning in a region where there is considerable variation of precipitation through the year. Best- and worst-case predictions may mean crop failures in bad years and underexploitation of the land in good years. It is better, whenever possible, to give a statistical distribution of the range of values and the confidence with which the range is held to be true. Sometimes it is possible to calculate the probability of a given impact using probability risk assessment (Carpenter, 1995). Direct measurement may be supplemented or replaced by modelling of various forms (e.g. Holling, 1978: 66), extrapolative exercises (i.e. projection of past trends to try to indicate future ones) and normative approaches (i.e. working backwards from a desired development outcome to try to see whether it would cause impacts). Normative approaches are useful in the difficult task of identifying impacts that lie well into the future.

Cumulative impacts

Perhaps the greatest challenge facing assessors during the impact identification stage is posed by cumulative impacts. These may be consequences of a single development, more than one development or a combination of development and natural effects (Gilpin, 1995: 31). The assessing of cumulative impacts, usually termed *cumulative effects assessment*, is generating a growing literature (Bain *et al.*, 1986; USACE, 1988; Lane and Wallace, 1988; Herson and Bogdan, 1991; Cocklin *et al.*, 1992; Contant and Wiggins, 1991, 1993; Davies, 1992; Smit and Spaling, 1994; Damman *et al.*, 1995). It is important not only to mainstream environmental impact assessment but also on a global scale because cumulative impacts pose serious threats to the Earth (for a review, *see* Orians, 1995; *see also* the section 'Global-scale and transboundary impact assessment' in Chapter 6 of this book, p. 171).

Graham Smith (1993: 27) tried to list the types of cumulative impact process, suggesting:

- *linear additive effects* – the result of incremental additions or deletions, each of which has a fixed effect (e.g. toxic pollution episodes affecting a lake). The time factor may be crucial: an impact sequence a + b + c might be coped with if it happens slowly, but might be a problem if sudden.
- *amplifying or exponential effects* – each increment/deletion has a greater effect than the one preceding it (e.g. emissions of CO_2 to the global atmosphere);
- *discontinuous effects* – no apparent consequence until a threshold is crossed, then sudden, perhaps catastrophic change; for example, various chemical 'time bombs' like the slow build-up of toxic compounds in soils which suddenly cease to bind them, perhaps as a consequence of acid pollution (biological 'time bombs' are also possible). Another example would be lake eutrophication;
- *structural surprises* – spread of effects to encompass a range of physical, social, economic factors; for example, reduced inflows leading to drying up and pollution of the Aral Sea which led to ecological loss, the failure of a fishing-based economy, community breakdown, health problems and political repercussions.

An alternative approach to classifying cumulative impacts was adopted by Beanlands *et al.* (1986), who recognized three groups of cumulative impact process:

1. repeated incremental insults to the system – additive effects;
2. actions that lead to system change or a dynamic process;
3. accumulation of impacts by cycling over space and time, including 'chemical and biological time bombs'.

A cumulative impact occurs when either a threshold is crossed; a feedback has taken place; a synergistic effect has occurred; there has been a 'surge' – a sudden release of, say, a pollutant; or there has been a cycling effect. To predict these types of scenario requires monitoring, and perhaps modelling. Cumulative effects assessment makes use of a diversity of techniques, including network analysis, spatial analysis, ecological modelling and expert opinion to try to assess systematically whether a cumulative impact will appear. These methods are discussed further in Chapter 5.

It is obvious that cumulative impacts are complex, so that, even with improvement, cumulative effects assessment will never be free of uncertainty. Cumulative impacts may be experienced after long delays and at a distance from their causes. Unfortunately, most decision-makers, governments and the public think on too local, too sectoral and too short-term a time-scale to appreciate such threats. To deal with cumulative impacts, institutions and ethics with at least a regional and ideally a global (and long-term) focus are needed (Goldrick and James, 1994; Orians, 1995: 7). It is not surprising that many impact assessments neglect cumulative impacts.

Strategic environmental assessment has been promoted as a means of dealing with cumulative impact assessment (*see* Chapter 3).

Interpretation and evaluation of impacts

Interpretation of the significance of identified impacts is not a precise art and the presentation of the final assessment is also likely to vary in standard. Increasingly, governments and professional bodies seek to ensure standards. Nevertheless, users of impact assessment should make themselves aware of the qualities of the techniques and methods used and of the standard of assessment in practice (Thompson, 1990), seeking independent advice if need be.

Having been identified and measured, impacts are usually displayed so that patterns of impacts can be seen, comparisons between development alternatives made, rankings decided, past and present developments compared, observed phenomena explained, and so on (Munn, 1979: 33; Lahlou and Canter, 1993). Interpretation and evaluation is the 'core' stage of impact assessment, the point at which the assessor determines what is *significant* (Duinker and Beanlands, 1986; Canter and Canty, 1993). Once the assessor understands what has been observed, an attempt is made to predict future trends and establish whether a given impact deserves attention, and what reliability the assessment has.

Each identified impact may have one or a number of attributes that assessors can consider (Parkin, 1992). Impacts may be tested for statistical significance, ecological significance, social significance, etc. Impact interpretation and evaluation usually leads to actions and strategies designed to counter unwanted and strengthen wanted impacts – so there is a feedback into the development process that can mean that impacts may not appear in their predicted form (Mihai, 1984). Techniques and methods adopted to identify impacts can often assist interpretation and evaluation.

Significance is not an easy thing to define, and to do so involves subjective judgement; there is some truth in the cynical remark of one assessor: 'no impact is significant unless the ''punters'' [customers] feel it is' (Fuggle, 1989: 36). However, the public are often poorly aware of threats and opportunities. Some ecologists, and possibly other specialists, might argue that any change is significant. The assessor usually asks questions like: Is the impact a permanent change? Is it reversible? Is it an unwanted change? How wide an area and how many people are affected? In short, the assessor 'trades off' various factors against each other to arrive at a (subjective) judgement of impact.

Interpretation and evaluation have tended to split into two camps: quantifiers, and those happy to use qualitative data. The separation is probably overstated and will, it is to be hoped, break down in the future (Lawrence, 1995). Whether impacts are expressed in a qualitative or a quantitative manner, they are still based on subjective judgements; the quantitative rankings, indices, scaled attributes, etc. give only an appearance of objectivity and application of science. There is a danger that attempts to quantify will obscure the way in which judgements have been arrived at, and it is not unknown for the process to be statistically or mathematically dubious. For example, ranking is less informative than rating; also, manipulations of nominal and ordinal scales can sometimes be statistically doubtful (Skutsch and Flowerdew, 1976). If various sets of data are given scores which are aggregated and perhaps weighted to form a 'grand index', the user may fail to note whether component sets have effects, and it may be impossible to judge strengths and weaknesses of alternatives. It is possible for quantification to be used to side-step scrutiny by decision-makers (Bisset, 1978: 54). There is another criticism of quantitative assessment methods: that in providing scores for each development alternative, they allow decision-makers (elected or whatever) to avoid responsibility; in effect, they can say 'the assessor chose', rather than those who commissioned and used the assessment – it should not be the job of the assessor to decide (Thompson, 1990).

Despite the warning about the risks of quantification, it would be very valuable if impacts could be satisfactorily converted into standard units or comparable indices, as, for example, cost–benefit analysis does (it does so by quantifying with monetary units). A well-established quantitative method (*see* Chapter 5), the Environmental Evaluation System, assigns value functions to quantify different impacts with the aim of allowing meaningful comparison.

Mitigation and avoidance of impacts recommendations

Impact assessors may be expected to suggest impact mitigation or avoidance measures. In doing this, care should be taken to outline the range of possible actions, together with costs and an indication of associated problems and advantages. Impact assessors, in this and in other aspects of assessment, must ensure that planners and decision-makers are responsible for choice of action. It is not the role of environmental impact assessment to make a choice or present a course of action in a way that effectively forces a decision. Planners and decision-makers must be responsible for development choices and environmental impact assessment should help reinforce their choices. Environmental impact analysis should communicate information as objectively and as clearly as possible.

There are often weak measures in force to check whether selected mitigation or avoidance measures are followed. The US Department of Energy is an example of an agency that has, since the early 1990s, checked whether measures are complied with.

Communication of impact assessment

Before trying to communicate information about impacts, it is necessary to ask, 'With what person or group of people am I trying to communicate?' It could be the general public, decision-makers, technical advisers, a minister, a dictator, or all of these. Often the impact assessment is done by consultants hired by the developer or by government employees, and even if the quality is good, objective communication can be a problem: the assessors may communicate what they feel the commissioning body wants to hear and support what is proposed. The solution is to use approved (ideally, accredited) consultants, or an independent group and have an independent agency vet the results.

Environmental impact statements

The impact assessment may be communicated by report, statement or public meeting. The most common form is an *environmental impact statement,* sometimes called an *environmental effects statement* or *environmental impact report,* and in the UK 'environmental statement' is often used to mean the same thing. Sometimes a full-length environmental impact statement is presented in the form of a précis or short report; for example, the Commonwealth of Australia issues Public Environment Reports to the public, these being a sort of summary of the full environmental impact statement). Whatever method of communication is adopted, it must be accurate, concise and clear, and as far as possible predictive. In the USA there is usually a period of at least 3 to 4 months between publication of an environmental impact statement and the start of on-the-ground development. Some reports and environmental impact statements are unwieldy, taking up metres of shelf space, but sometimes the opposite is the case, and they are too circumspect or neglect things such as social impacts. James (1994: 7) argued for environmental impact statements to include economic impacts as well as environmental and social ones. Gallagher and Jacobson (1993), reviewing 150 US environmental impact statements, found poor standards of presentation (especially typography), to the extent that in 12 per cent of cases they deemed the product 'unreadable'. Often circulation is restricted; a ministry or company may not be keen to release results to too many people, or the costs involved may be too great. One effect of such 'grey literature' (material that fails to circulate freely) is that planners and decision-makers have difficulty in learning from hindsight, or fail to do so altogether. Hindsight knowledge should be vast: in the USA alone there had been over 23 500 environmental impact statements by 1985 (Hyman and Stiftel, 1988); whether it is accessible or even sought out is another matter.

Environmental impact statements are *aids* to decision-making, not the decision-making process itself. In the USA the use of independent mediators to improve their content and use is quite common. The USA and Canada have established regulations concerning environmental impact statements. The European Union has regulations outlining how they should be presented (part of EC Directive 85/337), but often there are no firm guidelines on how communication should be handled. The literature on environmental impact statements is full of hints and tips. Communication should:

- be objective and unemotional;
- meet any requirements or regulations in force;
- use consistent units and descriptions;
- avoid being vague;
- not duplicate information that is already reasonably available;
- show sources of information and give adequate scientific referencing;
- avoid jargon and cliché;
- list what has been excluded from study at the scoping stage;
- avoid unnecessary statistical analysis;
- use the same clear terminology throughout.

(Canter, 1977: 233–6; Gilpin, 1995: 16; Wood, 1995: 146; Kreske, 1996). Coverage should include:

- proposed actions;
- details of the environment affected;
- information on how environment and people would fare if the development did not take place;
- a forecast of significant impacts if action(s) are taken;
- consideration of likely impacts of alternative developments; information on mitigation measures;
- a non-technical summary

Some authorities ask for indication of irreversible impacts (if sustainable development is a goal, it is vital that irreversible impacts be flagged), and some ask for environmental impact statements or reports to adopt a phase-by-phase approach, setting out what happens during initiation and construction, what happens after completion, etc. An environmental impact statement should give guidance for ongoing monitoring. In many respects its terms of reference are decided, or should be, at the scoping stage.

Review of impact assessment

Impact assessment review is not a precise term; it most usually refers to checking the progress of an impact assessment, as opposed to establishing whether it was effective. It can take the form of monitoring, surveillance or audit (*see* the following sections). The impact assessment review process should be controlled by legislation and established procedures, and seek to ensure that impartial and proper consideration takes place, without undue delay, and that the environmental impact statements are adequately acted upon.

A review may or may not involve the public. If it is decided that it should, there are two broad ways: use of a mediator or mediators; and use of a review panel, or similar review body. Some countries have panels, boards, statutory commissions or committees that check the implementation of assessment and the final environmental impact statements. These may be a tight-knit group of officials or composed of selected members of the public, the judiciary, university staff and other representatives. Their effectiveness depends on how much power and independence they wield as well as their composition; some are primarily advisory rather than being effective overseers. Canada has operated an effective review process since the late 1970s: the Federal Environmental Assessment Review Process, which had provisions for mediation introduced in 1992. A number of countries, including some developed countries have unsatisfactory review processes; for example, the UK was being criticized in the mid-1990s for lacking an independent review panel.

The boundary between review and post-environmental impact assessment audit is not precise. An environmental impact statement may be sent by authorities to all agencies involved in the development, with a demand that they report back within a set time. If this allows modification of the environmental impact statement it is basically a review process; if it gathers information on an already finished environmental impact statement it is a post-environmental impact assessment audit approach.

Making use of impact assessment

It makes little sense, other than for data collection, to carry out assessments unless they are used to avoid, mitigate or remedy problems or grasp opportunities identified. There have been attempts to investigate how assessment is used. For instance, Sager (1995) examined how decision-makers use impact assessments in Norway. Novek (1995) examined the appliçation of environmental impact assessment to large pulp and paper-mill projects, and concluded that authorities, in the contradictory position of being both development promoters and environmental regulators (as is often the case), have approached it as a means of 'political legitimation and settlement of social claims'.

Mitigation attempts, and the scoping and impact assessment processes, may have impacts. Suppose that an environmental impact assessment warns that a dam will flood an area of forest; developers might accordingly clear the vegetation and in doing so pollute the region with herbicide. Or, perhaps, a flood protection bank built in response to environmental impact assessment warnings might provide a useful road link. There may be more than one mitigation or avoidance response to an assessed impact, one of which is likely to involve fewer side-effects than the others. Should the environmental impact statement identify mitigation options, or is that the responsibility of its users?

Mitigation may take the form of compensatory action. For example, if a river is contaminated with salts as a consequence of irrigation development before flowing into another country, to mitigate this the developer might choose to desalinate some of the river water rather than stop the irrigation development. Unfortunately, mitigation is often not implemented or lapses. It might be that a holiday camp visible from an upland scenic area is required, as a condition of planning permission, to have its roofs painted in neutral colours to reduce visual impact. A few years later they are repainted in bright colours but the change is not 'noticed' by local officials (Hollick, 1981a).

In the past, perhaps less so nowadays, it was felt that impact assessment ceased with the release of the environmental impact statement or completion of the development implementation. Now it is increasingly accepted that impacts can continue and that post-implementation monitoring is needed. Mitigation measures should also be monitored to ensure that they remain effective.

The 'heart' of the assessment process should be the option to modify or halt development or change a policy or programme. Too often assessment takes place when there is no longer an opportunity to do this. As is often the case with environmental issues, the law relating to assessment tends to lag behind perceived needs (Sheate, 1993).

ESTABLISHING WHETHER IMPACT ASSESSMENT HAS WORKED

Surprisingly, until recently the question 'has impact assessment worked?' was seldom asked. It may be that authorities would rather not have their planning problems disclosed, or that expertise was unavailable, or that there was little or no budget for quality control and auditing after expenditure on impact assessment, or that consultants prefer not to publicize their errors.

Before considering post-environmental impact assessment audits, it should be stressed that there are other ways of evaluating the adequacy of impact assessment: court actions and official inquiries. Early in the development of environmental impact assessment in North America this frequently happened. For example, in the USA there was the Calvert Cliffs case of 1971, which established a requirement for an environmental impact statement by an independent body on US Atomic Energy Commission nuclear facility proposals (*Calvert Cliffs Coordinating Committee* v. *Atomic Energy Commission* – Court of Appeal, the District of Columbia), and in Canada the Berger Inquiry of the 1970s (Ditton and Goodale, 1972: 23).

Although the terminology is somewhat confused, 'review' tends to refer to a process begun early on to check the effectiveness of assessment as it progresses, whereas 'audit' implies examination after completion of an assessment to determine whether it was adequate. 'Monitoring' may be used in either sense. 'Evaluation' is sometimes used instead of 'audit', but may indicate a less objective procedure, and probably excludes checks on compliance (that is, whether the environmental impact statement is acted upon; Sadler, 1988: 130; UNECE, 1990). There are, however, some countries where 'audit' means review of a draft environmental impact statement and it may be applied to the field of environmental auditing and eco-auditing (Bisset, 1980; Ortolano *et al.*, 1987; Ross, 1987; Buckley, 1991b; Lee and Brown, 1992; Wood, 1995: 197–211). And, confusingly, 'monitoring' (Wood and Jones, 1991) or 'post-project analysis' has been applied to impact assessment review (UNECE, 1990). 'Post environmental impact assessment audit' would seem to be the best term to adopt.

Post-environmental impact assessment audit

Buckley (1991a) described post-environmental impact assessment audit as 'systematic comparison of predicted and actual impacts'. The stress is on *hindsight* audit of environmental impact assessment. Some authors talk of 'post-development audit', (Rigby, 1985), but this implies a focus on development in general, rather than on the effectiveness of the impact assessment. Since the 1980s interest in impact assessment auditing has grown, and in some countries it is now well-established.

The success of impact assessment depends on a large number of factors being favourable as well as an effective procedure, so there are many ways in which it can fail: First, the assessment can be side-stepped (often for 'strategic' reasons or by powerful special-interest groups), or the body required to conduct an assessment manages to get the authorities to modify the assessment regulations ('regneg') so as to make it easier to satisfy them or avoid them.

Second, deliberate attempts to distort the impact assessment process are not unknown. A 'cosmetic' assessment may hide problems by using jargon, manipulating statistics or manipulating the presentation of data. Alternatively, a 'straw man' approach is adopted, whereby a bad alternative is stressed in the hope that it will draw attention from one that is favoured (but which may be unwise).

Third, as discussed earlier, changes to environments, societies and economies are difficult to predict, so assessment deals in uncertainties, and predictions are often more or less hypothetical. Assessment takes place but impacts are 'missed' or 'distorted', or alternatives are inadequately considered, or the communication of results may be poor. A common fault is that assessment considers things over too limited a time span (a 'snapshot' view is taken).

Fourth, the environmental impact statement may not be adequately heeded or agreed, with the result that mitigation measures are not taken. One example of this, ironically, is the case of the Linha Vermelha, a highway connecting Rio de Janeiro's Galileo Airport with downtown Rio. The agency responsible for assessment prior to construction is reported to have been pressured to exempt it from environmental impact assessment so that it could be opened in time for the 1992 UN Conference on Environment and Development – the 'Earth Summit' (Ortolano and Shepherd, 1995: 14).

Fifth, post-environmental impact assessment audits suggest that some environmental impact statements present information on impacts in a very vague way, making it difficult to do a before-and-after comparison – and raising the question, 'How did the developer make an objective decision?' (Fairweather, 1989: 143).

The problem of environmental impact statements presenting a distorted picture could be eased if better bias-checking procedures were applied to presented data and used to check what had been omitted. Better monitoring of the assessment process should lead to its improvement (Canter and

Fairchild, 1986). In fact, without ongoing monitoring, auditing is difficult (Bisset and Tomlinson, 1988).

Conflict over assessment has been reviewed by a number of workers (e.g. Covello *et al.*, 1985: 3; Rickson *et al.*, 1989). There seem to be three common reasons for it: local people oppose the action proposed or the mitigation measures advised by assessment; or second, one level of government or a department may not co-operate with another (local versus federal conflict is common); or, third, different sectors oppose assessment or its recommendations; for example, there may be opposition from unions, non-governmental organizations or employers.

What is tested and appropriate in, say, the USA may not work in another country. Assessment often has to be adapted and tested before it runs well for a given country or type of development. There are often pressures where funds are short to delay assessment until a development looks as if it will proceed – 'why waste money assessing an uncertainty?' – and that is too late for considering alternatives. Some assessments run into problems because they were done as a sort of 'snapshot', and at a later point the environment and/or the development, programme or policy changed. A solution to this last difficulty is adaptive assessment, with assessment accepted as *investigation*, not *determination*, of impacts; *see* Holling, 1978.)

To help ensure that an environmental impact statement is acted upon requires there to be enforcing mechanisms. These may take the form of legal requirements or rules and a sufficiently independent and powerful overseeing body, such as the Council on Environmental Quality in the USA. Alternatively, as in Australia, citizens may be given the legal backing to take developers to court if an environmental impact statement is not satisfactorily acted upon (Ortolano and Shepherd, 1995: 20). Wood (1995: 8) felt that environmental impact assessment functions best where there are specific legal requirements for it and where the authorities are responsible for considering the results.

Impact assessment audits may consider:

- the effectiveness of the impact assessment in assessing impacts;
- how well it is used;
- whether it can be modified to adapt to unforeseen developments;
- the relationship of the impact assessment with environmental management and planning.

Impact assessment audits can focus on techniques, procedures, process and the assessment's management and administration (for details, *see* Environment Canada and Transport Canada, 1988). It is important that the auditing be planned early on, not as an afterthought, and that the auditors are independent, competent and able to speak their views. Ideally, the audit results should be published somewhere accessible; however, there are still countries where auditing is not open to public scrutiny, and relatively few countries involve public representatives in the actual assessment process or consult the public.

The nature of impact assessment hinders auditing: Bisset and Tomlinson (1988) reported attempts with a number of UK environmental impact statements which proved difficult because only a small percentage of the predicted impacts could be properly studied. However, in the cases where studies were possible, an accuracy of 50 to 60 per cent was found. Making comparative reviews of various assessment exercises and systems is not easy because each case is unique (though there may be generic similarities). One solution might be to use computer techniques. Wood (1995) tried to overcome this problem, examining and comparing several environmental impact assessment systems (those of the USA, the UK, the Netherlands, Canada, Australia and New Zealand). Impact assessment often relies on expert opinion, rather than empirical evidence, and seldom is this opinion checked. Yee (1981) tried to assess the reliability of expert opinions on environmental impact relevant to foreign policy-making, and Culhane (1987) tried to establish the precision and accuracy of a sample of environmental impact statements produced in the USA. Case-studies of environmental impact assessment and the environmental management systems it served from 11 countries were audited by the UNECE (1990). In the UK, the Environmental Impact Assessment Centre of Manchester University carried out impact assessment audits for the Department of the Environment on environmental impact assessments conducted between 1988 and 1989 (EIA Centre, Manchester, 1992).

Assessment audit or review can focus on component parts of a case such as effectiveness of impact prediction or effectiveness of implementing the recommendations of an environmental impact assessment (*compliance auditing*), on the whole individual case, on several related cases, or on assessment systems. The focus may be on post-auditing or on checking on the progress of assessment at successive stages as it proceeds ('review'- which allows 'policing' and reaction to problems). Different modes of enquiry into assessment have been examined by Serafin *et al.* (1992).

Audits or reviews can be carried out by the developer, the assessor(s), an independent body or researcher(s), a government agency or a combination of one or more of these. The criteria checked are likely to be: whether predictions were accurate and comprehensive; whether there were gaps in prediction; whether mitigation measures were developed and adopted with satisfactory results; whether assessment led to useful feedbacks in the development process; whether the costs of the assessment were justified; and whether the process could be speeded up or otherwise improved. Wood (1995: 200) stressed that audits should not confine themselves to the sort of questions just listed, because one of the values of assessment is to provoke thought about development. There is also a need to check whether an assessment results in appropriate management action. Audit and review of assessment is important if the process is to learn from its mistakes and successes (Bisset, 1984; Tomlinson, 1984, 1987; Anon., 1986; Bailey and Hobbs, 1990; EIA Centre Manchester, 1992; Culhane, 1993).

Post-environmental impact assessment auditing suggests that a good deal of assessment is 'unsatisfactory' (Westman, 1985: 16–18; Bisset and Tomlinson,

1988: 125; Lee and Colley, 1990; Lee *et al.*, 1994; Gilpin, 1995: 21). Jones *et al.* (1991) suggested that about one-third of environmental impact assessments undertaken in the UK between 1988 and 1990 had shortcomings. Lee and Dancey (1993) surveyed environmental impact statements in the Republic of Ireland and the UK between 1988 and 1992 and concluded that 60 per cent were unsatisfactory, 20 per cent were good and the trend seemed to be one of improvement. Buckley (1991a, 1991b) found the average accuracy of Australian environmental impact assessment predictions he tested to be 44 per cent (± 5 per cent standard error). The faults reported by auditors frequently include the non-technical nature of the summary; missed impacts, especially complex cumulative impacts; expert judgement that is subjective and difficult to scrutinize; and vague predictions that are difficult to check for accuracy. Graham Smith (1993: 17) flagged four common faults:

- the adoption of a comprehensive, but superficial ('shotgun') approach – which could be solved by better scoping;
- the taking of too temporally limited a view (a 'snapshot');
- the placing of too much focus on baseline studies;
- the placing of insufficient emphasis on social impact assessment.

Assessment might be improved if there were better access to previous environmental impact statements and audits. A list of publications (up to 1985) is given by Wood and Gazidellis (1985: 152–7). Wood (1995: 245) observed that the UK had no official listing of environmental impact statements carried out (although copies were sent to appropriate government departments). The USA, in contrast, has made a large number available.

Clearly, having better regulations and a code of ethics, together with better monitoring of assessors, would help. In recent years there has been a move, at least in developed countries, towards accreditation (certification) or licensing of impact assessors and eco-auditors by governments, professional bodies or independent regulators (Moy, 1983; Thomson *et al.*, 1995). In the UK, the Institute of Environmental Assessment has undertaken registration and accreditation of assessors since the early 1990s.

MONITORING

'Monitoring' is not a precise term, there is some confusion between it and *review*, *post-environmental impact assessment audit* (both discussed earlier) and *surveillance* – repetitive or ongoing measurement, but of an observational and qualitative nature. *Monitoring* is repetitive or ongoing measurement of a more quantitative nature (possibly more focused). Monitoring may be started before, during or after development. Ideally it should begin as early as possible and continue as long as possible (Canter, 1993). Monitoring should be an integral part of the impact assessment process. Monitoring undertaken before development (baseline monitoring) should relate to the development proposal and established ongoing environmental management monitoring. Some countries or states, including California, require monitoring as an environmental impact assessment progresses. Monitoring can feed data to impact assessment and vice versa.

Effects monitoring concentrates on what has resulted from an activity or an impact assessment. *Post-development monitoring* focuses on compliance with assessment recommendations or effects; that is, with ongoing impacts (Mareus, 1979). Specialist fields of monitoring – environmental, biological, social and economic monitoring – are well-defined and have developed considerably in recent years (Karr, 1987; Spellerberg, 1991). Factors assisting this improvement include research and practice, better microcomputers, better measuring instruments, remote sensing and automatic ground instrumentation. There are still gaps, however, especially in less developed countries. Impact assessment often adopts a 'snapshot' approach (i.e. viewing things at a single point in time) and makes little provision for ongoing monitoring. *Post-environmental impact assessment monitoring* refers to monitoring initiated after completion of an environmental impact statement. For a discussion of its purposes and premisses, *see* Canter and Fairchild (1986).

Future development of the impact assessment process

Graham Smith (1993: 1) noted, 'Despite two decades of evolution and a myriad of techniques, present practice appears unable to help in preventing environmental disasters . . . or poor resource management . . . or the demise of species'. His opinion was that the blame lay with 'a flawed conceptualization of impact assessment and its role in development planning and resource management'. How then must impact assessment be improved? There is clearly a need and a strong trend to move beyond project application (with environmental impact assessment as a 'stand-alone' tool) towards closer integration with planning, policy-making and environmental management. Impact assessment might be the way to better link environmental management and planning. Such a linkage will require a wider adoption of strategic environmental assessment and a more holistic approach.

Too much impact assessment is responsive or reactive, undertaken after a development has been initiated; there is a need for a more proactive approach.

If the predictions made by impact assessment were subjected to follow-up monitoring to test their accuracy, there might be improvement (Fairweather, 1989: 142).

There was interest in impact assessment at the UN Conference on Environment and Development 1992 (the Rio 'Earth Summit'). However, mention of environmental impact assessment in *Agenda 21* (Keating, 1993) was added seemingly as an afterthought. Gilpin (1995: 158–61) reviewed the shortcomings of environmental impact assessment, concluding that there were the following needs:

- better public participation and circulation of information;
- reduction of poverty so that environmental impact assessment can be given a higher priority;
- a firm legislative formulation;
- a clear schedule of what is to be subject to environmental impact assessment;
- better data, planning and environmental management;
- participant integrity;
- adequate avenues of appeal if there are problems.

A problem which, although not restricted to less developed countries, seems increasingly common in those countries is the appropriation of common resources by individuals, special-interest groups, multinational corporations or central government. Impact assessment could be used to help to control this trend and safeguard the traditional rights of local peoples (Graham Smith, 1993).

Research has been focusing on the challenge of improving the assessment of cumulative impacts; on strategic environmental assessment; on the contribution of impact assessment to sustainable development; and in coping with transboundary impacts and urban problems. Human populations are increasingly urban and face worsening problems, especially in rapidly expanding cities in less developed countries, and an issue of *Built Environment* (1980: **6(2)**) focused on urban impact analysis. Improvements in computer hardware and software and the development of graphical information systems offer new opportunities for impact assessment, monitoring, data retrieval and storage, and modelling. Remote sensing can feed into environmental impact assessment and could be more widely used (Khanna and Kondawar, 1991).

Production of environmental impact statements is not enough; impact assessment must not be just a 'planning tool', it has to be better integrated into planning and administration to improve the quality of decision-making (Biswas and Agarwala, 1992; Ortolano and Shepherd, 1995). MacDonald and Brown (1995: 486) suggested that impact assessment should move beyond environmental impact analysis by focusing on cumulative impact assessments, involving the community, and linking policy, planning and assessment. The role of impact assessment in policy-making must be strengthened (for an examination of this point, *see* Bartlett, 1989).

There is a need to further adapt and improve the use of assessment in developing countries (discussed in Chapter 7). There should be co-ordination of impact assessment between local and project levels, regional and programme levels, national and policy and global levels. Already, multinational groupings such as, the European Union have impact assessment, environmental management and eco-audit legislation, guidelines and institutions that are starting to foster co-ordination. Growing acceptance that impacts often result from multiple causes leads to awareness that impact assessment focused on physical and biological factors is not enough and what is required is a move to *total impact assessment* that also considers socio-economic factors. In the future, impact assessment looks set to become a broader, holistic process, more reliant on information technology and adaptive in approach.

References

Anon. 1986: EIA auditing: an overview. *American Society of Civil Engineers* **112(4)**, 638–46.

Bailey, J. and **Hobbs, V.** 1990: A proposed framework and database for EIA auditing. *Journal of Environmental Management* **31(2)**, 163–72.

Bain, M.B., Irving, J.S., Olsen, R.D., Shill, E.A. and **Witmer, G.W.** 1986: *Cumulative impact assessment: evaluating the environmental effects of multiple human developments*. Washington DC: US Department of Energy.

Bartlett, R.V. (ed.) 1989: *Policy through impact assessment: institutionalized analysis as a policy strategy*. Toronto: Greenwood Press.

Beanlands, G.E. 1988: Scoping methods and baseline studies in environmental impact assessment. In Wathern, P. (ed.), *Environmental impact assessment: theory and practice*. London: Unwin Hyman, 33–46.

Beanlands, G.E., Erckmann, W.J., Orians, G.H., O'Riordan, T., Policansky, D., Sadar, M.H. and **Sadar, B.** (eds) 1986: *Cumulative environmental effects: a binational perspective.* (Canadian Environmental Assessment Research Council/US National Research Council). Ottawa: Ministry of Supply and Services.

Bisset, R. 1978: Quantification, decision-making and environmental assessment in the UK. *Journal of Environmental Management* **7(1)**, 43–58.

Bisset, R. 1980: Problems and issues in the implementation of EIA audits. *Environmental Impact Assessment Review* **1(4)**, 379–96.

Bisset, R. 1984: Post-development audits to investigate the accuracy of environmental impact predictions. *Zeitschrift für Umweltpolitik* **7(4)**, 463–84.

Bisset, R. and **Tomlinson, P.** 1988: Monitoring and auditing of impacts. In Wathern, P. (ed.), *Environmental impact assessment: theory and practice*. London: Unwin Hyman, 117–28.

Biswas, A.K. and **Agarwala, S.B.** (eds) 1992: *Environmental impact assessment for developing countries*. Oxford: Butterworth-Heinemann.

Buckley, R.C. 1991a: Auditing the precision and accuracy of environmental impact predictions in Australia. *Environmental Monitoring and Assessment* **18(1)**, 1–23.

Buckley, R.C. 1991b: How accurate are environmental impact predictions? *Ambio* **XX(3–4)**, 161–2.

Cairns, J. Jr and **Niederlehner, B.R.** 1993: Ecological function and resilience: neglected criteria for environmental impact assessment and ecological risk analysis. *Environmental Professional* **15(2)**, 116–24.

Canter, L.W. 1977: *Environmental impact assessment*, 1st edn. New York: McGraw-Hill.

Canter, L.W. 1993: The role of environmental monitoring in responsible project management. *Environmental Professional* **15(1)**, 76–87.

Canter, L.W. 1996: *Environmental impact assessment*, 2nd edn. New York: McGraw-Hill.

Canter, L.W. and **Canty, G.A.** 1993: Impact significance determination: basic considerations and a sequenced approach. *Environmental Impact Assessment Review* **13(4)**, 275–95.

Canter, L.W. and **Fairchild, D.M.** 1986: Post-EIS monitoring. *Impact Assessment Bulletin* **4(3–4)**, 265–85.

Carpenter, R.A. 1995: Risk assessment. *Environmental Assessment* **13(2)**, 153–87.

Clark, B.D., Gilad, A., Bisset, R. and **Tomlinson, P.** (eds) 1984: *Perspectives on environmental impact assessment*. Dordrect: D. Reidel.

Cocklin, C., Parker, S. and **Hay, J.** 1992: Notes on cumulative environmental change I: Concepts and issues. *Journal of Environmental Management* **35(1)**, 31–49.

Contant, C.K. and **Wiggins, L.L.** 1991: Defining and analysing cumulative environmental impacts. *Environmental Impact Assessment Review* **11(4)**, 297–309.

Contant, C.K. and **Wiggins, L.L.** 1993: Towards defining and assessing cumulative impacts: practical and theoretical considerations. In Hildebrand, S.G. and Cannon, J.B. (eds), *Environmental analysis: the NEPA experience*. Boca Raton, FL: Lewis Publishers, 336–56.

Covello, V.T., Mumpower, J.L., Stallen, P.J. and **Uppuluri, V.R.R.** (eds) 1985: *Environmental impact assessment, technology assessment, and risk analysis: contributions from the psychological and decision sciences*. Berlin: Springer-Verlag.

Culhane, P.J. 1987: The precision and accuracy of US environmental impact statements. *Environmental Monitoring and Assessment* **8(3)**, 217–38.

Culhane, P.J. 1993: Post-EIS environmental auditing: a first step to making rational environmental assessment a reality. *The Environmental Professional* **15(1)**, 66–75.

Culhane, P.J., Friesema, H.P. and **Beecher, J.A.** (eds) 1987: *Forecasts and environmental decision-making: the contents and productive accuracy of environmental impact statements*. Boulder, CO: Westview.

Damman, D.C., Cressman, D.R. and **Husain, S.M.** 1995: Cumulative effects assessment: the development of practical frameworks. *Impact Assessment* **13(4)**, 433–54.

Davies, K. 1992: *Cumulative environmental effects: a sourcebook*. Hull, Quebec: Federal Environmental Assessment Review Office.

De Jongh, P. 1988: Uncertainty in EIA. In Wathern, P. (ed.), *Environmental impact assessment: theory and practice*. London: Unwin Hyman, 62–84.

Ditton, R.B. and **Goodale, T.L.** (eds) 1972: *Environmental impact analysis: philosophy and methods*. Madison, WI: University of Wisconsin, Sea Grant Program, Sea Grant Publications Office.

Duinker, P.N. and **Baskerville, G.L.** 1986: A systematic approach to forecasting in environmental impact assessment. *Journal of Environmental Management* **23(3)**, 271–90.

Duinker, P.N. and **Beanlands, G.E.** 1986: The significance of environmental impacts: an exploration of the concept. *Environmental Management* **10(1)**, 1–10.

EIA Centre, Manchester 1992: *Reviewing environmental impact statements*. EIA Centre Leaflet Series 11. Manchester: University of Manchester, Department of Planning and Landscape.

Environment Canada and **Transport Canada** 1988: *Environmental monitoring and audit: guidelines for post-project analysis of development impacts and assessment methodology*. Manuscript Series. Ottawa: Environment Canada, Environmental Impact Systems Division.

Erickson, P.A. 1979: *Environmental impact assessment: principles and applications*. New York: Academic Press.

Fairweather, P.G. 1989: Environmental impact assessment: where is the science in EIA? *Search* **20(5)**, 141–4.

Fortlage, C.A. 1990: *Environmental assessment: a practical guide*. Aldershot: Gower Technical.

Fuggle, R.F. 1989: Integrated environmental management: an appropriate approach to environmental concerns in developing countries. *Impact Assessment Bulletin* **8(1-2)**, 31–45.

Gallagher, T.J. and **Jacobson, W.S.** 1993: The typography of environmental impact statements: criteria, evaluation, and public participation. *Environmental Management*, **17(1)**, 99–109.

Gilpin, A. 1995: *Environmental impact assessment (EIA): cutting edge for the twenty-first century*. Cambridge: Cambridge University Press.

Glasson, J., Thérivel, R. and **Chadwick, A.** 1994: *Introduction to environmental impact assessment: principles and procedures, process, practice and prospects*. London: University College London Press.

Goldrick, G. and **James, D.** 1994: Assessing cumulative impacts of aluminium smelting in the Hunter Valley, NSW, Australia. In James, D. (ed.), *The application of economic techniques in environmental impact assessment*. Dordrecht: Kluwer, 275–98.

Graham Smith, L. 1993: *Impact assessment and sustainable resource management*. Harlow: Longman.

Herson, A.I. and **Bogdan, K.M.** 1991: Cumulative impact analysis under NEPA. *Environmental Professional* **13(2)**, 100–6.

Hipel, K.W. 1988: Non-parametric approaches to environmental impact assessment. *Water Resources Bulletin* **24(3)**, 487–92.

Hollick, M. 1981a: Enforcement of mitigation measures resulting from environmental impact assessment. *Environmental Management* **5(6)**, 507–13.

Hollick, M. 1981b: The role of quantitative decision-making methods in environmental impact assessment. *Journal of Environmental Management* **12(1)**, 65–78.

Holling, C.S. 1978: *Adaptive environmental planning and management*. Chichester: Wiley.

Hyman, E.L. and **Stiftel, B.** 1988: *Combining facts and values in environmental impact assessment: theories and techniques*. Boulder, CO: Westview.

James, D. (ed.) 1994: *The application of economic techniques in environmental impact assessment*. Dordrecht: Kluwer.

Jones, C.E., Lee, N. and **Wood, C.M.** 1991: *UK environmental statements 1988–1990: an analysis*. EIA Centre Occasional Paper 29. Manchester: University of Manchester, Department of Planning and Landscape.

Julien, B., Feneves, S.J. and **Small, M.J.** 1992: An environmental impact identification system. *Journal of Environmental Management* **36(3)**, 167–84.

Karr, J.R. 1987: Biological monitoring and environmental assessment: a conceptual framework. *Environmental Management* **11(2)**, 249–56.

Keating, M. 1993: *The Earth Summit's agenda for change: a plain language version of Agenda 21*. Geneva: Centre for Our Common Future (Palais Wilson, 52 rue des Pâquis, CH-1201, Geneva, Switzerland).

Kennedy, A.J. and **Ross, W.A.** 1992: An approach to integrate impact scoping with environmental impact assessment. *Environmental Management* **16(4)**, 475–84.

Khanna, P. and **Kondawar, V.K.** 1991: Application of remote sensing techniques for environmental impact assessment. *Current Science* **61(3–4)**, 252–6.

Kosko, B. 1994: *Fuzzy thinking: the new science of fuzzy logic*. London: Flamingo (HarperCollins).

Kozlowski, J.M. 1989: *Integrating ecological thinking into the planning process: a comparison of the EIA and UET concepts*. Paper N. FS-II-89–404. Berlin: Wissenschaftszentrum Berlin für Sozialforschung.

Kreske, D. 1996: *Environmental impact statements: management and preparation*. Chichester: Wiley.

Kung, H.T., Ying, L.G. and **Liu, Y.C.** 1993: Fuzzy clustering analysis in environmental impact assessment: a complement tool to environmental quality index. *Environmental Monitoring and Assessment* **28(1)**, 1–14.

Lahlou, M. and **Canter, L.W.** 1993: Alternatives evaluation and selection in development and environmental remediation projects. *Environmental Impact Assessment Review* **13(1)**, 37–61.

Lane, P.A. and **Wallace, R.R.** 1988: *A user's guide to cumulative effects assessment in Canada*. Ottawa: Canadian Environmental Assessment Research Council.

Lawrence, D.P. 1993: Quantitative versus qualitative evaluation: a false dichotomy? *Environmental Impact Assessment Review* **13(1)**, 3–11.

Lee, N. 1982: The future development of environmental impact assessment. *Journal of Environmental Management* **14(1)**, 71–90.

Lee, N. and **Brown, D.** 1992: Quality control in environmental impact assessment. *Project Appraisal* 7(1), 41.

Lee, N. and **Colley, R.** 1990: *Reviewing the quality of environmental statements*. Manchester: University of Manchester, Department of Planning and Landscape.

Lee, N. and **Dancey, R.** 1993: The quality of environmental impact statements in Ireland and the United Kingdom: a comparative analysis. *Project Appraisal* **8(1)**, 31–6.

Lee, N., Walsh, F. and **Reeder, G.** 1994: Assessing the performance of the EA process. *Project Appraisal* **9(3)**, 161–72.

MacDonald, G.T. and **Brown, L.** 1995: Going beyond environmental impact assessment: environmental input to planning and design. *Environmental Impact Assessment Review* **15(6)**, 483–96.

Maclaren, V.W. 1987: The use of social surveys in environmental impact assessment. *Environmental Impact Assessment Review* **7(4)**, 363–75.

Mareus, L.G. 1979: *A methodology for post-EIS monitoring*. US Geological Survey Circular 782. Washington DC. Government Printing Office.

Marsden, D.M., Oakley, P. and **Pratt, B.** (eds) 1994: *Measuring the process: guidelines for evaluating social development*. Oxford: Intrak (PO Box 563, Oxford).

Matthews, W.H. 1975: Objective and subjective judgements in environmental impact analysis. *Environmental Conservation* **2(20)**, 121–31.

Mihai, V. 1984: Some contributions to the estimation of effects in impact studies. *Impact Assessment Bulletin* **3(3)**, 18–24.

Morris, P. and **Thérivel, R.** (eds) 1995: *Methods of environmental impact assessment*. London: University College London Press.

Moy, P.J. 1983: Environmental impact assessment consultants: the case against self-regulation. *Journal of Environmental Management* **17(4)**, 393–401.

Munn, R.E. 1979: *Environmental impact assessment: principles and procedures*, 2nd edn. SCOPE Report 5. Chichester: Wiley.

Nijkamp, P. 1986: Multiple criteria analysis and integrated impact analysis. *Impact Assessment Bulletin* **4(3–4)**, 226–61.

Novek, J. 1995: Environmental impact assessment and sustainable development: case-studies of environmental conflict. *Society and Natural Resources* **8(2)**, 145–59.

OECD 1991: *Environmental indicators*. Paris: Organisation for Economic Co-operation and Development.

Orians, G.H. 1995: Thought for the morrow: cumulative threats to the environment. *Environment* **37(7)**, 6–14, 33–6.

Ortolano, L., Jenkins, B. and **Abracosa, R.** 1987: Speculations on when and why EIA is effective. *Environmental Impact Assessment Review* **7(4)**, 285–92.

Ortolano, L. and **Shepherd, A.** 1995: Environmental impact assessment: challenges and opportunities. *Impact Assessment* **13(1)**, 3–30.

Ott, W.R. 1979: *Environmental indices: theory and practice*. Ann Arbor, MI: Ann Arbor Science.

Parkin, J. 1992: A philosophy for multiattribute evaluation in environmental impact assessment. *Geoforum* **23(4)**, 467–75.

Rickson, R.E., Hundloe, T. and **Western, J.S.** 1989: Impact assessment in conflict situations: World Heritage Listings of Queensland's northern tropical rainforests. *Impact Assessment Bulletin* **8(1–2)**, 179–90.

Rigby, B. 1985: Post-development audits in environmental impact assessment. In Maclaren, V.W. and Whitney, J.B. (eds), *New directions in environmental impact assessment in Canada*. Toronto: Methuen, 179–220.

Ross, W.A. 1987: Evaluating environmental impact statements. *Journal of Environmental Management* **25(2)**, 137–47.

Sadler, B. 1988: The evaluation of assessment: post-EIS research and process development. In Wathern, P. (ed.), *Environmental impact assessment: theory and practice*. London: Unwin Hyman, 129–42.

Sager, T. 1995: From impact assessment to recommendation: how are the impact assessment results presented and used? *Environmental Impact Assessment Review* **15(4)**, 377–97.

Save the Children 1995: *Toolkits: a practical guide to assessment, monitoring, review and evaluation*. Save the Children Development Manual 5. London: Save the Children.

Schindler, D.W. 1976: The impact statement boondoggle. *Science* **192(4239)**, 509.

Serafin, R., Nelson, G. and **Butler, R.** 1992: Post hoc assessment in resource management and environmental planning: a typology and three case studies. *Environmental Impact Assessment Review* **12(4)**, 271–94.

Sheate, W. 1993: *Making an impact: a guide to EIA law*. London: Cameron May.

Skutsch, M.McC. and **Flowerdew, R.T.N.** 1976: Measurement techniques in environmental impact assessment. *Environmental Conservation* **3(3)**, 209–17.

Smit, B. and **Spaling, H.** 1994: Methods for cumulative effects assessment. *Environmental Impact Assessment Review* **15(1)**, 81–106.

Spellerberg, I.F. 1991: *Monitoring ecological change*. Cambridge: Cambridge University Press.

Spellerberg, I.F. and **Minshul, A.** 1992: An investigation into the nature and use of ecology in environmental impact assessment. *British Ecological Society Bulletin* **23(1)**, 38–45.

Thompson, M.A. 1990: Determining impact significance in EIA: a review of 24 methodologies. *Journal of Environmental Management* **30(3)**, 235–50.

Thomson, D.R., Bacon, R.A., Tarling, J.P. and **Baverstock, S.J.** (eds) 1995: *The EARA register of environmental auditors*. London: Earthscan.

Tomlinson, P. 1984: EIA audits. *Environmental Impact Assessment Worldletter* **1(5)**, 6.

Tomlinson, P. 1987: Environmental audits – special edition of *Environmental Monitoring and Assessment* **8(3)**, 183–261.

UNDP 1991: *Human development report 1991*. Oxford: Oxford University Press.

UNECE 1990: *Post-project analysis in environmental impact assessment*. Environmental Series 3. Geneva and New York: United Nations Economic Commission for Europe.

USACE 1988: *Methodology for analysis of cumulative impacts of Corps. Permit activities* (US Army Corps of Engineers). Washington DC: Institute for Water Resources Policy.

Warner, M.L. and **Preston, E.H.** 1973: *A review of environmental impact assessment* *methodologies*. Washington DC: Office of Research and Development, US Environmental Protection Agency.

Wathern, P. (ed.) 1988: *Environmental impact assessment: theory and practice*. London: Unwin Hyman.

Westman, W. 1985: *Ecology, impact assessment and environmental planning*. Chichester: Wiley.

Wood, C.M. 1995: *Environmental impact assessment: a comparative review*. Harlow: Longman.

Wood, C.M. and **Gazidellis, V.** 1985: *A guide to training materials for environmental impact assessment*. Occasional Paper 4. Manchester: University of Manchester, Department of Town and Country Planning.

Wood, C.M. and **Jones, C.** 1991: *Monitoring environmental assessment and planning*. London: HMSO.

Yee, H. 1981: Reliability and validity of expert ratings on environmental impact of foreign policy decision making. *Political Methodology* **7(1)**, 1–25.

5

TECHNIQUES AND METHODS

INTRODUCTION

This chapter outlines the techniques and methods of environmental impact assessment, whereas Chapter 3 considers impact assessment approaches and strategies. Chapter 8 outlines techniques and methods of social impact assessment. In this chapter and in Chapter 8 the assumption is that readers are seeking an introduction to impact assessment or are 'consumers' or 'commissioners'; I do *not* to seek to train in the practice of impact assessment or related fields. The focus is on how techniques and methods function, and on their strengths and weaknesses.

Literature on impact assessment tends to use terms such as 'techniques' and 'methods' in an imprecise way; often they are treated as synonymous. Conceptually, *methods* can be distinguished from *techniques*; one of the few authors in the field to trouble to do so is Bisset (1983: 169; 1988: 47). The former are concerned with various aspects of assessment such as the identification or description of likely impacts (and may incorporate means to scale, weigh and compare impacts), and aid the collection and classification of impact data; the latter provide the data, which are organized in accord with the operational principles of particular methods (*see also* Canter, 1996: 56). So, a technique provides data on some parameter (e.g. past and present noise levels and those likely to be generated by a development); those data are then utilized by a method which might present and evaluate them. A technique may be used to evaluate or present information, etc., but is basically a 'building-block' for a method. An environmental impact assessment, social impact assessment, risk assessment or eco-audit method may use more than one technique. The application of methods is usually controlled by published guidelines or rules.

Change of an environmental, social or economic parameter is often apparent only if an assessor recognizes and monitors a suitable indicator. Ideally, such indicators and indices are widely applicable and everywhere convey the same meaning. Perhaps there is even some agreed and published standard.

An *index* may combine information from one or on more than one parameter, presenting it as a single value.

The main tasks that need to be addressed in impact assessment are, first, impact identification and measurement; second, impact assessment (which involves prediction, interpretation and evaluation – that is, how 'significant', how likely, when?); and third, impact communication (to decision-makers and perhaps the public). There may well be additional tasks (such as monitoring, and enforcement of an environmental impact statement) and subdivisions of those just listed. Before impact identification can proceed, baseline data are required and methods are needed to collect and assemble those data. Often, development could take place by way of one of a number of possible actions. Ways will be needed to compare these; the assessor may have to suggest problem mitigation or avoidance strategies; and techniques may be needed to audit and compare assessments. Tasks may be difficult to separate, and particular techniques and methods may well cover more than one task.

There is no single comprehensive approach capable of doing well all that is required for impact assessment, and those with a limited focus are frequently far from perfect even at their specialized task (Warner and Preston, 1973; Skutsch and Flowerdew, 1976; Canter, 1983; Bisset, 1980, 1981, 1987). Even straightforward comparison of likely environmental conditions if there were to be no development with those predicted if development proceeds is difficult, because the environment is complex and dynamic and human behaviour is changeable (Clark *et al.*, 1984: 242).

There have been attempts to compare and assess various impact assessment methods. Most textbooks make a general attempt within their first few chapters. For a concise review, *see* Shopley and Fuggle (1984). Lapping (1975) provided a critique of early environmental impact assessment methods, and Atkins (1984) examined several methods. Hyman and Stiftel (1988: 155–223) reviewed 14 environmental impact assessment methods, seeking to evaluate them all according to the same criteria. Duke *et al.* (1994) recognized over 50 impact assessment methodologies, and argued the need for synthesis of what is available for a given task. This chapter considers the more common techniques and methods.

The choice of technique(s) or method(s) used in an assessment depends on the time and the resources available; what goals the assessment is required to meet (e.g. is it to brief planners or public and planners?); what criteria are to be assessed; and what personnel comprise the assessment team (Heer and Hagerty, 1977: 185). Study may be through one or more of the following: desk-research; field visits; experiments *in situ*; laboratory studies; and modelling or simulation exercises.

It is easier to refer to approach or organizing principles than to divide methods and techniques according to the tasks they address, and the former approach is the one adopted in this chapter and Chapter 8, and in much of the literature. On the basis of approach and organizing principles, the following groups of techniques and methods can be recognized: *ad hoc* procedures; overlay techniques; checklists; matrices; networks and systems diagrams; quantitative methods; modelling; manuals; cumulative impact assessment methods; and miscellaneous techniques (some of which have been discussed in Chapter 2 – for example, orchestrated expert opinion of the form used in the Delphi technique or expert systems applications). Figure 5.1 presents an evaluation of some of the common assessment methods.

Type of method	Adressses impact								Ease of application			Resource requirement			
	Identification		Measurements		Interpretation		Evaluation					Staff needed		Computer needed	
	Yes	No	Yes	No	Yes	No	Yes	No	Difficult	Moderately difficult	Not difficult	Highly skilled	Moderately skilled	Desired	Not needed
Ad hoc	X			X		X		X			X		X		X
Overlays	XX			X		X		X		X		X			X
Checklists	XXX			X		X		X			X		X		X
Matrices	XXX		X		X		XX			X			X		X
Networks	XXX		XXX		XXX		XXX		X			X		X	

X = Fair XX = Good XXX = Excellent

Figure 5.1 Evaluation of selected assessment methodologies (for methods suitable for developing countries, *see* Chapter 7)
Source: Based on various sources, including Jain *et al.* (1977)

Difficulties faced in applying methods

Sometimes assessment may be limited to *ad hoc* discussion and use no other techniques. Things are seldom static; an environment, society or economy may change, and so methods that worked at one point in time may not be as effective later or in another situation. Apparently similar situations may in reality be markedly different. Ideally, techniques and methods will assist an assessor to identify impacts, not just aid documentation of obvious impacts (Julien *et al.*, 1992). Many of the techniques used in assessment were developed for other tasks and have been adapted. Methods and techniques for measuring ecosystem function are less developed than those for measuring its structure (Treweek, 1995: 296). Methods and techniques should be standardized and consistent, replicable and adaptable. To this list might be added accurate and reliable, cheap, not too demanding of manpower and expertise, and fast. 'Black-box' techniques should be avoided, with the approach being made as transparent as possible to ensure that statistical manipulations do not obscure the character of the original data or the way they have been transformed. There is a trend for impact assessment to be subjected to greater legal and scientific scrutiny. Increasingly, especially in developed countries, developers are being held more accountable for their actions. Impact assessment must therefore avoid errors and 'gaps', and stand up to cross-examination in court if those responsible for commissioning assessments are to accept them.

Ad hoc Methods

'*Ad hoc*' methods were the first to evolve and are 'unsophisticated'. They provide minimum guidance for the assessor, limiting themselves to suggesting broad areas of possible impacts. They give no indication of specific parameters to investigate. It is therefore possible for assessors to miss one or more impacts. *Ad hoc* methods are of limited value for scoping and, as one-off approaches, are likely to be difficult to transfer (Rau, 1980: 8.1). Approaches that pre-dated impact assessment, and that today may still feed into it, include the public inquiry or commission of inquiry. (A commission of inquiry is less open to public scrutiny.) Parliamentary democracies like the UK have long used public inquiries and commissions to assess the likely effects of certain developments or to determine why problems have arisen. An example is the Windscale Inquiry into proposals by British Nuclear Fuels to construct a nuclear reactor fuel reprocessing plant in Cumbria (Parker, 1978).

Checklist methods

Simple checklists are based on a priori judgements, and list, perhaps in some sort of hierarchy, the factors to be considered when assessing (Westman, 1985; Hyman and Stiftel, 1988: 155–223; Graham Smith, 1993: 18–29). Their task is thus primarily one of impact identification and listing, but there may be some assessment of the character and nature of the impact: for example, an entry may be made against a particular impact stating whether it is adverse or beneficial, significant or insignificant, will have long- or short-term impact.

Checklists mainly serve to:

- order thought;
- aid data-gathering;
- help ensure that the assessor does not overlook a possible impact;
- assist the assessor to screen large amounts of data so that impact assessment can be focused.

Simple checklists can help describe impacts and give some measurement and prediction. More sophisticated checklists may apply scaling or weighting techniques to try to give some measure of impact or a utility function (i.e. indicate the relative merit of various alternatives). An example is the Adkins–Burke method (*see* Box 5.1). It is also possible to combine checklists with land capability analysis (Galloway, 1978).

Box 5.1 Adkins–Burke method (checklist) for three alternative developments

Factor	Definition/explanation	Alternatives 1	2	3
		Ratings*		
Air pollution	Rule or threshold	–1	–3	+1
Drainage	Chance of floods changed	+2	+1	–1
Noise	Increased noise	–1	0	0
Visual impact	Unpleasant view	0	0	–1
	Total impact	0	–2	–1

Source: Based on Canter (1977: 204)

Note: In practice there are likely to be far more factors under consideration. This checklist rates identified impacts on a relative scale (a subjective judgement) of –5 to +5, which permits an impacts total to be obtained for each proposed development.

More sophisticated checklists may be descriptive or organized by development phases: planning and design phase, construction phase, management phase, etc. (Canter, 1996: 86). Questionnaire checklists are composed of a series of questions posed to the assessor relating to possible development impacts. USAID, the United States Agency for International Development, used questionnaire checklists in the early 1980s (Clark *et al.*, 1984: 203–4; Bisset, 1987: 16).

Uncritical use of checklists can result in a blinkered approach to assessment, whereby what is not on a checklist is ignored. The 'pigeon-hole' approach may also encourage a simplistic view of environmental systems: because checklists do not show interactions between impacts, an impact may be double-counted under different categories, or even counted more than twice. Checklists can become long and unwieldy, and often make no indication of the likelihood of an impact and do not prioritize impacts (Rau, 1980: 8.5). A checklist of critical environmental thresholds could be a useful tool for the assessor, particularly with sustainable development a goal. Some checklists take the form of a dichotomous key that guides the assessor from one finding on to the next relevant questions.

Overlay methods

Overlay methods are largely derived from town and country planning, landscape architecture and land capability assessment. They generally rely on a set of maps, sometimes referred to as 'sieve maps' (Fortlage, 1990: 126) of a locality where a development is proposed, each map showing physical, biological, social, economic, natural hazards, etc. By overlaying these maps a systematic, composite picture can be obtained which can aid in screening sites or routes for impacts. Early examples pre-date the appearance of environmental impact assessment, and include the work of McHarg (1968, 1969), who mapped environmentally sensitive factors at urban, metropolitan and river basin levels and uses in route planning (Skutsch and Flowerdew, 1976: 215; Bisset, 1987: 22–4). Originally maps were drawn on transparent sheets, but this approach becomes inaccurate if too many sheets are used (the maximum is about 10) because inaccuracies of scale increase and the numerous layers become insufficiently transparent. Today things have been made faster and easier and greater precision provided by the use of computer databases. It is also possible to go well beyond 10 'overlay sheets' when using computer techniques, if need be embarking on complex multiple comparisons.

Overlay methods do not separate direct and indirect impacts, give no indication of the probability of an impact's occurring and do not do much to show causal relationships. A problem that still remains in spite of computing is that of defining 'boundaries'. Air photos, satellite data or fieldwork seldom provide maps with clear, sharp boundaries; there is overlap and fuzziness (Clark *et al.*, 1984: 217). Because nature seldom provides completely distinct boundaries, methods of utilizing fuzzy data may be appropriate.

An extension of the McHarg overlay method is the metropolitan landscape planning model (METLAND) method, which uses computers, draws on more variables and has a more regional focus. Neither the McHarg nor the METLAND approach does much to assist public involvement (Hyman and Stiftel, 1988: 163).

Overlay techniques are useful when considering linear developments such as the building of roads, railways and canals. They assist in planning a route involving the lowest cost, lowest risk and least disruption. They are also appropriate for land use planning and industrial siting. A closely related approach is *sieve mapping*, used by town and country planners to overlay land use and hazard maps to identify the best areas for development (Fortlage, 1990: 136).

GEOGRAPHICAL INFORMATION SYSTEMS

Geographical information systems allow regular or even real-time (i.e. constant and immediate) updating of the data input, and 'fine tuning' of what is overlaid on what (Schaller, 1990, 1992). They may help impact assessment become more proactive by allowing regular updating of data, and can also play a part in assessing cumulative impacts (Parker and Cocklin, 1993; Eedy, 1995; and *see* the section 'Dealing with cumulative impacts' later in this chapter). Geographical information systems make it possible to quantify incremental or cumulative impacts on a locality or region (Treweek, 1995: 297). They also allow impact assessment to contribute to and combine with predictive modelling (Siegel and Moreno, 1993; Fedra, 1994). They can be updated regularly or in real time by remote or other forms of sensing (Ross and Singhroy, 1985; Moreno and Heyerdahl, 1990). Sweden integrates data from environmental impact assessment and geographical information systems to support planning decisions (Bowman *et al.*, 1994).

MATRIX METHODS

Matrices can be used for screening and for impact assessment, especially impact identification. A simple matrix is the combination of two checklists, one describing potential impacts of the proposed action (columns) and the other listing environmental, including socio-economic, conditions that might be affected (rows). They thus list potential impacts of a development's effects, showing simple causal relationships. Simple matrices may go beyond identifying impacts and rank them too. More complex matrices can identify indirect impacts. Simple matrices generally do relatively little to help in interpretation so so they may give no indication of whether impacts are delayed or instantaneous, long-term or short-term.

Goals-achievement matrix

A goals-achievement matrix, permits some evaluation of alternative development options, and was one of the earliest methods to be used in impact assessment, slightly pre-dating 'mainstream' environmental impact assessment (Hill, 1968). This type of matrix is oriented towards a set of multiple community objectives. A good matrix acts as a heuristic model; that is, it asks questions of the assessor.

Leopold matrix

A number of matrix approaches are fast, cheap and well-tested (Barber *et al.*, 1990). One of the best-known is the Leopold matrix (*see* Fig. 5.2). This is an open-cell interaction matrix originally developed by the US Geological Survey for use on US construction projects (Leopold *et al.*, 1971; Westman, 1985: 133; Glasson *et al.*, 1994: 96–100). The Leopold matrix links environmental factors with development activities, and by using a standardized form helps ensure that no potential impact is overlooked. A subjective judgement, preferably by an expert, produces a ranking for each impact's magnitude and importance, so a cell indicates how widespread and how significant an impact is judged to be. Values in cells are rankings (the scales are ordinal, and impacts should not be added, subtracted, multiplied or divided), therefore comparison of cells and attempts to add or mathematically transform these ordinal scores are best avoided. For example, a score of 10 in one cell is of higher rank than a score of 5 in another, but it is not valid to conclude that the former is twice as important, nor is it acceptable to add cells for one

PROPOSED ACTIONS →

EXISTING CHARACTERISTICS AND CONDITIONS OF THE ENVIRONMENT ↓

INSTRUCTIONS

1. Identify all actions (located across the top of the matrix) that are part of the proposed project.
2. Under each of the proposed actions, place a slash at the intersection with each item on the side of the matrix if an impact is possible.
3. Having completed the matrix, in the upper left-hand corner of each box with a slash, place a number from 1 to 10 which indicates the MAGNITUDE of the possible impact; 10 represents the greatest magnitude of impact and 1, the least (no zeroes). Before each number place + if the impact would be beneficial. In the lower right-hand corner of the box place a number from 1 to 10 which indicates the IMPORTANCE of the possible impact (e.g. regional versus local): 10 represents the greatest importance and 1, the least (no zeroes).
4. The text which accompanies the matrix should be a discussion of the significant impacts, those columns and rows with large numbers of boxes marked and individual boxes with the larger numbers.

SAMPLE MATRIX

	a	b	c	d	e
a		2/1			8/5
b		1/2	3/8	3/1	6/7

A. MODIFICATION OF REGIME

			a. Exotic flora or fauna introduction	b. Biological controls	c. Modification habitat	d. Alteration of ground cover	e. Alteration of ground water hydrology	f. Alteration of drainage	g. River control and flow modification	h. Canalization	i. Irrigation	j. Weather modification	k. Burning	l. Surface or paving	m. Noise and vibration
A. PHYSICAL AND CHEMICAL CHARACTERISTICS	1. EARTH	a. Mineral resources													
		b. Construction material													
		c. Soils													
		d. Land form													
		e. Force fields and background radiation													
		f. Unique physical features													
	2. WATER	a. Surface													
		b. Ocean													
		c. Underground													
		d. Quality													
		e. Temperature													
		f. Recharge													
		g. Snow, ice and permafrost													
	3. ATMOSPHERE	a. Quality (gases, particulates)													
		b. Climate (micro, macro)													
		c. Temperature													
	4. PROCESSES	a. Floods													
		b. Erosion													
		c. Deposition (sedimentation, precipitation)													
		d. Solution													
		e. Sorption (ion exchange, complexing)													
		f. Compaction and settling													
		g. Stability (slides, slumps)													
		h. Stress–strain (earthquakes)													
		i. Air movements													

Rows = development actions
Columns = impacts
Cells: top-left = magnitude (1–10)
bottom-right = importance (1–10)

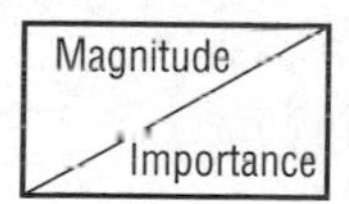

FIGURE 5.2 Leopold matrix (a portion of)
Source: Leopold *et al.* (1971); Bisset (1983: 170, Fig. 1); Hyman and Stiftel (1988)

development proposal and compare the total with that obtained for another development proposal. A Leopold matrix is thus of little value for evaluating beneficial or harmful impacts *in toto* or for comparing alternative developments because there is no standardized way of assigning the scores, nor is there a means of assigning weights to different impacts to determine relative importance (Bisset, 1983: 171).

A matrix should be accompanied by a report describing how values in cells are arrived at. The user should look for patterns: columns where there are numerous impacts.

Ideal for preliminary assessments and when there is little time available, a matrix is also relatively cheap and production is quite rapid. The Leopold matrix has also been the basis for other matrix methods. A standard Leopold matrix lists 100 actions (columns) and 88 environmental factors (rows), which generates 8800 cells – which can become unwieldy. The rows are usually grouped into physical and chemical factors, and biological and ecological ones. It is also possible to consider social, economic and cultural impacts. The assessor indicates where impacts are likely by a slash across a cell, and then (as a subjective judgement) assigns a magnitude (extent) and importance (significance) value in the upper and lower half of each cell respectively (and should explain in the accompanying report or key how these values are determined).

The Leopold matrix can give a good visual summary. If it is too cluttered, a précis version can be extracted or, if it lacks detail, it is possible to expand the checklists. This summary can serve as a sort of simple environmental impact statement. Colour shading of cells can be used to ease visual interpretation: green for a beneficial impact, red for an adverse, some other colour for unknown or neutral/benign. Cells are sometimes divided into four to show magnitude, importance and other characteristics, such as confidence of prediction or reversibility. Matrices have some interpretive and measurement value. Leopold matrices can be applied at different phases in a development – construction, operation, post-operation – and at various spatial levels, such as local/site, regional/off-site, even national or global.

The Leopold matrix relies heavily on subjective evaluation by experts and does not facilitate public involvement. As with other simple matrices that list only direct impacts, there is a risk that the layperson (who one hopes is not the decision-maker) will gain a false sense of security, when in reality, indirect impacts might be generated and are not shown. It is also possible to double-list an impact; that is, list it in each of two cells. It must also be remembered that a matrix is likely to be an individual's viewpoint, or that of a small assessment team, as there is little or no public input. For a non-uniform population (consisting perhaps of distinct cultures and age groups) it may make sense to construct a Leopold matrix for each group.

Alternative types of matrix

The failure of simple matrices to identify indirect impacts and their inability to support total or compare ranking scores have helped promote the development of a plethora of alternative types of matrix (Fortlage, 1990: 127). Some are simply impact rankings, or weighted rankings, for different alternatives that allow a mean or total impact index for each development

Box 5.2 Sphere matrix (developed by Sphere Associates)

Impacts	Sites:	a	b	c	d
	dust	1	1	1	0
	noise	1	2	4	6
	flood	0	0	1	4

Note: In practice there are likely to be many more impacts to consider

alternative, such as the *Sphere impact matrix* (Box 5.2) or the *optimum-pathway matrix*. Others seek to improve on the range of impact information presented, even showing indirect relationships between development and environment or society. An example of this type is the *Saratoga Associates matrix* (*see* Fig. 5.3). First developed in Boston, this consists of two matrices per development. The first lists on one axis activities likely to cause impacts (about 100), and on the other all the aspects likely to be affected (about 50). Each cell is divided into four by means of diagonal lines, and each of the four areas portrays a different attribute of the impact, such as whether the impact is major

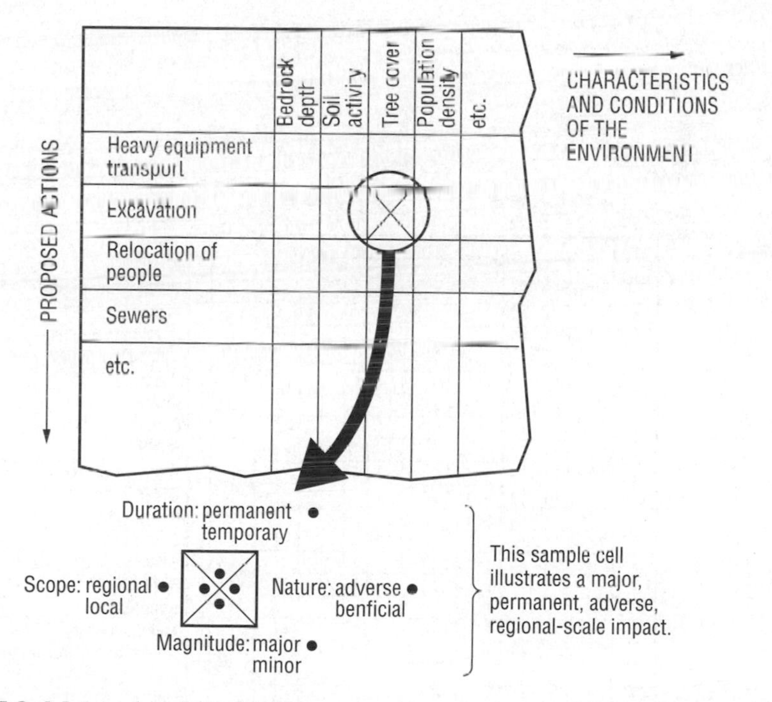

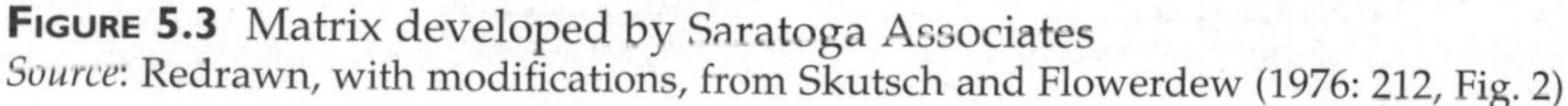

Figure 5.3 Matrix developed by Saratoga Associates
Source: Redrawn, with modifications, from Skutsch and Flowerdew (1976: 212, Fig. 2)

or minor, regional or local. The second matrix is compiled in the same way but the cells log four different criteria describing the *management* of each impact: for example, whether the impact is dealt with by a single agency or more than one. These latter four criteria are dichotomous, and a dot is entered or not entered.

The Saratoga matrix sorts causal factors and impact types. It is not a good visual summary but is useful as an index to more detailed description in a report (Skutsch and Flowerdew, 1976: 212).

The *component interaction matrix* can identify indirect impacts, using matrix algebra. It was developed by Environment Canada in 1974 (Environment Canada, 1974; Clark *et al.*, 1984: 214; Westman, 1985: 138). The same ecosystem components are listed horizontally and vertically (*see* Fig. 5.4a), and direct dependencies between the components are identified by scoring a cell (1 for first-order, 2 for second-order impacts, and so on) if the horizontal component (row) *depends* on the vertical (column). For example, songbirds are food for birds of prey, not vice versa. These are first-order impact relationships. Multiplying the matrix by itself (squaring) using matrix multiplication (developed from linear algebra in network analysis and cost–benefit analysis) can be used to determine second-order (indirect) relationships (two–link chains). Multiplying this second matrix by the first (cubing) gives three–link chain relationships (third-order impacts), and so on. This process could be pursued, provided there are sufficient data, time and funds, to iden-

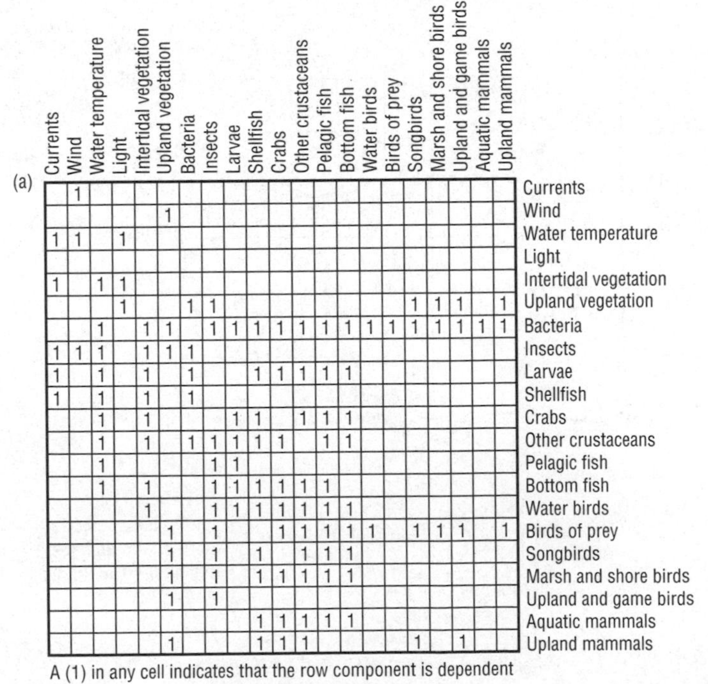

(a)

	Currents	Wind	Water temperature	Light	Intertidal vegetation	Upland vegetation	Bacteria	Insects	Larvae	Shellfish	Crabs	Other crustaceans	Pelagic fish	Bottom fish	Water birds	Birds of prey	Songbirds	Marsh and shore birds	Upland and game birds	Aquatic mammals	Upland mammals
Currents		1																			
Wind						1															
Water temperature	1	1		1																	
Light																					
Intertidal vegetation	1		1	1																	
Upland vegetation				1			1	1									1	1	1		1
Bacteria			1		1	1		1	1	1	1	1	1	1	1	1	1	1	1	1	1
Insects	1	1	1		1	1	1														
Larvae	1		1		1		1			1	1	1	1	1							
Shellfish	1		1		1		1														
Crabs			1		1				1	1		1	1	1							
Other crustaceans			1		1		1	1	1	1	1		1	1							
Pelagic fish			1					1	1												
Bottom fish			1		1			1	1	1	1	1	1								
Water birds					1			1	1	1	1	1	1	1							
Birds of prey						1		1			1	1	1	1	1		1	1	1		1
Songbirds						1		1		1		1	1	1							
Marsh and shore birds						1		1		1	1	1	1	1							
Upland and game birds						1		1													
Aquatic mammals										1	1	1	1	1							
Upland mammals						1				1	1	1					1		1		

A (1) in any cell indicates that the row component is dependent on the column component

(b)

	Currents	Wind	Water temperature	Light	Intertidal vegetation	Upland vegetation	Bacteria	Insects	Larvae	Shellfish	Crabs	Other crustaceans	Pelagic fish	Bottom fish	Water birds	Birds of prey	Songbirds	Marsh and shore birds	Upland and game birds	Aquatic mammals	Upland mammals
Currents	4	1	4	3	4	2	3	3	4	4	4	4	4	4	4	4	3	3	3	4	3
Wind	3	3	3	2	3	1	2	2	3	3	3	3	3	3	3	3	2	2	2	3	2
Water temperature	1	1	4	1	4	2	3	3	4	4	4	4	4	4	4	4	3	3	3	4	3
Light	0	0	0	0	0	0	0	0	0	0	0	0	0	0	0	0	0	0	0	0	0
Intertidal vegetation	1	2	1	1	5	3	4	4	5	5	5	5	5	5	5	5	4	4	4	5	4
Upland vegetation	2	2	2	1	2	2	1	1	2	2	2	2	2	2	2	2	1	1	1	2	1
Bacteria	2	2	1	2	1	1	2	1	1	1	1	1	1	1	1	1	1	2	1	1	1
Insects	1	1	1	2	1	1	1	2	2	2	2	2	2	2	2	2	2	2	2	2	2
Larvae	1	2	1	2	1	2	1	2	2	1	1	1	1	1	2	2	2	3	2	2	2
Shellfish	1	2	1	2	1	2	1	2	2	2	2	2	2	1	2	2	2	3	2	2	2
Crabs	2	2	1	2	1	3	2	2	1	1	2	1	1	1	3	3	3	4	3	3	3
Other crustaceans	2	2	1	2	1	2	1	1	1	1	1	2	1	2	2	2	2	3	2	2	2
Pelagic fish	2	2	1	2	2	2	2	1	1	2	2	2	2	2	3	3	3	3	3	3	3
Bottom fish	2	2	1	2	1	2	2	1	1	1	1	1	1	2	3	3	3	3	3	3	3
Water birds	2	2	2	2	1	2	2	1	1	1	1	1	1	1	3	3	3	3	3	3	3
Birds of prey	2	2	2	2	2	1	2	1	2	2	1	1	1	1	1	3	1	1	1	3	1
Songbirds	2	2	2	2	2	1	2	1	2	1	2	1	2	1	3	3	2	2	2	3	2
Marsh and shore birds	2	2	2	2	2	1	2	1	2	1	1	1	2	1	3	3	2	2	2	3	2
Upland and game birds	2	2	2	2	2	1	2	1	3	3	3	3	3	3	3	3	2	2	2	3	2
Aquatic mammals	2	3	2	3	2	3	2	2	2	1	1	1	1	1	3	3	3	1	3	3	0
Upland mammals	2	3	2	2	2	1	2	2	2	1	1	1	2	2	3	3	1	2	1	3	2

(c)

	Currents	Wind	Water temperature	Light	Intertidal vegetation	Upland vegetation	Bacteria	Insects	Larvae	Shellfish	Crabs	Other crustaceans	Pelagic fish	Bottom fish	Water birds	Birds of prey	Songbirds	Marsh and shore birds	Upland and game birds	Aquatic mammals	Upland mammals
Currents		${}^{0}_{0}0^{0}_{0}$																			
Wind						${}^{0}_{1}1^{1}_{1}$															
Water temperature	${}^{1}_{1}1^{1}_{1}$	${}^{2}_{1}1^{1}_{1}$		${}^{2}_{2}1^{1}_{1}$																	
Light																					
Intertidal vegetation	${}^{3}_{2}1^{2}_{1}$		${}^{1}_{1}1^{1}_{1}$	${}^{3}_{3}3^{3}_{3}$																	
Upland vegetation				${}^{0}_{3}3^{3}_{3}$			${}^{0}_{1}1^{1}_{1}$	${}^{0}_{1}1^{1}_{1}$									${}^{0}_{1}1^{1}_{1}$	${}^{0}_{1}1^{1}_{1}$	${}^{0}_{1}1^{1}_{1}$		${}^{0}_{1}1^{1}_{1}$
Bacteria			${}^{1}_{1}1^{1}_{1}$		${}^{2}_{2}2^{2}_{1}$	${}^{0}_{2}2^{2}_{2}$		${}^{1}_{1}1^{1}_{1}$	${}^{1}_{1}1^{1}_{1}$	${}^{1}_{1}1^{1}_{1}$	${}^{1}_{1}1^{1}_{1}$	${}^{1}_{1}1^{1}_{1}$	${}^{1}_{1}1^{1}_{1}$	${}^{1}_{1}1^{1}_{1}$	${}^{1}_{1}1^{1}_{1}$	${}^{0}_{1}1^{1}_{1}$	${}^{0}_{1}1^{1}_{1}$	${}^{1}_{1}1^{1}_{1}$	${}^{0}_{1}1^{1}_{1}$	${}^{1}_{1}1^{1}_{1}$	${}^{0}_{1}1^{1}_{1}$
Insects	${}^{1}_{0}0^{0}_{0}$	${}^{1}_{1}1^{1}_{1}$	${}^{1}_{1}1^{1}_{1}$		${}^{1}_{1}1^{1}_{1}$	${}^{0}_{1}1^{1}_{1}$	${}^{1}_{1}1^{1}_{1}$														
Larvae	${}^{2}_{2}2^{2}_{2}$		${}^{2}_{2}2^{2}_{2}$		${}^{3}_{3}3^{3}_{3}$		${}^{2}_{2}2^{2}_{2}$			${}^{2}_{2}2^{2}_{2}$	${}^{2}_{2}1^{2}_{0}$	${}^{2}_{2}2^{2}_{2}$	${}^{2}_{2}2^{2}_{3}$	${}^{2}_{2}2^{2}_{2}$							
Shellfish	${}^{2}_{2}2^{2}_{2}$		${}^{2}_{2}2^{2}_{2}$		${}^{3}_{3}3^{3}_{3}$		${}^{2}_{2}2^{2}_{2}$														
Crabs			${}^{2}_{1}0^{2}_{0}$		${}^{2}_{2}0^{2}_{0}$				${}^{2}_{2}0^{2}_{0}$	${}^{2}_{2}0^{1}_{0}$		${}^{2}_{2}0^{2}_{0}$	${}^{1}_{1}0^{1}_{0}$	${}^{2}_{2}0^{1}_{0}$							
Other crustaceans			${}^{2}_{2}2^{2}_{2}$		${}^{2}_{2}2^{2}_{2}$		${}^{1}_{1}1^{1}_{1}$	${}^{1}_{1}1^{1}_{1}$	${}^{2}_{2}1^{1}_{1}$	${}^{2}_{2}2^{2}_{2}$	${}^{1}_{1}0^{1}_{0}$		${}^{1}_{1}1^{1}_{1}$	${}^{1}_{1}1^{1}_{1}$							
Pelagic fish			${}^{2}_{2}2^{2}_{2}$					${}^{2}_{2}2^{2}_{2}$	${}^{3}_{3}3^{2}_{3}$												
Bottom fish			${}^{2}_{2}2^{2}_{2}$		${}^{2}_{2}2^{2}_{2}$			${}^{2}_{2}2^{2}_{2}$	${}^{3}_{3}3^{3}_{3}$	${}^{1}_{1}0^{1}_{0}$	${}^{1}_{1}0^{1}_{0}$	${}^{3}_{3}3^{3}_{3}$	${}^{1}_{2}2^{2}_{2}$								
Water birds					${}^{0}_{2}1^{1}_{1}$			${}^{1}_{1}0^{0}_{0}$	${}^{2}_{2}0^{1}_{0}$	${}^{3}_{2}1^{1}_{0}$	${}^{3}_{2}0^{1}_{0}$	${}^{3}_{2}1^{1}_{0}$	${}^{2}_{2}1^{1}_{1}$	${}^{2}_{2}1^{1}_{0}$							
Birds of prey						${}^{0}_{3}3^{3}_{3}$		${}^{0}_{1}1^{1}_{1}$			${}^{1}_{1}0^{0}_{0}$	${}^{1}_{1}0^{1}_{0}$	${}^{1}_{1}1^{1}_{1}$	${}^{1}_{1}0^{1}_{0}$	${}^{2}_{2}1^{1}_{1}$		${}^{0}_{2}2^{2}_{2}$	${}^{0}_{1}0^{1}_{0}$	${}^{0}_{1}1^{1}_{1}$		${}^{0}_{2}2^{2}_{2}$
Songbirds						${}^{0}_{3}3^{3}_{2}$		${}^{1}_{3}3^{3}_{2}$		${}^{1}_{1}0^{1}_{0}$		${}^{2}_{3}1^{2}_{1}$	${}^{1}_{1}1^{1}_{0}$	${}^{1}_{1}1^{1}_{1}$							
Marsh and shore birds						${}^{0}_{2}2^{2}_{2}$		${}^{0}_{1}1^{1}_{0}$		${}^{1}_{1}0^{1}_{0}$	${}^{1}_{1}0^{1}_{0}$	${}^{2}_{2}2^{2}_{2}$	${}^{1}_{1}1^{1}_{1}$	${}^{3}_{3}1^{3}_{1}$							
Upland and game birds						${}^{0}_{3}3^{3}_{3}$		${}^{0}_{1}1^{1}_{1}$													
Aquatic mammals										${}^{1}_{1}1^{1}_{0}$	${}^{2}_{2}0^{2}_{0}$	${}^{2}_{2}2^{2}_{2}$	${}^{2}_{2}2^{2}_{2}$	${}^{2}_{2}2^{2}_{2}$							
Upland mammals						${}^{0}_{3}3^{3}_{3}$				${}^{0}_{2}2^{3}_{2}$	${}^{0}_{1}0^{1}_{0}$	${}^{0}_{3}3^{3}_{3}$					${}^{0}_{1}1^{1}_{1}$		${}^{0}_{1}1^{1}_{1}$		

A (1) in any cell indicates that the row component is dependent on the column component

FIGURE 5.4 (a) Component interaction matrix, (b) Minimum-link matrix and (c) disruption matrix
Source: Redrawn from Clark *et al.* (1984: 214, 216, Figs 1 and 2)

tify longer-link chains; Environment Canada (1974) went as far as identifying five–link chains (fifth-order impacts). The cells are filled in with a number indicating the number of links in the chain – for example, songbirds are three links away from aquatic mammals in terms of dependency – and a *minimum-link matrix* (Fig. 5.4b) is derived.

The minimum-link matrix shows dependency relationships – links in the shortest chains of dependency between pairs of components; in effect, indirect impacts. Assessors can thus see likely chains of causation and response (Clark *et al.*, 1984: 217). This matrix approach is quite complicated and probably most useful for showing biogeophysical relationships. As is also true of component interaction and minimum-link matrices, the diagram presented is some steps removed from raw data, which may obscure how value judgements were made.

It is possible to computerize matrix manipulations, and some efforts have been made to develop component interaction matrix techniques so that they can be applied more widely (Shopley *et al.*, 1990).

Once direct dependencies have been identified with the aid of a component interaction matrix these could then be scored on a 'disruption level' scale of 0–3 to form a *disruption matrix*. The cells in the example matrix shown in Fig. 5.4c show five alternative developments, each with a disruption level score.

The *Moore impact matrix* (Moore *et al.*, 1973) was developed in the USA for investigating the relationship between manufacturing and regional impacts. The *goals achievement matrix* was developed before environmental impact assessment appeared in 1970 and seeks to show how groups involved in a development benefit or lose (Hill, 1968) (Box 5.3). The *compatibility matrix* lists spatial strategy objectives along rows and columns, and a record is made in the cells as to whether the sets of objectives are compatible or incompatible. Another kind of matrix, the *policy impact matrix*, lists policies on one axis and environmental criteria on the other (DoE, 1993: 19–32). Still another, the *three-dimensional matrix*, allows biophysical and socio-economic impacts to be charted against development activity.

Box 5.3 Goals Achievement Matrix (in million $)

Producer(s)	**Plan 1**	**Plan 2**	**Plan 3**
Airline	4.3	4.5	3.6
Consumer(s)			
Air traveller	?	2.5	3.0
Nearby public	0	–1.3	0
Wildlife	0	1.0	?
National trade	20	35.0	30.0

Unknown effect shown by?
Negative value shown by –

Source: Based on Westman (1985: 156)

Multi-attribute utility theory-based methods

Multi-attribute utility theory-based methods seek to recognize particular attributes of a development option, determine how they will behave, and establish the desirability of such change. Through this process an attempt is made to compare and trade off attributes between various options. The method depends on the subjective opinion of experts (Bisset, 1988: 50–3).

Network diagrams, stepped matrices, systems diagrams, linear graphs and networks

There are a number of assessment approaches based on cause-and-effect networks (network analysis) or systems diagrams, some virtually hybrids with matrices (e.g. the stepped matrix), that seek to recognize the complex web of relationships present in the real world (Shopley and Fuggle, 1984; Canter, 1996: 69–86). A *network* uses multiple matrices to account for temporal interactions of impacts and to show the full range of potential impacts. It is possible to examine the structure and function of a system to be developed or to trace the characteristics of a development. Stepped matrices can be useful for showing the public the nature of impacts because they show relationships – chains of causation – clearly.

Sorenson network

The Sorensen network (a linear graph or stepped matrix) approach, developed in the 1970s in the USA, was one of the first to explore dredging impacts and is one of the earliest systems approaches to environmental impact assessment (Sorensen, 1971, 1974; Bisset, 1983: 172, 1987: 24–27). It combines a matrix format with a network diagram to show what interactions occur and how they are produced, and the dependency relations. Thus it is a 'cause–condition–effect network', and shows both the direct and indirect linkages. A network can show second and higher-order impacts of proposed development actions. It integrates into one display a matrix for the identification of impacts, a network of consequent impacts, and proposals for actions to avoid or mitigate adverse consequences (Clark *et al.*, 1984: 244).

Network diagrams are unlikely to give information on impact probability, relative importance or magnitude. The Sorensen network identifies impacts but does little or nothing to evaluate them. Additional information on mitigation measures, and so on, may be added to the diagram.

The Sorensen network is a well-known approach, and is increasingly computer-generated (Clark *et al.*, 1984: 221). It is probably best applied to ecological impacts and is useful for presenting or clarifying relationships. It is reputedly difficult to apply Sorensen networks to socio-economic impact assessment. Networks can become complex and difficult to follow, but relatively simple ones can provide a good visual presentation (Fig. 5.5). Once

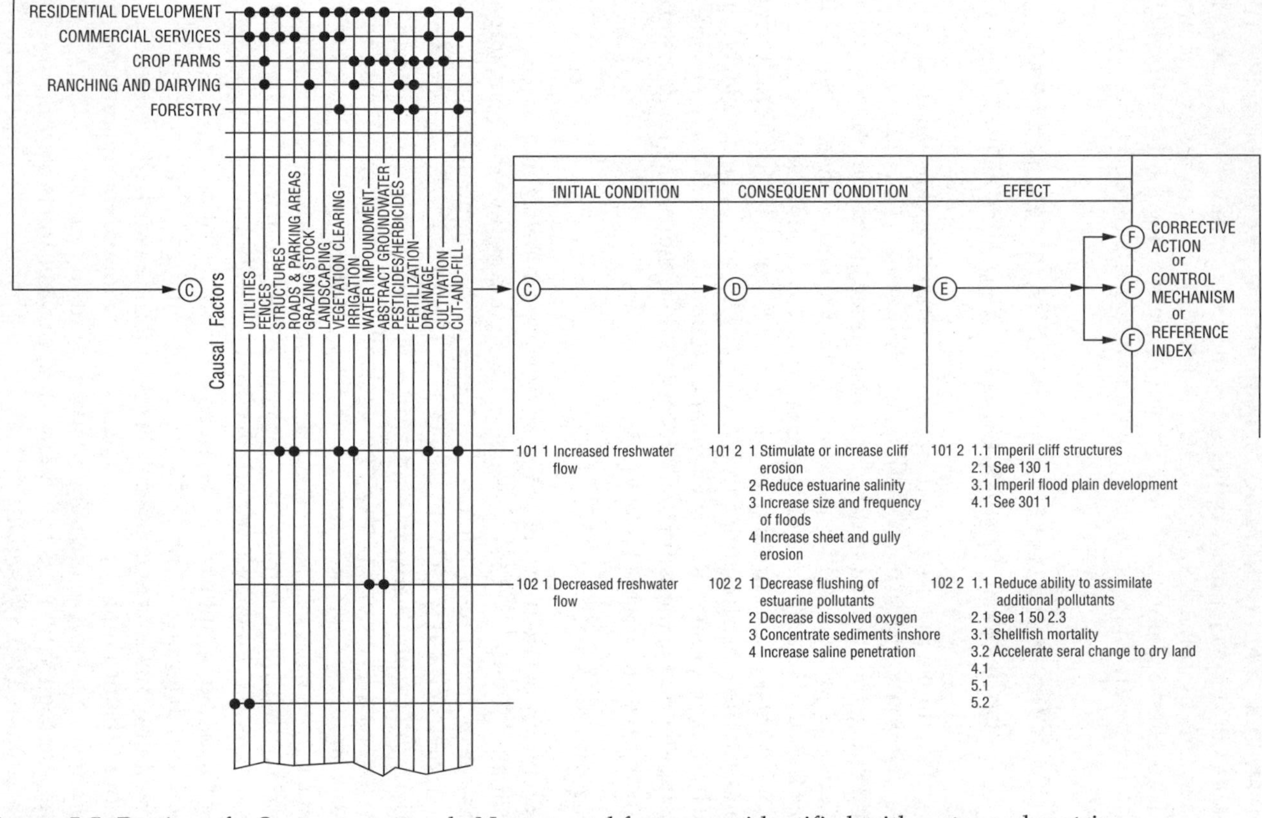

FIGURE 5.5 Portion of a Sorensen network. Note: causal factors are identified with a stepped matrix.
Source: Based on various sources, all ultimately deriving from Sorensen (1971: 15)

developed and 'tuned' for a particular type of project, they become quicker to prepare and quite cheap and simple to carry out. Networks can point out direct and indirect impacts for further study, but generally do not show which impacts are important. Comparison of Sorensen networks for different sites can be difficult.

Systems diagrams

A 'system' can be defined as a set of logically interconnected models, a 'model' being a set of equations used to predict the behaviour of variables in a particular topic area. Systems analysis has been applied to impact assessment to explore impacts or the nature of the ecosystem impacted (DeSouza, 1979). The approach can be qualitative – simple synergistic models – or quantitative.

Systems diagrams developed out of work in ecological energetics by Odum (1972). Ecological or socio-economic components are linked in a diagram to show energy flows between environmental components with their magnitude indicated (appropriate units such as calories, decibels or curies, etc. are used). These diagrams show ecological impacts, but they are less good at mapping ecological relations that are not dependent on energy flows, so that shelter needs, nesting sites, etc. would be missed. Ecological and economic issues could both be dealt with if a common unit such as coal-equivalents were used (Clark *et al.*, 1984: 221).

Ecosystems may have stable subsystems, but linkages are often unclear and difficult to predict, and this limits systems and network approaches.

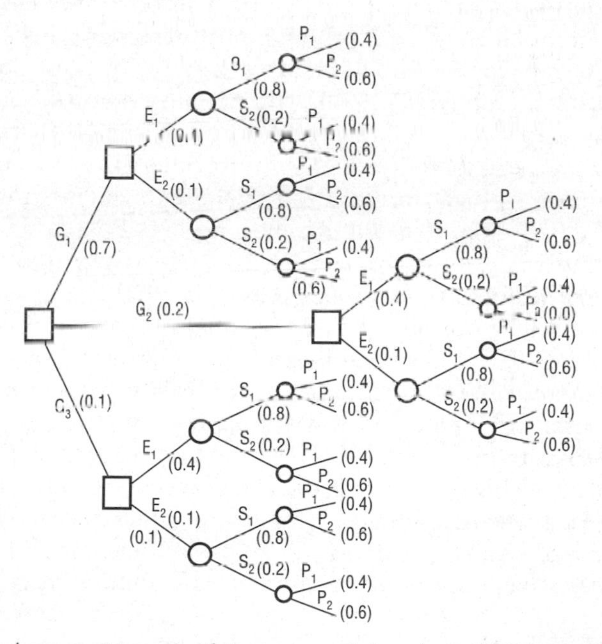

FIGURE 5.6 An event tree, in this case used for air quality assessment
Source: Redrawn from Hyman and Stiftel (1988: 37, Fig. 2–2)

Their value is largely restricted to prediction of ecological impacts, but they do not provide visual summaries useful to the public (Gilland and Risser, 1977; Bisset, 1983: 175; 1988: 53–57; Graham Smith, 1993: 23).

EVENT TREES

A way of analysing problems that involve an ordered sequence of decisions and chance outcomes that depend on earlier decisions and chance outcomes is the event tree. The event tree shows subjective probabilities and (judged by experts) impact magnitudes of outcomes (Fig. 5.6; *see* Whyte and Burton, 1980: 49–50; Hyman and Stiftel, 1988: 35). Event trees are ideal for complex technology assessments, such as the checking of nuclear power-stations. Their preparation is normally computerized.

USE OF COMPUTERS AND EXPERT SYSTEMS

Environmental impact analysis by computer

There have long been attempts to computerize impact identification and assessment (for early examples, *see* Krauskopf and Bunde, 1974; Sorensen, 1974; Thor *et al.*, 1978). More recent studies are provided by Guariso and Page (1994) and Benoît (1995: 421–6). Benoît and also Canter (1996: 45) argue that as impact assessment gets more complex and laborious (in order to be more holistic and adaptive), computerization becomes more important. Computer techniques have been used for interpreting impacts (*see* Baumwerd-Ahlmann *et al.*, 1991). The development of better microcomputers and improved software has made it easier to run impact assessment methods, expert systems, environmental information systems, models, geographical information systems, systems analysis, etc. (Jialin *et al*, 1989; Julien *et al.*, 1992; Fernandes, 1993). Nevertheless, progress has still been limited by lack of user-friendly programs and by the relatively low number of assessors within the field of environmental impact assessment who are skilled in the application of computers (Guariso and Page, 1994). There has been interest in integrating environmental impact assessment, monitoring and graphical information systems through computer use, and in applying computing to social impact assessments (Leistritz *et al.*, 1995).

Guariso and Page (1994) warn that the application of computing to environmental impact assessment should be 'transparent' to reduce the risk of accidental or deliberate errors, unauthorized disclosure, etc. Accidents like the Chernobyl disaster have prompted a number of countries to co-operate in developing joint rapid-impact assessment and data exchange systems. This is vital for coping with transboundary problems, like airborne pollution or radioactive fallout. The European Community (now the European Union) has gone part way to developing such a system for radioactive fallout by establishing the EC Urgent Radiological Information Exchange (ECURIE) in 1987.

Expert systems

Expert systems or 'knowledge-based systems' software have been applied to impact assessment, regional environmental impact analysis, environmental planning, environmental assessment (Lein, 1988; Loehle and Osteen, 1990; Fedra, 1991; Antunes and Camara, 1992; Geraghty, 1993; Wright *et al.*, 1993; Canter, 1996: 46), eco-auditing and environmental management (Benoît and Podesto, 1995) since the 1980s. These systems perform problem-solving based on a database of expert knowledge; they draw on heuristic ('rule-of-thumb') reasoning and act as 'advisers' or provide decision support. Expert systems may be slow and difficult to develop, but once they have been developed they may offer great potential for impact assessment and eco-auditing and for data management (Berka and Jirku, 1995).

Gray and Stokoe (1988) reviewed the potential and limitations of expert systems for impact assessment and environmental management, one of their hopes being that they could help achieve consistent quality of impact assessments. Schibuola and Byer (1991) proposed a knowledge-based system for environmental impact assessment that could facilitate public participation (*see* Chapter 2). Mercer (1995), recognizing that impact assessment increasingly uses qualitative methods of assessment, tried to develop an expert system capable of coping with this. Expert systems methods have been applied to environmental impact assessment for regional planning by Burde *et al.* (1994). Tucker and Richardson (1995) applied an expert system in a sort of impact assessment to screen introduced plants in order to evaluate which ones might invade native vegetation.

Use of expert systems could enable a diversity of planners and managers to consider impacts. For example, engineers could call on it (Mercer, 1995). This would help to improve development proposals, but it should not be a substitute for thorough assessment by independent specialists.

QUANTITATIVE METHODS: SCALING, WEIGHTING, INDICES

The characteristics and relative merits of quantitative and qualitative data are discussed in Chapter 4. Interest in quantitative approaches is in part driven by the desire to make comparisons of different impacts identified by an assessment and to facilitate comparison of different assessments. This has encouraged the use of scaling, weighting, standardizing and aggregation of impacts to produce composite indices, which offer single, apparently objective, benchmarks (Ott, 1979; Hollick, 1981; Bisset, 1983: 175; Westman, 1985: 135; Ying and Liu, 1995). The goal of quantitative methods is to give a comparable unit, although in some cases this gives a false sense of scientific precision. Kung *et al.* (1993) have applied fuzzy logic to weighting intended to derive a representative environmental quality index. For a critique of index methods, *see* Bisset (1988: 53, 60–1).

Weighing and other statistical transformations may not be as objective as they seem. Often they are just the 'value judgements of experts', and conceal the original (raw) data. Undistorted qualitative data may well be preferable to transformed quantitative data. Weighting should be applied only to interval or ratio-scale data.

Environmental valuation system

The environmental evaluation system (Whitman *et al.*, 1971; Dee *et al.*, 1973; Battelle Columbus, 1977; Heer and Hagerty, 1977: 165; Bisset, 1983: 176; Clark *et al.*, 1984: 201–3; Westman, 1985: 150–252; Glasson *et al.*, 1994: 102) is a complex scaled checklist, the weighting procedure having originally been developed in 1972 by Battelle Memorial Laboratories, Columbus, Ohio, for the USA Bureau of Land Reclamation for application to water resources development projects (Dee *et al.*, 1973). It is useful for comparing alternative development options but could miss impacts. It is seemingly a quantitative method but uses value judgements (for a critique, *see* Ortolano and Shepherd, 1995: 8).

The approach collects data on a quantitative, aggregated index of 78 environmental attributes grouped into four accounts. For example, these could be ecology, environmental pollution, aesthetics and human interest, but alternatives are possible, such as environmental quality, regional development, social well-being and national economic development (*see* Fig. 5.7). At least one measurement technique is used per attribute. The result is an expression of impacts, whether physical, biological, social, or whatever, in commensurate units; that is, negative and positive impacts are expressed as common units by transforming attributes using *value functions*. The value functions are developed by a panel of experts and are therefore subjective measurements. The measurement for each attribute is converted into a 0 (bad) to 1 (good) scale of environmental quality. Experts assign weights to each attribute using a ranked pair-wise comparison technique. The weighted list is then subjected to what is basically the Delphi technique to obtain a consensus. Each environmental quality value is then multiplied by a weighting to give an index value. The final environmental impact value is the difference between the sum of impacted values with and without development taking place (*see* Clark *et al.*, 1984: 244).

The environmental evaluation system method has been adapted for use in a variety of development situations including applications to social impact assessment. It can identify indirect impacts, but does not show the interrelations between impacts. Hyman and Stiftel (1988: 25) warn that this, like the Water Resources Assessment Methodology (the latter is dealt with below), may involve the dubious practice of multiplying ordinal numbers and that value functions can 'blur facts and values'. Bisset (1988) outlined modified environmental evaluation system methods developed in Thailand which use techniques derived from cost–benefit analysis.

Water Resources Assessment Methodology

Developed by the US Army Corps of Engineers in the 1970s to assess the environmental, economic and social impacts of water resources development and flood control projects, the Water Resources Assessment Methodology uses a checklist of potential impacts. This checklist is divided into four accounts: environmental quality; national economic development; regional development; and social well-being. Scaling, especially with non-quantitative data, may involve pair-wise comparisons. Weighting can also use pair-wise comparisons. The accounts are aggregated and weighted according to the views of the authority managing the development. There is some flexibility as to how the scaled values are produced. The 'comparable' impact values entered into these accounts are obtained by scaling in a manner similar to that described for the environmental evaluation system; indeed, some practices from that system were incorporated into Water Resources Assessment Methodology.

The Water Resources Assessment Methodology is useful for comparing alternative development options (Solomon *et al.*, 1977; Richardson *et al.*, 1978). The importance of each variable relative to all others can be seen in a ranking table, and the method can cope with qualitative data; that is, values are displayed as an 'account', not aggregated as a single value as in the environmental evaluation system. The Water Resources Assessment Methodology approach does not address indirect or cumulative impacts and, if it involves the multiplication of ordinal scale values, this should be treated with caution.

Sondheim method

The Sondheim method is another scaled checklist approach, developed in 1978 in the USA (Sondheim, 1978) and updated in 1983 (Bisset, 1988: 49). It gives a total score that allows comparison of different development alternatives. Four steps are involved. First, expert panels assess the impacts of alternative development proposals using appropriate techniques. Each expert produces a set of ratings on an ordinal scale. Second, normalization is carried out. Each impact is weighted so as to be 'comparable': the public may feel, say, that noise is less unpleasant than smell, and weightings reflect such attitudes. Third, scores obtained from the weighting process are placed in a matrix: the rows are the development alternatives, the columns are components of the environment. Fourth, a 'weighting panel', which can include representatives of the public, may adjust scores in the matrix. The result is a numerical preference ranking of alternatives.

Assessors are still using and developing procedures that rely on value functions (e.g. Beinate *et al.*, 1994). Quantitative methods have advantages, but these may be overshadowed by the drawbacks of cost and delay, and by their obscuring of raw data, making it difficult to see how impact significance has been judged. Ideally, impact assessment should be based on the best use of data whether these be quantitative, qualitative or a combination.

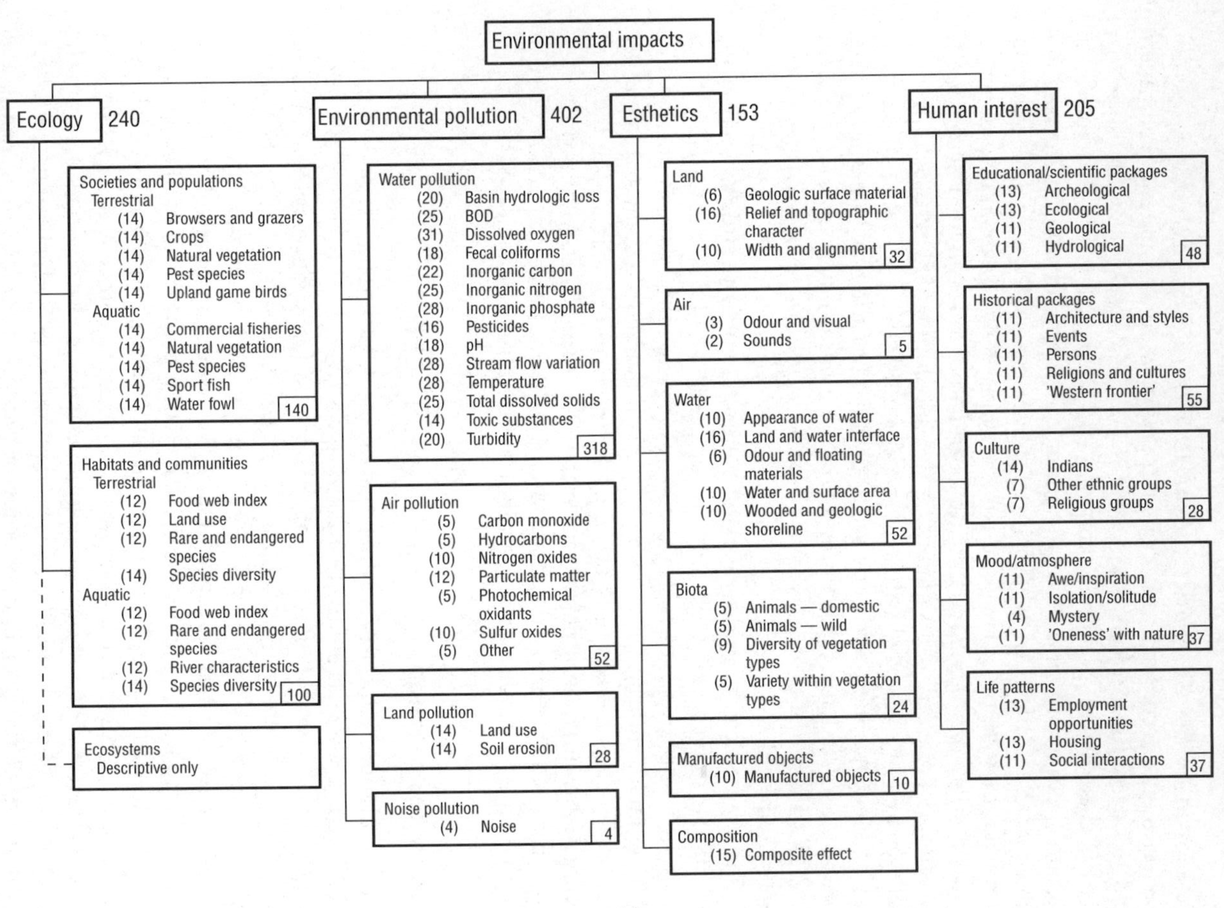
Environmental impacts
Ecology 240
Societies and populations
Terrestrial
(14) Browsers and grazers
(14) Crops
(14) Natural vegetation
(14) Pest species
(14) Upland game birds
Aquatic
(14) Commercial fisheries
(14) Natural vegetation
(14) Pest species
(14) Sport fish
(14) Water fowl
140
Habitats and communities
Terrestrial
(12) Food web index
(12) Land use
(12) Rare and endangered species
(14) Species diversity
Aquatic
(12) Food web index
(12) Rare and endangered species
(12) River characteristics
(14) Species diversity
100
Ecosystems
Descriptive only
Environmental pollution 402
Water pollution
(20) Basin hydrologic loss
(25) BOD
(31) Dissolved oxygen
(18) Fecal coliforms
(22) Inorganic carbon
(25) Inorganic nitrogen
(28) Inorganic phosphate
(16) Pesticides
(18) pH
(28) Stream flow variation
(28) Temperature
(25) Total dissolved solids
(14) Toxic substances
(20) Turbidity
318
Air pollution
(5) Carbon monoxide
(5) Hydrocarbons
(10) Nitrogen oxides
(12) Particulate matter
(5) Photochemical oxidants
(10) Sulfur oxides
(5) Other
52
Land pollution
(14) Land use
(14) Soil erosion
28
Noise pollution
(4) Noise
4
Esthetics 153
Land
(6) Geologic surface material
(16) Relief and topographic character
(10) Width and alignment
32
Air
(3) Odour and visual
(2) Sounds
5
Water
(10) Appearance of water
(16) Land and water interface
(6) Odour and floating materials
(10) Water and surface area
(10) Wooded and geologic shoreline
52
Biota
(5) Animals — domestic
(5) Animals — wild
(9) Diversity of vegetation types
(5) Variety within vegetation types
24
Manufactured objects
(10) Manufactured objects
10
Composition
(15) Composite effect
Human interest 205
Educational/scientific packages
(13) Archeological
(13) Ecological
(11) Geological
(11) Hydrological
48
Historical packages
(11) Architecture and styles
(11) Events
(11) Persons
(11) Religions and cultures
(11) 'Western frontier'
55
Culture
(14) Indians
(7) Other ethnic groups
(7) Religious groups
28
Mood/atmosphere
(11) Awe/inspiration
(11) Isolation/solitude
(4) Mystery
(11) 'Oneness' with nature
37
Life patterns
(13) Employment opportunities
(13) Housing
(11) Social interactions
37

(a)

FIGURE 5.7

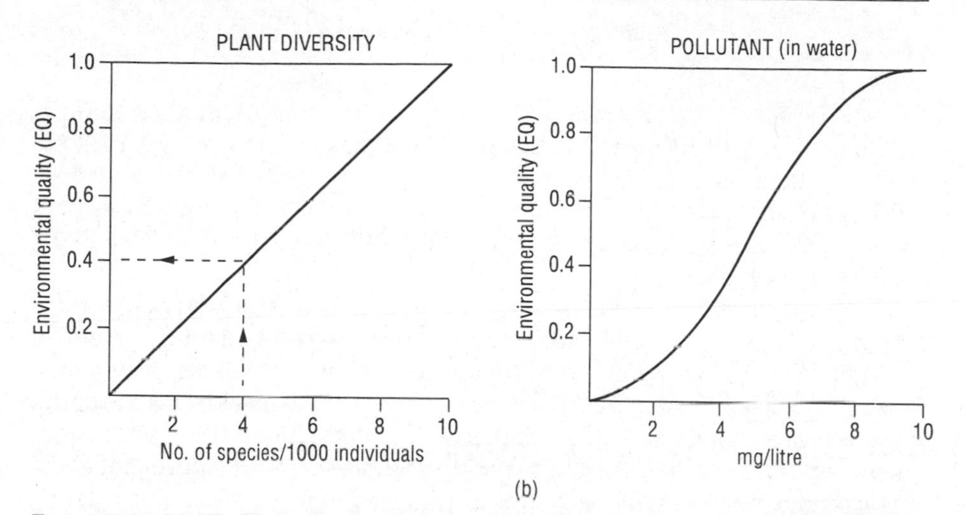

FIGURE 5.7 (a) The Battelle environmental evaluation system. It uses a checklist of 78 environmental, social and economic parameters that may be affected by the proposed development. Data on each of these are collected – not allowing for development; that is, to determine the baseline situation. These data are converted to environmental quality (EQ) scale values (in all, 78 of them). These scale values are multiplied by their individually assigned parameter importance unit (PIU); this is basically a subjective judgement by assessors. EQ × PIU gives a composite score for the environment without the project. For each development alternative the assessors predict the change in environmental attribute/parameter. For each development alternative the change in the environmental attribute/parameter is predicted. Using the change in the parameter values, the EQ and PIU for each parameter and each alternative are determined. The EQ values for each alternative are multiplied by each PIU, and the information is aggregated to give a total composite score. Finally a composite score is produced for each development option. The numbers in parentheses are PIUs, the numbers in boxes composite scores for each account section. BOD = biochemical oxygen demand. (b) In practice, the determination of the EQ scale requires the construction of 78 value function graphs, of which two examples are shown. These are constructed by experts and show the lowest likely value (zero) to highest likely to be encountered for a parameter on the horizontal axis, and a 0 to 1.0 value function scale on the vertical axis (EQ scale). For example, if a field botanist recorded 4 species per 1000 plants surveyed, the EQ would be 0.4 (left hand graph).

Predicted impacts for proposed developments are 'scaled' using these graphs. For example, a 'good' level of oxygenation in a river would be equated to value function 1.0, and a 'good' species diversity, as one might expect, would also be 1.0. Value functions (or scalars) are common units on 0 to 1.0 scale (no negative numbers). Note that the vertical EQ scale (which is notional and arbitrary) is constant for all 78 graphs.

Modelling methods

Various types of models (computer, analogue, conceptual models, simulation modelling, word models, etc.) have been applied in ecology, regional planning, land use change, climatology, hydrology, pollution studies, social studies, economics, and many other disciplines, as well as impact assessment (Bisset, 1988: 57–60; Hyman and Stiftel, 1988: 198; Whitehead, 1992a, 1992b; Physick *et al.*, 1995).

Conceptual models are widely used to see what needs study, to help formulate and check hypotheses, to organize ideas and to identify errors of thought. *Simulation models,* developed for impact assessment (Graham Smith, 1993: 24–6), can be useful for management once development is under way. The prediction of dispersal patterns for pollutants in the atmosphere, rivers and water bodies like the sea and the assessment of future global climatic change (increasingly with supercomputers running general circulation models) is vital and often depends on simulation modelling. These models are often far from accurate or rely on imperfect data and understanding of the processes being modelled.

Modelling, unless it has already been developed for similar activities, is generally demanding of expertise, time and possibly data. The main use has been on larger and more complex projects. Modelling can be an attempt to simplify reality so that things may be described and understood and, if possible, predictions made. There are mechanistic (analogue) models – for example, tidal estuary models that can be flooded or drained and monitored; mathematical models; workshop-based models; role-playing games; world models; and word models. Modelling can facilitate the consistent testing of alternatives (Hyman and Stiftel, 1988: 16).

An advantage that is often stressed is the ability of models to reflect the dynamic character of the environment and human activity. When used for impact assessment this can allow for uncertainty. For example, the adaptive environmental assessment and adaptive environmental assessment and management approaches (*see* Chapter 3) include modelling workshops and project or regional modelling, including input–output modelling. However, modelling has tended to have little provision for public involvement. Some models deal with one or a limited number of factors, but recently the trend has been toward more complex models, even ones modelling an entire ecosystem.

Walters (1993) examined the relationship of modelling and experiment – extreme positions, which on their own each yield inadequate results. He advocated 'wise modelling'; that is, application of modelling to things on which it works with experimentation to support and complement it. Holling (1978) argued for an adaptive, 'less static' quick modelling approach to assessment and environmental management: adaptive environmental assessment and management (Graham Smith, 1993: 25). The Holling approach to modelling relies on expert workshops to draw up the model, and seeks a close integration of analysis and evaluation that allows an adap-

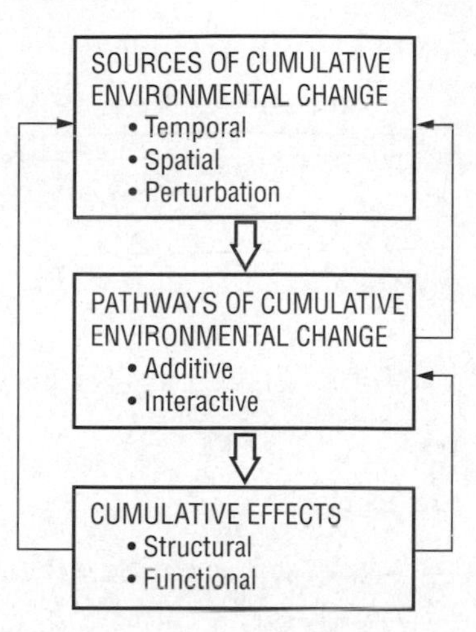

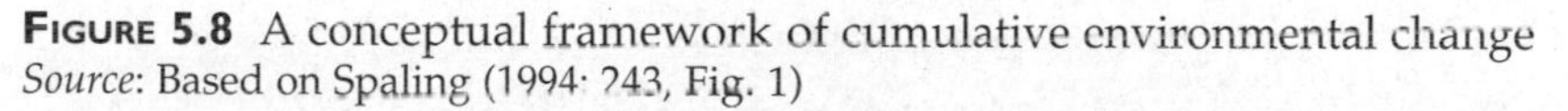

FIGURE 5.8 A conceptual framework of cumulative environmental change
Source: Based on Spaling (1994: 243, Fig. 1)

- synergistic impacts;
- impact occurs when a threshold is passed as a consequence of some 'trigger effect' (e.g. 'chemical time bomb' or 'biological time bomb');
- irregular 'surprise effects';
- impacts triggered by a feedback process ('antagonistic' – feedback which reinforces a trend; or 'ameliorative' – feedback which counters a trend).

Any one of these cumulative impact scenarios may result in a 'runaway' process that exceeds some critical threshold and may be difficult to remedy. For example, global warming leads to uncontrollable releases of 'greenhouse gases' from various sinks resulting in a strong, perhaps 'runaway', positive feedback – uncontrollable warming. Impact assessment needs to be watchful for such interactions. There may have been some interaction of impacts *before* a development takes place.

Table 5.1 suggests a typology of cumulative effects.

In practice, accurate cumulative effects assessment is difficult. Nevertheless, there are techniques and methods available, for example the component interaction matrix and minimum-link matrix (*see* Figs 5.4a and 5.4b), layered matrices and combined networks. There are a variety of cumulative effects assessment methods (*see* Cocklin, *et al.*, 1992; Spaling and Smit, 1993; Smit and Spaling, 1994). A Swedish retrospective study which tried to assess accumulated environmental impact over the period between 1940 and 1990 for cadmium heavy metal pollution made use of trade and manufacturing statistics, fertilizer use records and other 'natural and societal weathering rate' data. The end result was a comparison of human versus natural emissions of cadmium for an extended period (Bergbeck *et al.*, 1994).

Some workers have tried to assess cumulative impacts by adopting a regional (Dixon and Montz, 1995) or strategic stance (*see* the discussion of

TABLE 5.1 A typology of cumulative effects

Type	Main characteristics	Example
Time crowding	Frequent and repetitive impacts on an environmental system	Forest harvesting rate exceeds regrowth
Time lags	Delayed effects	Exposure to carcinogens
Space crowding	High spatial density of impacts on an environmental system	Pesticides in streams from non-point sources
Trans-boundary movement	Impacts occur away from this source	Acid rain deposition
Fragmentation	Change in landscape pattern	Fragmentation of wetlands
Compounding effects	Effects arising from multiple sources or pathways	Synergism among pesticides
Indirect effects	Secondary impacts	Release of methyl mercury in reservoirs
Triggers and thresholds	Fundamental changes in system behaviour or structure	Global climate change

Source: Spaling (1994: 247 – table 2)

strategic environmental assessment in Chapter 3), others have tried to view things at the project level (Lawrence, 1994). Rees (1995) argued that it is now necessary to assess global as well as local and regional cumulative effects, because global stability and even 'life-support systems' are increasingly shaped by cumulative impacts and global impacts affect local and regional systems.

Smit and Spaling (1994: 83) suggest that cumulative effects assessment methods can be grouped into primarily analytical approaches (e.g. spatial analysis, including geographical information systems; network analysis; interactive matrices; ecological modelling) and primarily planning approaches (e.g. expert opinion; multi-criteria evaluation; programming models, such as linear programming; land suitability evaluation). Geographical information systems have considerable promise as a method for cumulative effects assessment, although there are problems incorporating consideration of impact accumulation processes and no differentiation of additive and interactive processes (Johnson *et al.*, 1988). Indirect impact assessment is a complex process, and for best results a mix of different methods is likely to be required (Smit and Spaling, 1994: 101).

COMMUNICATING IMPACT ASSESSMENT RESULTS

Some of the techniques and methods discussed in this chapter serve to communicate the assessment to the 'user', rather than or as well as identifying and perhaps assessing impacts. However, the assessor must usually prepare supportive information or present the results from less communicative techniques or from an assemblage of techniques. The law may require a certain form of communication, and those commissioning the impact assessment generally lay down certain requirements (terms of reference). The communication may take the form of a report, a précis diagram, a public presentation, or a combination of these. This 'end-product' for environmental impact assessment is usually termed an environmental impact statement.

It is not uncommon for otherwise adequate assessments to be seriously flawed at the communication stage by careless or deliberately poor presentation. To minimize the risk of substandard impact assessment communication, the first step should be to establish the motives for impact assessment. It is likely to be undertaken for one or more of the following reasons:

- to select from alternative developments;
- to justify a decision;
- to facilitate public participation;
- to forewarn the developer of problems;
- to protect the developer against a lawsuit for negligence or damaging decisions;
- to design mitigation measures, impact avoidance or contingency plans.

It is unlikely that an impact assessment statement will meet all the demands made upon it adequately. In any case, the assessor is often limited by regulations that state what shape the communication should take. Whether the statement is aimed at decision-makers, the public or any other group, it is vital that it be clear and concise. Excessive 'clutter' may hide things and make publication more costly – yet it is not unknown for communication to be in the form of several volumes of technical reports that may not even have a reasonable summary.

Overview of methods

There is no 'universal' method or combination of methods. Hyman and Stiftel (1988: 221) felt that few, if any, impact assessment methods they had reviewed satisfactorily elicited and incorporated information from a broad enough array of groups in affected communities, and it was most important to remedy this. Figure 5.1 is an attempt to provide a summary of method qualities in terms of seven criteria. It is not, however, valid to assume that a method is better than another because it satisfies more of the criteria than another; methods must be selected for a particular situation, and there may be a need to use two or more methods that complement each other.

REFERENCES

Antunes, M.P. and **Camara, A.** 1992: HyperAIA: an integrated system for environmental impact assessment. *Journal of Environmental Management* **35(2)**, 93–111.

Atkins, R. 1984: Analysis of the utility of EIA methods. In Clark, B.D., Gilad, A., Bisset, R. and Tomlinson, P. (eds), *Perspectives on environmental impact assessment*. Dordrecht: D. Reidel, 241–52.

Barber, W.F., Bartlett, R.V. and **Dennis, C.** 1990: Matrix organization theory and environmental impact assessment. *Social Science Journal* **27(3)**, 235–52.

Battelle Columbus 1977: *Environmental evaluation in project planning*. Contract 6–07–DR-50/50. Columbus, OH: Battelle Columbus Laboratories.

Baumwerd–Ahlmann, A., Jaschek, P., Kalinski, J. and **Lehmkuhl, H.** 1991: Using KADS for generating explanations in environmental impact assessment. In Bullinger, H.J. (ed.), *Human aspects in computing*. Proceedings of the HCI International, 1–6 September, 1991. Stuttgart, Amsterdam: Elsevier, 866–76.

Beinate, E., Nijkamp, P. and **Rietveld, P.** 1994: Value functions for environmental pollutants: techniques for enhancing the assessment of expert judgements. *Environmental Monitoring and Assessment* **30(1)**, 9–23.

Benoît, R. 1995: Current and future directions for structured impact assessments. *Impact Assessment,* **13(4)**, 403–32.

Benoît, R. and **Podesto, M.** 1995: Environmental assessment: transition of a decision support system to the environmental management of projects. *Impact Assessment* **13(2)**, 117–33.

Bergbeck, B., Anderberg, S. and **Lohm, U.** 1994: Accumulated environmental impact: the case of cadmium in Sweden. *Science of the Total Environment* **145(1–2)**, 13–28.

Berka, P. and **Jirku, P.** 1995: An expert system for environmental data management. *Environmental Monitoring and Assessment* **34(2)**, 185–95.

Bisset, R. 1980: Methods for environmental impact analysis: recent trends and future prospects. *Journal of Environmental Management* **11(1)**, 27–43.

Bisset, R. 1981: Recent EIA methods: a review. In Breakall, M. and Glasson, J. (eds), *Environmental impact assessment: from theory to practice*. Oxford: Department of Town Planning, Oxford Polytechnic, 39–54.

Bisset, R. 1983: A critical survey of methods for environmental impact assessment. In O'Riordan, T. and Turner, R.K. (eds), *An annotated reader in environmental planning and management*. Oxford: Pergamon, 168–86.

Bisset, R. 1987: Methods for environmental impact assessment: a selective survey with case studies. In Biswas, A.K. and Qu Geping (eds), *Environmental impact assessment for developing countries*. London: Tycooly International, 3–64.

Bisset, R. 1988: Developments in EIA methods. In Wathern, P. (ed.), *Environmental impact assessment: theory and practice*. London: Unwin Hyman, 47–61.

Bowman, J., Olsson, M. and **Sundholm, M.** 1994: SAMBA: a way to deal with GIS. In Guariso, G. and Page, B. (eds), *Computer support for environmental impact assessment* IFIP Transactions B: Applications in technology – Conference on Computer Support for EIA, CSEIA 93, Como, Italy, 6–8 October 1993. Amsterdam: North-Holland, 1–11.

Burde, M., Jackel, T., Dieckmann, R. and **Hemker, H.** 1994: Environmental impact assessment for regional planning with SAFRAN. *IFIP Transactions B: Applications in Technology* **16**, 245–56.

Canada 1986: *Proceedings of the Workshop on Cumulative Environmental Effects:a binational perspective* (the Canadian Environmental Assessment Research Council and the US National Research Council). Ottawa: Supply and Services Canada.

Canter, L.W. 1977: *Environmental impact assessment*, 1st edn. New York: McGraw-Hill.

Canter, L.W. 1983: Methods for environmental impact assessment: theory and application. In University of Aberdeen, Project and Development Control Unit (ed), *Environmental impact assessment*. The Hague: Martinus Nijhoff, 165–234.

Canter, L.W. 1996: *Environmental impact assessment*. 2nd edn. New York: McGraw-Hill. (1st edn 1977)

Canter, L.W. and **Kamath, J.** 1995: Questionnaire checklist for cumulative impacts. *Environmental Impact Assessment Review* **15(4)**, 321–39.

Clark, B.D., Gilad, A., Bisset, R. and **Tomlinson, P.** (eds) 1984: *Perspectives on environmental impact assessment*. Dordrecht: D. Reidel.

Cocklin, C., Parker, S., and **Hay, J.** 1992: Notes on cumulative environmental change II: a contribution and methodology. *Journal of Environmental Management* **35(1)**, 51–67.

Contant, C.K. and **Wiggins, L.L.** 1991: Defining and analysing cumulative environmental impacts. *Environmental Impact Assessment Review* **11(3)**, 297–309.

Contant, C.K. and **Wiggins, L.L.** 1993: Towards defining and assessing cumulative impacts: practical and theoretical considerations. In Hildebrand, S.G. and Cannon, J.B. (eds), *Environmental analysis: the NEPA experience*. Boca Raton, FL: Lewis Publishers, 336–56.

Dee, N., Baker, J.K., Drobney, N.L., Duke, K.M., Whitman, I. and **Fahringer, D.** 1973: An environmental evaluation system for water resource planning. *Water Resources Research* **9(3)**, 523–35.

DeSouza, G.R. 1979: *System methods and environmental impact analysis*. Lexington, MA: Lexington Books.

Dixon, J. and **Montz, B.E.** 1995: From concept to practice: implementing cumulative impact assessment in New Zealand. *Environmental Management* **19(3)**, 445–56.

DoE 1993: *Environmental appraisal of development plans* (UK Department of the Environment). London: HMSO.

Duke, K.M., Cornaby, B.W. and **Velagaleti, R.R.** 1994: Technology transfer of environmental impact assessment methodologies to developed and developing countries. *Journal of Scientific and Industrial Research* **53(8)**, 609–18.

Eedy, W. 1995: The use of GIS in environmental assessment. *Impact Assessment* **13(2)**, 199–206.

Environment Canada 1974: *An environmental assessment of Nanaimo Port alternatives*. Ottawa: Environment Canada.

Fedra, K. 1991: *Expert systems for environmental screening*. Laxenburg: International Institute for Applied Systems Analysis.

Fedra, K. 1994: *GIS and environmental modelling*. Research Reprint 94–2. Laxenburg: International Institute for Applied Systems Analysis.

Fernandes, J.P. 1993: Ecogis-Ecosad: a methodology for the biogeophysical environmental assessment within the planning process. *Computers, Environment and Urban Systems* **17(4)**, 347–54,

Fortlage, C.A. 1990: *Environmental assessment: a practical guide*. Aldershot: Gower Technical.

Galloway, G. 1978: *Assessing man's impact on wetlands*. Water Resources Research Institute Report 78–136. Raleigh, NC: University of North Carolina.

Geraghty, P.J. 1993: Environmental assessment and the application of expert systems: an overview. *Journal of Environmental Management* **39(1)**, 27–38.

Gilland, M.W. and **Risser, P.G.** 1977: The use of systems diagrams for environmental impact assessment: procedures and an application. *Ecological Modelling* **3(3)**, 183–209.

Glasson, J., Therivel, R. and **Chadwick, A.** 1994: *Introduction to environmental impact assessment: principles and procedures, process, practice and prospects*. London: University College London Press.

Graham Smith, L. 1993: *Impact assessment and sustainable resource management*. Harlow: Longman.

Gray, A. and **Stokoe, P**. 1988: *Knowledge-based or expert systems and decision support tools for environmental assessment and management: their potential and limitations*. Final Report for the Federal Environmental Assessment and Review Office. Halifax, Nova Scotia: Dalhousie University, School for Resource and Environmental Studies.

Guariso, G. and **Page, B.** (eds) 1994: *Computer support for environmental impact assessment*. IFIP Transactions B: Applications in Technology. Conference on Computer Support for EIA - CSEIA 93, Como, Italy, 6–8 October 1993. Amsterdam: North-Holland.

Heer, J.E. Jr and **Hagerty, D.J.** 1977: *Environmental impact assessment*. New York: Van Nostrand Reinhold.

Hill, M. 1968: A goals-achievement matrix for evaluating alternative plans. *Journal of the American Institute of Planners* **34(1)**, 19–28.

Hollick, M. 1981: The role of quantitative decision-making methods in environmental impact assessment. *Environmental Management* **12(1)**, 65–78.

Holling, C.S. 1978: *Adaptive environmental assessment and management*. Chichester: Wiley.

Horberry, J. 1983: *Status and application of environmental impact assessment for development*. Gland: Conservation for Development Centre, International Union for Conservation of Nature and National Resources.

Hyman, E.L. and **Stiftel, B.** (eds) 1988: *Combining facts and values in environmental impact assessment: theories and techniques*. Boulder, CO: Westview.

Jain, R.K., Urban, L.V. and **Stacey, G.S.** 1977: *Environmental impact assessment: a new dimension in decision making*. New York: Van Nostrand Reinhold (2nd edn, 1981).

Jialin, X., Wei, Q. and **Huadong, W.** 1989: A regional environmental information system for analysing and predicting the impacts of nonferrous mining on an agricultural environment. *Environmental Conservation* **13(2)**, 259–69.

Johnson, C.A., Detenbeck , N.E., Bonde, J.P. and **Niemi, G.J.** 1988: Geographic information systems for cumulative impact assessment. *Photogrammetric Engineering and Remote Sensing* **54(11)**, 1609–15.

Julien, B., Fenves, S.J. and **Small, M.J.** 1992: An environmental impact identification system. *Journal of Environmental Management* **36(3)**, 167–84.

Krauskopf, T.M. and **Bunde, D.C.** 1974: Evaluation of environmental impact through a computer modelling process. In Ditton, R.B. and Goodale, T.L. (eds), *Environmental impact analysis: philosophy and methods*. Madison, WI: University of Wisconsin, Sea Grant Program, Sea Grant Publications Office, 107–25.

Kung, H.T., Ying, L.G. and **Liu, Y.C.** 1993: Fuzzy clustering analysis in environmental impact assessment: a complement tool to environmental quality index. *Environmental Monitoring and Assessment* **28(1)**, 1–14.

Lapping, M.B. 1975: Environmental impact assessment methodologies: a critique. *Environmental Affairs* **4(1)**, 123–34.

Lawrence, D.P. 1994: Cumulative effects assessment at the project level. *Impact Assessment* **12(3)**, 253–73.

Lein, J.K. 1988: An expert systems approach to environmental impact assessment. *Journal of Environmental Studies* **33(1)**, 13–27.

Leistritz, L.F., Coon, R.C. and **Hamm, R.R.** 1995: A microcomputer model for assessing socio-economic impacts of development projects. *Impact Assessment* **12(4)**, 373–84.

Leopold, L.B., Clarke, F.E., Hanshaw, B.B. and **Balsley, J.R.** 1971: *A procedure for evaluating environmental impact*. US Geological Survey Circular 645 – N71–36757. Washington DC: US Department of the Interior, Geological Survey.

Loehle, C. and **Osteen, R.** 1990: IMPACT: an expert system for environmental impact assessment. *AI Applications in Natural Resource Management* **4(1)**, 35–43.

McHarg, I. 1968: *A comprehensive highway route selection method*. Highway Research Record 246. Washington DC: Highway Research Board, 1–15.

McHarg, I. 1969: *Design with nature*. New York: Doubleday (Natural History Press).

Mercer, K.G. 1995: An expert system utility for environmental impact assessment in engineering. *Journal of Environmental Management* **45(1)**, 1–23.

Moore, J.L., Manty, P.B., Cheney, J.L. and **Rhuman, J.L.** 1973: *A methodology for evaluating environmental impact statements on proposed manufacturing developments in Delaware's coastal zone*. Report to the State of Delaware. Columbus, OH: Battelle Memorial Institute.

Moreno, D.D. and **Heyerdahl, L.A.** 1990: Advanced GIS modelling techniques in environmental impact assessment. *Conference on Geographical Information Systems, 7–10 November 1990* (GIS/LIS 90 Proceedings), vols 1 and 2. Anaheim, CA.

Morris, P. and **Thérivel, R.** (eds) 1994: *Methods of environmental impact assessment*. London: University College London Press.

Odum, H.T. 1972: Use of energy diagrams for environmental impact statements. In *Tools for Coastal Management*. Proceedings of the 1972 Conference of the Marine Technology Society. Washington DC: Marine Technology Society, 197–213.

Ortolano, L. and **Shepherd, A.** 1995: Environmental impact assessment: challenges and opportunities. *Impact Assessment* **13(1)**, 3–30.

Osemans, G., Krtzschmar, J., Janssen, L. and **Maes, G.** 1995: The third workshop on environmental impact assessment model intercomparison exercise. *International Journal of Environment and Pollution* **5(4–6)**, 785–98.

Ott, W.R. 1979: *Environmental indices: theory and practice*. Ann Arbor, MI: Ann Arbor Science.

Parker, The Honourable Mr Justice. 1978: *The Windscale Inquiry* (3 vols). London: HMSO.

Parker, S. and **Cocklin, C.** 1993: The use of geographical information systems for cumulative environmental effects assessment. *Computers, Environment and Urban Systems* **17(5)**, 393–407.

Physick, W.L., Hurley, P.J. and **Manins, P.C.** 1995: Environmental impact assessment of industrial development at Gladstone, Australia. *International Journal of Environment and Pollution* **5(4–6)**, 548–56.

Rau, J.G. 1980: Summarization of environmental impact. In Rau, J.G. and Wooten., D.C. (eds), *Environmental impact analysis handbook*. New York: McGraw-Hill, 8–1–8–32.

Rees, W. 1995: Cumulative environmental assessment and global change. *Environmental Impact Assessment Review* **15(4)**, 295–309.

Richardson, S.E., Hansen, W.J., Solomon, R.C. and **Jones, J.C.** 1978: *Preliminary field test of the Water Resources Assessment Methodology (WRAM): Tensas River, Louisiana*. Miscellaneous Paper Y-78–1. Vicksburg, MS: US Army Corps of Engineers.

Ross, D.I. and **Singhroy, V.** 1985: The application of remote sensing technology in the environmental assessment process. In Maclaren, V.W. and Whitney, J.B. (eds), *New directions in environmental impact assessment in Canada*. Toronto: Methuen, 144–78.

Schaller, J. 1990: Geographical information system applications in environmental impact assessment. In Scholten, H.J. and Stillwell, J.C.H. (eds), *Geographical information systems for urban and regional planning*. Kluwer: Dordrecht, 107–18.

Schaller, J. 1992: GIS: applications in environmental planning and assessment. *Computers, Environment and Urban Systems* **16(4)**, 337–53.

Schibuola, S. and **Byer, P.H.** 1991: Use of knowledge-based systems for the review of environmental impact assessments. *Environmental Impact Assessment Review* **11(1)**, 11–27.

Shopley, J.B. and **Fuggle, R.F.** 1984: A comprehensive review of current environmental impact assessment methods and techniques. *Journal of Environmental Management* **18(1)**, 25–47.

Shopley, J.B., Sowman, M. and **Fuggle, R.F.** 1990: Extending the capability of component interaction matrix as a technique for addressing secondary impacts in environmental assessment. *Journal of Environmental Management* **31(3)**, 197–213.

Siegel, M.S. and **Moreno, D.D.** 1993: Geographical information systems: effective tools for siting and environmental impact assessment. In Hildebrand, S.G. and Cannon, J.B. (eds), *Environmental analysis: the NEPA experience*. Boca Raton, FL: Lewis Publishers, 178–86.

Skutsch, M.McC. and **Flowerdew, R.T.N.** 1976: Measurement techniques in environmental impact assessment. *Environmental Conservation* **3(3)**, 209–17.

Smit, B. and **Spaling, H.** 1995: Methods for cumulative effects assessment. *Environmental Impact Assessment Review* **15(1)**, 81–106.

Solomon. C., Colbert, B.K., Hansen, W.J., Richarson, S.E., Canter, L. and **Vlachos, E.C.** 1977: *Water Resources Assessment Methodology (WRAM): impact assessment and alternative evaluation*. Technical Report Y-77–1. Vicksburg, MS: US Army Corps of Engineers, Waterways Experiment Station.

Sondheim, M.W. 1978: A comprehensive method for assessing environmental impact. *Journal of Environmental Management* **6(1)**, 27–42.

Sorensen, J.C. 1971: *A framework for the identification and control of resource degradation and conflict in the multiple use of the coastal zone*. Berkeley, CA: University of California, Department of Landscape Architecture.

Sorensen, J.C. 1974: Some procedures and programs to assist environmental impact statement process. In Ditton, R.B. and Goodale, T.I. (eds), *Environmental impact analysis: philosophy and methods*. Madison, WI: University of Wisconsin, Sea Grant Program, Sea Grant Publications Office, 97–106.

Spaling, H. 1994: Cumulative effects assessment: concepts and principles. *Impact Assessment* **12(3)**, 231–51.

Spaling, H. and **Smit, B.** 1993: Cumulative environmental change: conceptual frameworks, evaluation approaches, and institutional perspectives. *Environmental Management* **17(5)**, 587–600.

Thor, E.C., Elsner, G.H., Travis, M.R. and **O'Loughlin, K.M.** 1978: Forest environmental impact analysis: a new approach. *Journal of Forestry* **76**, 723–5.

Treweek, J. 1995: Ecological impact assessment. *Impact Assessment* **13(3)**, 289–316.

Tucker, K.C. and **Richardson, D.M.** 1995: An expert system for screening potentially invasive alien plants in the South African fynbos. *Journal of Environmental Management* **44(4)**, 309–38.

Vlachos, E. 1985: Assessing long-range cumulative impacts. In Covello, VT: Mumpower, J.L., Stallen, P.J.M. and Uppuluri, V.R.R. (eds), *Environmental impact assessment, technology assessment and risk analysis: contributions from the psychological and decision sciences*. NATO ASI series G, Ecological Science, vol. 4. Berlin: Springer-Verlag, 49–80.

Walters, C.J. 1993: Dynamic-models and large-scale field experiments in environmental impact assessment and management. *Australian Journal of Ecology* **18(1)**, 53–61.

Warner, M.L. and **Preston, E.H.** 1973: *A review of environmental impact assessment methodologies*. Washington DC: Office of Research and Development, US Environmental Protection Agency.

Westman, W.E. 1985: *Ecology, impact assessment and environmental planning*. New York: Wiley.

Whitehead, P. 1992a: Examples of recent models in environmental impact assessment. *Journal of the Institution of Water and Environmental Management* **6(4)**, 475–84.

Whitehead, P. 1992b: The role of models in environmental impact assessment. In Falconer, R.A. (ed.), *Water quality modelling*. Brookfield, VT: Ashgate Publishing, 53–68.

Whitman, I.L., Dee, N., McGinnnis, J.T., Fahringer, D.C. and **Baker, J.K.** 1971: *Design of an environmental evaluation system*. Columbus, OH: Battelle Columbus Laboratories.

Whyte, A.V. and **Burton, I.** (eds) 1980 *Environmental risk assessment*. SCOPE Report 15. Chichester: Wiley.

Wright, J.R., Wiggins, L.L., Jain, R.K. and **Kim, T.J.** (eds) 1993: *Expert systems in environmental planning*. Berlin: Springer-Verlag.

Ying, L.G. and **Liu, Y.C.** 1995: A model of objective weighting for EIA. *Environmental Monitoring and Assessment* **36(2)**, 169–82.

6

THE ORIGINS OF IMPACT ASSESSMENT AND ITS SPREAD

This chapter examines the origins of impact assessment and its spread from the USA to developed countries and areas of international interest such as the Arctic and Antarctica. The use of impact assessment in developing countries and by international organizations involved in development is examined in Chapter 7.

Origin of impact assessment

Commissions charged to examine environmental impacts were set up from time to time throughout history and in various countries. For example, Fortlage (1990: 1) noted some similarities between the activities of a Royal Commission (of 3 months' duration) investigating the Weald iron mills and furnaces in southern England in AD 1548 and modern environmental impact assessment. However, nothing quite like environmental impact assessment had appeared before the late 1960s, although the *Report of the Volta Preparatory Commission* (HMSO, 1956) has been recognized as an effective proactive development assessment, despite its failure to ease the resettlement of up to 84 000 people adequately (Chambers, 1970). The UK made use of commissions of inquiry to assess impacts and, to some extent, to keep the public informed. Some of these were held well before 1970, but they generally took a good deal of time and were applied in an *ad hoc* manner in response to ministerial or popular concern, involved the public in a limited way and were not as systematic as modern impact assessment. One of the best-known examples was the Roskill Commission on the third London airport (Roskill: Chairman, 1971).

Environmental impact assessment evolved from fields including land use planning, cost–benefit analysis, multiple-objective analysis and modelling and simulation, and was primarily stimulated by a piece of US legislation, the National Environmental Policy Act 1969, which came into force on 1 January 1970 (*see* Box 6.1, and Flamm, 1973; Caldwell, 1982; Ditton and Goodale, 1974: 145–51). That Act was prompted by various factors, including the 'media and information revolution' which was under way by the late 1950s; concern on the part of the public and non-governmental organizations for the environment; the development of assessment techniques; developments in planning theory; and the activities of the environmental movement. What was new about the environmental impact assessment approach was its *systematic* assessment and presentation of results on impacts, alternatives and mitigation possibilities. Graham Smith (1993: 8) noted that environmental impact assessment grew up in an era dominated by a 'technocratic perspective on problem solving' and with an emphasis on biophysical impacts. This may help explain why socio-economic impacts and social impact assessments have received less support (although, as mentioned in Chapter 8, activities very similar to social impact assessments pre-date environmental impact assessment).

McHarg's (1969) *Design with nature* stressed the value of proactive and systematic consideration of environmental limits, development impacts and alternatives, and is seen by many as a forerunner of the environmental impact assessment approach. Gilbert White also came close to proposing environmental impact assessment in the 1960s (White, 1968). The first true environmental impact assessment appears to have been commissioned in February 1967 into copper mining in Puerto Rico (Mayda, 1993; Gilpin, 1995: 115).

The National Environmental Policy Act, known as NEPA, has not been the only environmental impact assessment initiative in the USA. Sixteen of the country's 50 states, plus the District of Columbia and Puerto Rico, had passed similar laws by 1991. Collectively these are often called 'little NEPAs'. Some of them add to the National Environmental Policy Act or replace it. For example, in 1970 California passed the California Environmental Quality Act, which in practice has a broader coverage than the National Environmental Policy Act, applying to virtually all developments in California. Vermont (Act 250), Washington and New York are among those states that have their own equivalent of the Act (Bass, 1990; Wood, 1995b: 27); for a list of 'little NEPAs', *see* Canter (1996: 20).

Discretionary guidelines were issued to improve the National Environmental Policy Act in 1971 and 1973, and the Act was reviewed in 1976. In 1977 President Carter issued an Executive Order (119911) which effectively amended the original environmental impact assessment–environmental impact statement provisions of the Act to strengthen environmental impact assessment procedures and, in 1978, to establish the Council on Environmental Quality and give regulations more force. The council revised the guidelines, aiming at more cost-effective and better-quality environmental impact assessments with less delay. These changes, notably the new Council on Environmental Quality regulations, made 1978 something of a watershed

Box 6.1 The US National Environmental Policy Act 1969

President Theodore Roosevelt had called for foresight in respect to pollution control during his 1908 Conference on Conservation, but it was not until the second half of the century that effective legislation was enacted. Discussions leading to the National Environmental Policy Act (NEPA) began in the early 1960s, when the need was perceived for the USA to have a declaration of national environmental policy and an action-forcing provision (Ditton and Goodale, 1974; Glasson *et al.*, 1994: 25–47; Canter, 1996: 1–35). The US government was largely reacting to public opinion that conventional planning did not adequately take account of the environment. Before NEPA the USA had little effective federal control over the environment and lacked land use regulations of the kind found in countries like the UK or France (Wood, 1995b: 16).

In December 1969 NEPA was passed by Congress and signed into US law on 1st January 1970 by President Nixon. It was initiated not as a policing action but to reform federal policy-making and with the intent to influence the private sector – the hope being to transform and reorientate values (for reviews, *see* Heer and Hagerty, 1977; Caldwell, 1982; Dreyfus and Ingram, 1976). Originally it was intended that NEPA would change the nature of federal decision-making, however, over the years it has become more of a procedural requirement (Wood, 1995b: 75). Caldwell (1989) felt that, had it not happened in the USA, something similar would have appeared elsewhere.

NEPA required an environmental impact assessment prior to federally funded projects that 'significantly' affected the environment – a message to federal officials of 'look before you leap' (Cheremisinoff and Morresi, 1977). The crucial sections of NEPA included Section 101, which set regulations to protect the environment, Section 102(2)(c), which ensured that they were pursued, and Section 103, which included provision for inadequate environmental impact assessment statements to be challenged in court. Court challenges happened a lot at first because NEPA was untested and because expressions such as 'significant', 'federal action' and 'human environment' were less than rigidly defined in law (Wathern, 1988: 24; Hildebrand and Cannon, 1993). There was also some need to clarify what developments required assessment and how it was to be conducted. One of the court actions which tested NEPA was the 1971 Calverts Cliff case, through which the US Court of Appeals for the District of Columbia ruled that the US Atomic Energy Commission must undertake an *independent* evaluation of environmental impacts for a nuclear power-station proposed for Calverts Cliff, Maryland.

Legislation for environmental impact assessment (and virtually the first use of the expression) appears in Section 102(2)C of NEPA, where US federal agencies are required to prepare an environmental impact statement (bearing the costs against taxes and sending copies to federal and state agencies and to the public) using environmental impact assessment, *prior* to taking action. This is where use of these terms became established. For a list of the federal agencies involved, *see* Corwin *et al.*, 1975: 41).

NEPA required US federal agencies to prepare an environmental impact statement for major federal actions (i.e. federal-funded or -supported developments) that could significantly affect the quality of the human environment. There were three main elements in the Act. First, it announced a US national policy for the environment. Second, it outlined procedures for achieving the objectives of that policy. Third, provision was made for initiating the establishment of a US Council on Environmental Quality which was to advise the US President on the environment, review the environmental impact assessment process, review draft environmental impact statements and see that NEPA was followed by providing recommendations and co-ordination. The Council on Environmental Quality effectively administers environmental impact assessment legislation in the USA and issues the regulations that ensure that effective environmental impact statements are produced.

Also in 1970 the US government created the US Environmental Protection Agency. Its sole was to co-ordinate the attack on environmental pollution and to be responsible for the environmental impact assessment process. In effect, the agency is the overseer of impact assessment in the USA (Gilpin, 1995: 115–19).

Section 102(2)(c) of NEPA required an environmental impact statement to consider:

- environmental impacts (the intent was to include 'human' aspects, but in practice socio-economic assessment was less thorough than biophysical);
- adverse environmental impacts that could not be avoided;
- alternatives to the proposed development;
- the relationship between local, short-term impacts and larger-scale, long-term environmental quality;
- irreversible impacts or commitments.

NEPA created a more systematic, 'product-driven' process of environmentally informed decision-making. That is, the approach was largely determined by the aforementioned environmental impact statement requirements (Graham Smith, 1993: 9). NEPA was the first time that US law had really allowed for development to be delayed or abandoned for the long-term good of the environment, and also it made efforts to co-ordinate public, state, federal and local activities. Public participation is strongly written into NEPA, to the extent that it might be described as a 'corner-stone'. Overall, it represented a revolution in values in a country where state intrusion was anathema. For this reason, many see it as a sort of 'Magna Carta', although it stopped short of making a healthy environment a constitutional right, a failure that some campaigners have been seeking to correct (Yost, 1990).

NEPA is statutory law; that is, it was written out after deliberation and did not evolve from custom, practice or tradition. Consequently, like a charter it was not perfected. There were problems, especially delay, as litigation took place over the 'significance' of impacts, etc. Many felt that NEPA had been 'abducted by lawyers' (Caldwell, 1989: 78) and could become a bureaucratic delaying tactic hindering development.

By the late 1980s most of the initial weaknesses had been overcome and at least 30 other countries had adopted similar procedures. NEPA has been a seminal concept and catalyst for environmental impact assessment in other countries (Manheim, 1994), although bodies such as the Canadian Environmental Assessment Research Council and the International Association for Impact Analysis also deserve considerable credit for spreading and developing impact assessment.

year in the evolution of impact assessment. A number of US federal agencies have promoted environmental impact assessment since 1970, for example the US Food and Drug Administration.

There was less progress during the 1980s under Ronald Reagan's presidency. By the late 1980s there were calls for reform (Renwick, 1988). Caldwell

(1989), author of the original draft of National Environmental Policy Act, argued for better incorporation of social impacts into environmental impact assessment procedures and for the Act to be more strongly written into the US Constitution (something which has yet to happen). However, by the late 1970s US aid agencies had been forced, partly by citizen action, to apply environmental impact assessment to their activities outside the USA. An attempt in 1990 to amend the Act – to further increase its overseas application to developing countries and global change, biodiversity loss and transboundary pollution – failed.

SPREAD OF IMPACT ASSESSMENT BEYOND THE USA

Although it had issued guidelines on environmental assessment in 1974, the US Agency for International Development (USAID) had failed to implement these adequately and was sued in 1975 by a US public-interest group in an attempt to force it to prepare environmental impact statements on its grants and loans. Consequently, by 1976 USAID and some other bodies, notably lending banks and the US State Department, were applying environmental impact assessment to overseas investments and aid. The Foreign Assistance Act of 1979 effectively extended the National Environmental Policy Act to the USA's foreign aid activities. Within the USA by that time there was a trend away from concern for formal procedures towards concern for more effective environmental impact assessment that actually reduced impacts.

The National Environmental Policy Act attracted interest and was carefully assessed by many governments, less carefully by others. A number of countries quite rapidly copied its provisions, others delayed much longer (in the case of the UK until after 1988). There have been considerable local modifications to the procedures and processes involved in impact assessment (McCormick, 1993; Coenen, 1993), and the quality of impact assessment varies greatly. By the early 1990s over 40 countries had established environmental impact assessment programmes (Robinson, 1992), and by 1995 probably about half the world's nations required environmental impact assessment in some form.

Adoption of impact assessment has usually involved adaptation to techniques and procedures, because US experience may not be relevant in other countries and because approaches and techniques are constantly evolving (OECD, 1979; Kennedy, 1984; Hollick, 1986; Hawke, 1987; Conacher, 1988; Robinson, 1992). So far, the greatest progress outside the USA has probably been made in Australia, Canada (Prasad, 1993) and the Netherlands. The latter countries, together with Sweden and Norway, are probably the most progressive environmental impact assessment users nowadays. By 1988, Spain, New Zealand, Japan, the Republic of Ireland, Germany, France, Canada and Australia had adopted formal systems for environmental impact assessment, together with quite a number of developing countries (*see* Chapter 7 for an account of these).

In some developing countries, the introduction of impact assessment has been swifter than in a number of developed countries, a situation quite common with the spread of innovations. Often, developing countries had fewer established physical planning procedures, so both had a need for and could adopt impact assessment with less reorganization. A significant amount of developing countries' adoption of impact assessment has been prompted by aid organizations, which, since the mid-1970s, have insisted on it. (International bodies and aid agencies are considered in Chapter 7.)

There are a number of ways in which impact assessment can be adopted:

- Existing planning procedures can be adapted to incorporate it, as in Germany, Sweden, Denmark and the UK, for example.
- Impact assessment legislation can be created, as in the USA, Australia and Canada (for a list, *see* Gilpin, 1995: 3).
- Global impact assessment regulations and supportive institutions can be developed – something raised, but not agreed, at Rio during the World Conference on Environment and Development or 'Earth Summit' in 1992.

Global-scale and transboundary impact assessment

Increasing interest has centred on globally focused and transboundary impact assessment. Even though no binding agreement was reached at Rio in 1992, nor much funding made available, the 'Earth Summit' made clear that the field was seen to be important. A great deal of effort has been concentrated in predicting likely global environmental changes and the probable impacts of the forecast changes, especially those triggered by accelerated global warming (e.g. Mitchell *et al.*, 1992) and ozone damage (Wallington *et al.*, 1994). World trade developments such as GATT, the General Agreement on Tariffs and Trade, now the World Trade Organization), and NAFTA, the New Zealand and Australia Free Trade Agreement, and structural adjustment policies have also generated impact assessments.

Global change impact studies have been dominated by climate studies, especially those that forecast global warming. There has been too little attention to problems of soil degradation, global pollution and loss of biodiversity. Global climate change impact studies have been dominated by the work of the International Panel on Climate Change (IPCC), set up in 1988 by the World Meteorological Organization and United Nations Environment Programme to investigate global warming. This panel is a group of over 80 experts from over 80 countries. Global warming may well cause huge impacts, but at present the models used to predict impacts are far from perfect (Taplin and Braaf, 1995).

In 1991 the UN Economic Commission for Europe launched the Convention on Environmental Impact Assessment in a Transboundary Context. This was signed by 28 countries including the USA, plus the European Community (now the European Union) at Espoo, Finland. This Espoo Convention (often called the 'Espoo Treaty') is the first multilateral treaty on transboundary rights relating to proposed activities. It has yet to be fully adopted by signatories but potentially could improve transboundary consultation and assessment. The Espoo Convention provides for the notification of all affected parties likely to suffer an adverse 'transboundary impact' by a proposed development, and signatories undertake to give equal rights concerning impact assessment to all those affected by a development, even if they are citizens of different countries; they could therefore be represented in the developer nation's public inquiries, for example. Gilpin (1995: 77) listed the types of development to which the Espoo Convention applies. A European Union directive, the EEC Environmental Assessment Directive (85/337EEC of 1985), makes similar though less wide-ranging provisions, but goes further to permit affected parties to participate in impact assessment, if they so wish (Kennedy, 1982; Jorissen and Coenen, 1992). The Espoo Convention goes beyond making provisions for project-level impact assessment to encourage programme- and policy-level assessments.

It is important that there be greater harmonization of standards, so that countries can share data. Strategic environmental assessment offers a route to global environmental impact assessment able to deal with transboundary impacts (Thérivel *et al.*, 1992: 131).

Offshore impact assessment

Beanlands and Duinker (1984) examined the application of environmental impact assessment offshore. In a number of countries, impact assessment or environmental assessment has been applied to marine sewage disposal or fish farming (Thérivel *et al.*, 1992: 78), and the impacts of shipwrecks and pollution spills have also been examined (e.g. Leewis, 1991; Nyholm, 1992). The impacts of offshore oil-drilling have been assessed off California, the UK, and elsewhere since the 1970s (Hyland *et al.*, 1994), and there has been some study of aquaculture (e.g. Enell, 1995). Tiger-prawn aquaculture has expanded in a number of developing countries' coastal areas and has attracted attention from impact assessors. There is still no sign of an adequate 'Law of the Sea' in the near future, which makes it difficult to achieve standardized and enforceable environmental impact assessment procedures. There is clearly a need to improve application of environmental impact assessment and environmental assessment to 'enclosed seas' like the Japanese 'Inland Sea', the Mediterranean (especially the Adriatic and Aegean Seas), the Aral Sea, the Caspian Sea, the Black Sea, and the Baltic and North Seas, where pollution and other problems are already manifest, partly because tidal currents are less effective at dispersing pollutants (*see* Healey and Harada, 1991).

Multinational and transnational corporations and impact assessment

Many multinational and transnational corporations have developed impact assessment and eco-auditing policies and practices (Scuperholme, 1984;

Robinson, 1993). The advantage of the involvement of such corporations in impact assessment is that these bodies may have more resources and experience than some poor countries and could promote a more uniform, less *ad hoc* approach. The disadvantages are that it may be difficult for local authorities to monitor or apply quality control to impact assessments carried out by large corporations, and that these corporations are likely to have a vested interest in proposed developments' proceeding, and hence might play down adverse impacts.

Antarctica and impact assessment

All human activities in Antarctica have been subject to careful impact assessment since 1991 (Shears, 1992). The Antarctic environment is the largest and most pristine wilderness on Earth (British Antarctic Survey, undated) and offers unique opportunities for conservation and research, but it is highly vulnerable. There have already been marine oil spills, excessive sealing and whaling (prior to the mid-1960s), over-exploitation of fish, krill and squid stocks, and introduction of alien species of plants and animals (mainly on the sub-Antarctic islands). However, so far no minerals exploitation has taken place, and there has been agreement on a ban on such activity that runs until AD 2041). Environmental damage can be very slow or impossible to rectify because organisms grow and pollutants decompose slowly. Tourism has already led to damage to flora, fauna and geology through habitat disturbance, trampling and oil spillage from damaged cruise ships. Antarctica and the southern high latitudes are also affected by stratospheric ozone reduction, mainly caused by global pollution. The monitoring of atmospheric conditions in Antarctica in 1987 alerted the world to the global problem of ozone thinning and the value of monitoring.

Although various nations have long had bases or even commercial operations in Antarctica, there is no territorial possession under Article 4 of the Antarctic Treaty, which was first signed in 1959 and by 1993 had 25 signatory nations. In effect, Antarctica and surrounding seas are a nature reserve dedicated to peace and science. There is a Protocol on Environmental Protection in the Antarctic Treaty, which obliges all treaty signatories to apply environmental impact assessment to any activity likely to generate impacts (Gilpin, 1995: 80–2). The Antarctic nations have a history of data-sharing and co-operation; progress with impact assessment may therefore point the way to strategies that might be successful in other parts of the world.

The first impact assessment conducted in the Antarctic was carried out in the 1970s to evaluate the effects of a research drilling programme and to select the least vulnerable sites for it. It used a modified Leopold matrix (Parker and Howard, 1977; Keys, 1992). In 1981 France started to construct an airstrip to serve one of its Antarctic stations without conducting an environmental impact assessment. Controversy ensued and, during the building, Greenpeace (a non-governmental organization) conducted an impact assessment. Airstrip construction at Rothera Point (on Adelaide Island, alongside the Antarctic Peninsula) by the British Antarctic Survey became one of the first developments subject to thorough environmental impact assessment. Guidelines for a formal environmental impact assessment procedure to be used by all nations active in the Antarctic were initiated by the Scientific Committee on Antarctic Research in 1987 (Spellerberg, 1991: 272–3). An annex to the Protocol on Environmental Protection to the Antarctic Treaty (1991) requires the environmental monitoring of human activities in Antarctica. The object of such monitoring is to provide information that can be used by decision-makers to prevent or minimize environmental impacts. This annex to the protocol requires the use of environmental impact assessment at the planning stage and in the operation of all activities in Antarctica. The protocol provides a basis for nations active in Antarctica to co-ordinate and standardize impact assessment and monitoring (Walton and Shears, 1994).

The annex to the protocol lays down a three–stage procedure. First, there is a preliminary screening. Second, an initial environmental evaluation is carried out. If this indicates the chance of significant impacts, the third stage comes into effect, a comprehensive environmental evaluation which results in an environmental impact statement either published or circulated to Treaty nations and the Committee for Environmental Protection (of the Antarctic) as a report. The UK has released guidelines relating to its activities in Antarctica and the sub-Antarctic (FCO, 1995) and controls its developments by permit, granted only after an environmental impact assessment has been conducted and indicates minimal risk (British Antarctic Survey, 1991; Clark, 1994; *see also* the Antarctic Act 1994 and Antarctic Regulations 1995).

All the impact assessments conducted in Antarctica between the end of 1988 and mid-1995 (in all 49 – mainly related to the construction or dismantlement of research station facilities) were subjected to a post-environmental impact assessment audit (Benninghoff and Bonner, 1985; Bonner, 1989; Burgess *et al.*, 1992; Anon., 1995). The majority of these assessments have been prepared by the nations involved, some with the assistance of non-governmental organizations, some wholly by such organizations (e.g. Greenpeace), and a few by tour companies involved in Antarctic cruises or other tourist activity.

Arctic impact assessment

Impact assessment is important in harsh environments where, if there is disturbance, the recovery of soil, vegetation and animals may be slow and difficult. In addition to Antarctica, these environments include the Arctic, the high latitudes of the Northern and Southern Hemispheres, high mountains and deserts. The Arctic has been much more affected by human activity than the Antarctic. Airborne and sea pollution reaches it from Europe, Canada, the USA and the Commonwealth of Independent States. The exploitation of oil and minerals does take place in the Arctic, unlike in Antarctica. There is also increasing movement of shipping along the fringes of the Arctic (Rudback *et al.*, 1991), and the impact of technological and social changes, for example the spread of the motorized sledge ('skidoo') and modern cash economy and communications. Pollution is a problem where wildlife depends on limited ice-free areas among sea-ice (polynyas); an oil spill or other disturbance in these locations can have much wider consequences (Beanlands and Duinker, 1983: 119). Ice cover may prevent the breakdown of ocean or lake pollutants.

Impact assessment has focused on the impacts of mineral exploitation, especially natural gas and oil exploitation and transport (Hood and Burrell, 1976; Norton, 1979). Problems with nuclear waste disposal and fears that river development in Asia will alter the salinity of the Arctic Ocean have led to impact assessments and are likely to prompt future ones.

The eight nations with Arctic territories (Canada, Finland, Iceland, Norway, Russia, Sweden, the USA, Denmark/Greenland) were, at the time of writing, about to set up an international organization, the Arctic Council, under a treaty to deal with environmental and social issues. The Arctic Environmental Protection Strategy, signed in 1991, which was designed to deal with problems such as toxic pollutants and the survival of endangered species, will be incorporated into the council's terms, together with measures to cover issues such as sustainable development, the future of indigenous peoples and global environmental change. Representatives for indigenous peoples, and non-governmental organizations and countries with an interest in the Arctic (the UK, the Netherlands and Germany) will attend the council, it seems, as observers. One must hope that the council will strengthen environmental protection, assist co-ordination of impact assessment and monitoring, and give indigenous peoples and non-governmental organizations a chance to voice their concerns (*see New Scientist* 1996: **151(1512040)**, 8).

National experiences with impact assessment

Canada

Canada started to test environmental impact assessment in 1969, and in 1973 the federal Cabinet decided to adopt a federal-level impact assessment system. The federal Environmental Assessment Review Process was implemented in 1974 (and amended in 1977 and 1984) as a policy rather than law (Bowden and Curtis, 1988). In 1989 Canada passed an Environmental Protection Act and by the 1990s the following arrangements were in place:

- The Federal Environmental Assessment and Review Office (FEARO) of the Department of the Environment was established in 1973 to oversee Canadian impact assessment. This office issued environmental impact assessment guidelines in 1984, and has promoted impact assessment procedures and methods beyond Canada through its research and publications.
- Federal projects and federally funded developments likely significantly to affect the environment are subject to impact assessment.
- The federal government calls for environmental impact assessments.
- The 1984 Environmental Assessment Review Process established a

two–stage environmental impact assessment approach: stage 1, an initial self-assessment by the developer; stage 2, a public revue by an independent environmental impact assessment panel.
- The Federal Environmental Assessment and Review Office makes environmental impact assessments available to the public.
- For overseas aid, the Canadian International Aid Agency requires environmental impact assessment, but the recipient government is not required to publish environmental impact statements.

In 1992 the Environmental Assessment Act gave legal powers to federal bodies to conduct impact assessment. The Environmental Assessment Review Process was improved in 1995, the changes including a strengthening of its enforcement powers under the Canadian Environmental Assessment Act (Delicael, 1995).

Canada has been at or near the forefront of impact assessment adoption and the development of improved approaches, and the Canadian public has been increasingly involved in environmental impact assessment and social impact assessments (Beanlands and Duinker, 1983; FEARO, 1985; Maclaren and Whitney, 1985; Whitney and Maclaren, 1985; Gibson, 1993; Gilpin, 1995: 113; Wood, 1995b: 55–71).

The pipelines of Arctic Alaska and Canada have generated a number of seminal impact assessments (Sage, 1980).

Australia

Australia was one of the first countries beyond the USA to adopt an environmental impact assessment policy and has actively developed impact assessment techniques and approaches. The State of New South Wales pioneered Australia's adoption of environmental impact assessment in 1972 and the Commonwealth of Australia followed a few months later (Hollick, 1980). In 1974 an Environmental Protection (Impact of Proposals) Act was passed, extending environmental impact assessment legislation to Australia as a whole (Wood, 1995b: 3, 55–71). This Act sought to ensure that proposals likely to affect the environment significantly would be subjected to impact assessment. Under the Act the responsible Minister or authority decides whether a draft environmental impact statement is to be made public. This Act was amended in 1987 (*see* Fowler, 1982; Porter, 1984; Dunn, 1986; Formby, 1987; Thomas, 1987; Wathern, 1988: 227). (Social impact and the Aborigines of Australia are discussed in Chapter 8.)

Ortolano and Shepherd (1995: 13) noted that relatively few of Australia's developments are subject to environmental impact assessment, and that citizen action to sue the government is difficult in the Commonwealth of Australia.

New Zealand

New Zealand started to establish environmental impact assessment procedures in 1973 (by Cabinet directive) (Morgan, 1983, 1986a, 1986b; Wood,

1995b: 3, 55–71), but mandatory requirements were not introduced until 1991 (Morgan, 1983; Montz and Dixon, 1993; Wood, 1993; Gilpin, 1995: 124–9). Progress up to 1988 was by means of various laws relating to protection of the environment, conservation, pollution control, etc., rather than a single, specific environmental impact assessment law. Application has been to projects involving the government or requiring statutory approval, and there is a high degree of public involvement in impact assessment and eco-auditing (Wathern, 1988: 229; Dixon and Fookes, 1995). A Ministry of Environment was established in 1986, and environmental impact assessment was integrated into new resource management laws in the late 1980s, with procedures set out in the Environmental Protection and Enhancement Procedures. An improvement of impact assessment procedures took place after environmental impact assessment became required by law under the 1991 Resource Management Act. Montz and Dixon (1993) have reviewed environmental impact assessment law and practice in New Zealand (for a short review, *see* Anon., 1989: 12–13).

New Zealand has been particularly active in developing social impact assessments and is ahead of most countries in its application; for example, *see* the coverage of the Huntley Monitoring Project in Chapter 8.

Japan

The Japanese Cabinet decided to adopt environmental impact assessment in 1972 for all major projects; however, it was not until after 1984 that uniform rules for implementing it (for general application to large projects) were approved by the national government (Anon., 1989: 11–12; Gilpin, 1995: 134–6). In 1984 Japan's Environmental Agency promulgated Principles for Implementing Environmental Impact Assessment. In practice it would seem that there has been no real enforcement of national environmental impact assessment laws up to the mid-1990s (Gilpin, 1995: 134–6). The signs at the time of writing were that Japan would adopt a regional approach to impact assessment.

Barrett and Thérivel (1989, 1991) provided a critique of environmental policy and impact assessment in Japan, and Hamanaka (1981) examined technical adaptation of environmental impact assessment to meet Japanese needs. Impact assessment has been applied to large projects such as the Kansai International Airport (Maeda, 1991). Some large projects have generated stiff resistance from protest groups, and this has probably helped to make the authorities cautious about public involvement.

Public hearings are part of the Japanese environmental impact assessment process, and from the late 1980s draft environmental impact statements have had to be released for public review for 1 month (Wathern, 1988: 232). Nevertheless, Japanese decision-making is reputed to be autocratic and factionalized, and this has apparently presented challenges for effective public involvement. Barrett and Thérivel (1989) feared that impact assessment tended to be used to justify development decisions; they also concluded that economic growth appears to take precedence over concern for environmental quality. The situation seems to be one of not very effective environmental impact assessment, applied mainly at the local or regional authority level.

European Union

European Union-wide directives have helped accelerate adoption of impact assessment and eco-auditing. 'Accelerate' might not be the best word, for the gestation period before the environmental impact assessment directives had real effects was long: 1977 to 1988, in significant part because of UK objections based on a feeling that the UK's Town and Country Planning Acts were enough, but also as a consequence of repeated redraftings in Brussels.

In 1972 the Paris Declaration on the Environment noted the value of environmental impact assessment, and in 1979 a UN Economic Commission for Europe seminar recommended that it be used at policy, planning and programme levels in Europe. In 1979 the then European Economic Community issued a draft directive on environmental impact assessment. Adoption of environmental impact assessment within the European Union has been shaped by the Commission of the European Communities Council Directive 85/337/EEC (of 27 June 1985), the 'Environmental Assessment Directive' or 'EIA Directive' (Commission of the European Communities, 1985). This made it mandatory for member states each to have national impact assessment legislation applying environmental impact assessment, including cumulative impacts assessment, to 'major' problems by July 1988, but left the choice of methods and style of adoption to individual states (for a list of what was subject to environmental impact assessment under the 'EIA Directive', *see* Gilpin, 1995: 75). Some states, such as the UK and Belgium, failed to comply with the directive's deadline, taking until 1990 or 1992, although they did set out basic frameworks (Lee, 1978, 1983, 1995; Wathern, 1988: 192–209; Fortlage, 1990: 14–29; Lee and Wood, 1978; Wandesforde-Smith, 1979; Kennedy, 1982; Glasson *et al.*, 1994: 41–4; McHugh, 1994). The 'EIA Directive' lays down the rules for environmental impact assessment of major developments in European countries. Europe in general has incorporated environmental impact assessment into existing procedures; for a short review of the situation in various European Community (as the European Union was then called) countries in 1989 *see* Anon. (1989). For a comparison of European and USA procedures, *see* McHugh (1994), and for an outline of European Union policy on environmental impact assessment, *see* Lee (1995).

In 1974 the Nordic countries, Sweden, Norway, Denmark, Finland, now but not then within the European Union (except Norway, which is not an EU member), signed an environmental protection convention. This agreed that environmental impact assessment for a development should give the same weight to impacts likely to damage other Nordic countries as to impacts on the country commissioning the impact assessment. This convention helped prepare the way for later transboundary impact agreements – *see* discussion earlier in this chapter of transboundary environmental impact assessment and the Espoo Convention. There is a considerable degree of coordination between Nordic countries over environmental impact assessment procedures.

The Netherlands

Strong government interest in environmental impact assessment was expressed in 1980, and in 1981 legislation was sent to Parliament. Environmental impact assessment has been actively promoted in the Netherlands by the Ministry of Housing, Physical Planning and Environment (VROM). The Netherlands developed an independent commission, The Netherlands Commission for Environmental Impact Assessment (*c.* 200 members appointed by the Crown), as 'overseer' (for a short review of environmental impact assessment adoption, *see* Anon., 1989: 7–8). By the mid-1980s the Netherlands had developed an interim policy (which lasted from 1981 to 1987) for voluntary environmental impact assessment, had published an environmental impact assessment manual on the subject (Clark *et al.*, 1984: 57–68) and required post-environmental impact assessment audit as part of the impact assessment process.

The Dutch have approached environmental impact assessment in a deliberate and reasoned manner (Wathern, 1988: 197), testing and evaluating methods and procedures and making careful checks on other countries' experience. Environmental impact assessment in the Netherlands has been based more on Canadian experience than directly on that from the USA, and is used to promote sustainable development. The Netherlands has taken impact assessment further than other European countries (Gilpin, 1995: 100). At the time of writing (1996) a statutory strategic environmental assessment system was in place; it was set up in 1987 and subsequently strengthened (*see also* the discussion of strategic environmental assessment in Chapter 3).

Belgium

Belgium, like the UK, has treated environmental impact assessment with caution (*see* De Vuyst and Hens, 1991; De Vuyst *et al.*, 1993; Anon., 1989: 3), expressing interest as early as 1977, but deferring implementation until the early 1990s when the European Union's 'EIA Directive' forced it. Before the 1990s, concern for the environment was, in large part, devolved to the regions, and environmental impact assessment in Belgium has largely continued to be applied separately by Wallonia, Flanders and Brussels, rather than on a national scale. De Vuyst *et al.* (1993) reviewed environmental impact assessment procedures in Flanders (northern Belgium), concluding that there were administrative problems that reduced its effectiveness.

Germany

West Germany was the first European country to initiate an environmental impact assessment policy, in 1971. It put forward a Model Procedure (government guidelines, not a full system backed by law) in 1975. The adoption of environmental impact assessment in West Germany was a Cabinet decision and co-ordination has been through the German Federal Environmental

Agency. Kennedy (1980, 1981) reviewed the state of West German impact assessment at the end of the 1970s; the late-1980s situation is outlined by Wathern (1988: 195–6), Anon. (1989: 3–4) and Gilpin (1995: 94–6). In 1991 an environmental impact assessment Act was passed which made provision for environmental impact assessment in the whole of united Germany. The former East Germany had relied on a Governmental Environmental Inspectorate (district-based) to monitor local authority activities for impacts; monitoring was also undertaken by the District Hygiene Inspectorates (Gilpin, 1995: 97–8).

France

Like the UK, France had environmental planning measure long before environmental impact assessment emerged. In 1917 laws had been passed requiring public inquiries prior to approval of developments likely to be unhealthy, disturbing or dangerous. These were updated in 1976 and included provision for environmental impact assessment (Loi de Protection de la Nature, proposed 1976, operational 1978). The legislation required public or private developments that needed public authorization to have pre-development assessment (Sanchez, 1993). Environmental impact assessment became obligatory in France in 1980. For a review of the situation in the late 1980s, *see* Monbailliu (1984), Anon. (1989: 3) and ACE (1990). France depends a great deal on nuclear electricity generation and has applied environmental impact assessment to atomic power-station installation (Jammet and Madelmont, 1982). In France the development proponent prepares the environmental impact statement, which Sanchez (1993: 261) suggested might mean a tendency to understate negative impacts and domination by engineers, which should be countered by better monitoring.

Surprisingly, given a history of citizen involvement in government, there has been delay in getting adequate public participation in French impact assessment. The situation improved after 1983 with environmental impact statements being placed on public display and with the introduction of public inquiries (Wathern, 1988: 196; Gilpin, 1995: 93–4).

UK

From the late 1940s the UK had a well-developed, if somewhat *ad hoc* in application, land use planning system and used public inquiries, often chaired by a judge, to fulfil a purpose similar to that of environmental impact assessment. (The USA, before the National Environmental Policy Act, had nothing like this system; Wood, 1995b: 44–54). Town and Country Planning Acts (particularly those of 1947 and 1971 – the latter amended in 1977) empowered UK government Ministers to select which developments needed a public inquiry. When the inquiries were completed, a report went to a Minister, who then made decisions. A good example is the 1977 'Windscale Inquiry', which lasted about 6 months, into proposals by British Nuclear Fuels Ltd to construct a nuclear reactor fuel reprocessing plant.

With some environmental protection and pre-development inquiry measures already in force, the UK Department of the Environment (founded in 1970 to co-ordinate what had previously been dealt with by several ministries) resisted the formal adoption of environmental impact analysis, partly to assess the working of approaches adopted in other European countries and partly through a feeling that it was unnecessary (Breakell and Glasson, 1981: 15). In 1981 a manual on the subject was published by the UK Department of the Environment, preparing the way for legislation (DoE, 1981). In 1985 the European Community's 'EIA Directive' ended this debate by insisting on progress in adopting environmental impact assessment by 1988 (Bichard and Frost, 1988). As a consequence of the directive, laws requiring the use of impact assessment were passed in the UK in 1988.

There had been use of environmental impact assessment in the UK before 1988, at first mainly by consultants in relation to North Sea oil exploitation (Clark *et al.*, 1984: 69–79 review this, especially the contribution of the University of Aberdeen's Project and Development Control Unit, now the Centre for Environmental Planning and Management), and later, application to onshore hydrocarbon exploitation in the southern UK and to major construction works such as the Channel Tunnel (Clark *et al.*, 1981).

Several groups at universities and polytechnics were also active in promoting and developing impact assessment. One of these groups, the Project and Development Control Unit team of the University of Aberdeen's Department of Geography, was asked in 1973 to prepare a method to assist UK planners make a balanced assessment of major industrial activities, prompted by North Sea oil exploitation. The result was a widely read guide to impact identification and assessment (PADC, 1976). In 1974 Catlow and Thirlwall (1974) were asked to assess the desirability of the UK's adopting environmental impact assessment, and recommended that a system be adopted. The UK government was afraid that environmental impact assessment would increase bureaucracy and enlarge planning machinery, slowing development if adopted on a wide scale, and so moved slowly and with some reluctance (by the mid-1970s) to an *ad hoc* non-statutory and voluntary application and modification of existing planning practice. Under pressure from European legislation (the 'EIA Directive'), the UK began to implement legislative measures from the mid-1970s (Foster, 1984). The Town and Country Planning (Assessment of Environmental Effects – SI No. 1199) Regulations 1988 gave legal effect – partial adoption – to the European Community Directive 85/337/EEC. The Department of the Environment issued a guide to environmental impact assessment procedures soon afterwards (DoE, 1989), but it was not until the Environmental Protection Act 1990 came into force in 1992 that the UK fully implemented 'EIA Directive' rules (Wood and Jones, 1992). A White Paper, *This common inheritance*, indicates UK government environmental strategy to the end of the century. (DoE, 1990).

The UK authorities have used a different terminology to that generally adopted elsewhere. For example, 'environmental assessment' is used in place of 'environmental impact assessment' and 'environmental statement' instead of 'environmental impact statement'.

In the UK the developer is responsible for preparing an environmental impact statement and is obliged to collaborate with relevant bodies (a copy of the environmental statement must go to the Nature Conservancy Council, English Nature and the Countryside Commission or the Countryside Commission for Wales, relevant Scottish bodies, and in some cases to the HM Inspectorate of Pollution) and to release a non-technical summary of the environmental statement. Some critics have suggested that this procedure is too developer-oriented.

There is still much improvement needed to adequately support environmental impact assessment (Alder, 1993; Thérivel *et al.*, 1992: 18–19). In 1993 the UK was criticized for allowing various developments that required impact assessment under European Union regulations to slip through without assessment being carried out, especially where development has 'Crown immunity', meaning that it is exempted by the government from impact assessment. Other criticisms are that public participation takes place only after formal application for project or plan approval (i.e. well into the development decision-making process), and that some UK bodies commission and carry out impact assessments and are allowed to use their own environmental impact statements with little supervision; examples include the Forestry Commission and the Department of Transport. In 1996 the UK still had no official independent environmental impact assessment review body (the Institute of Environmental Assessment, a non-governmental organization, was available if called upon for review). The lack of adequate strategic and long-term planning in the UK is seen by some as a hindrance to environmental impact assessment and, especially, to adopting strategic environmental assessment (for a discussion of this latter topic, *see* Chapter 3).

Morris and Thérivel (1995: 348–9) provide a bibliography of publications on environmental impact assessment in the UK.

Republic of Ireland

New industrial projects have been assessed since 1970 and an Environmental Protection Agency promotes assessment. Statutory provisions for environmental impact assessment were added to the Local Government (Planning and Development) Act of 1976. These came into force between 1977 and 1980 and were improved in 1988 and 1990. At first, large private-sector projects were required to carry out impact assessment but local authorities could select which, among the developments for which they were responsible, should be subject to environmental impact assessment. In the late 1980s the 'EIA Directive' (85/337/EEC) was implemented: detailed impact assessments were now required before planning permission could be granted, and the environmental impact statements were made available to the public (Wathern, 1988: 198). Meehan (1989) reviewed the implementation of the European Community's Directive in Ireland (*see also* Anon., 1989: 3; O'Sullivan, 1990; Geraghty, 1996). Gilpin (1995: 99) felt that Ireland had been more active with impact assessment than the UK.

Luxembourg

The Act on Conservation of Nature and Natural Resources 1978 introduced Luxembourg's first legislation for environmental impact assessment – on developments likely to affect the natural environment (Clark *et al.*, 1984: 32; Wathern, 1988: 196). Gilpin (1995: 99) felt that by 1990 Luxembourg had made good progress at implementing the European Community's Directive and even upgrading from that. For a short review of Luxembourg's environmental impact assessment practice and legislation, *see* Anon (1989: 3).

Italy

Attempts were made to introduce formal environmental impact assessment procedures in 1984 but had limited success. Adoption of the 'EIA Directive' had not been good before 1989 (Bartlett, 1989: 27–36). The environmental impact assessment situation in Italy has been reviewed by Matarrese (1989); *see also* Anon. (1989: 3).

Spain

On joining the European Community, Spain moved toward implementing the 'EIA Directive', but before about 1986 the approach was piecemeal. The legal framework for environmental impact assessment in Spain was reviewed by Lee and Wood (1985) and Rocandio (1989) (*see also* Wathern, 1988: 198; Anon., 1989: 3; Gilpin, 1995).

Environmental assessment in Spain has been incorporated into land use planning (Rivas *et al.*, 1994). Efforts have been largely decentralized, and up to 1996 progress in achieving national legislation had been slow. Like the UK and Portugal, at least in the early 1990s, Spain had no official independent environmental impact assessment review body.

Portugal

The 1987 Portuguese 'Environmental Act' – Project of Law No. 79/iv – included provision for environmental impact assessment, but prior to the late 1980s there were no unified legislative or procedural measures. The European Community 'EIA Directive' was implemented from 1990 by means of various decrees that built upon the 1987 legislation. In 1992 Portugal still had no independent environmental impact assessment review body (Canelas, 1991). At the time of writing (late 1996), the following procedure operated: the developer carries out an environmental impact assessment; licensing authorities check it; it then goes to the Ministry for the Environment for checking and/or further work; finally, a decision is made. The impact assessment system is quite centralized. For short reviews of environmental impact assessment in Portugal, *see* Wathern (1988: 197–8), Anon. (1989: 3) and Gilpin (1995: 100).

Greece

Impact provisions in Greece, at least to 1986, were fragmentary, and some have claimed that the country's adoption of the European Community's 'EIA

Directive' has been poor (Bartlett, 1989: 27–36). Parliament introduced environmental impact assessment measures in 1986, which became law in 1990, for major projects or activities likely to cause adverse impacts (Gilpin, 1995: 99). Wathern (1988: 196) and Gazidellis (1989) have outlined the situation in the late 1980s; *see also* Anon. (1989: 3). The National University of Athens established a Centre for Environmental Impact Assessment in 1993.

Denmark

Denmark (together with Greenland and the Faeroes, autonomous territories of Denmark) had no specific environmental impact assessment legislation before the European Community's 'EIA Directive' prompted the environmental impact assessment Order of 1989, although the 1973 Environmental Protection Act made provision for assessment. Denmark produced a National Action Plan on Environment in 1988. Danish environmental impact assessment procedures include public hearings and are applied to development likely to affect the environment significantly. Some developments approved by Parliament can, however, escape impact assessment (Wathern, 1988: 195; Ministry of Environment, 1991; Gilpin, 1995: 77, 94–104; Anon., 1989: 3; Brorasmussen, 1992).

Sweden

Sweden has a strongly developed concern for environmental protection and has been at the forefront in applying environmental management, having established a National Environmental Agency in 1967 (i.e. well before the USA). Municipalities have master plans, and public participation is encouraged and can have considerable power. For example, a referendum initiated by citizens and non-government bodies worried after the Three Mile Island reactor accident in the USA prompted the alteration of Sweden's nuclear energy policy, at a time when atomic generation provided about half the national electricity supply, to non-nuclear, non-polluting renewable energy.

Environmental impact assessment has been done through an elaborate physical planning process and as part of the licensing of specific projects. It is applied to government policies, plans, programmes and foreign aid as well as projects. Procedures established by 1995 ensured that the developer pays for the assessment. The Swedish Board of Housing, Building and Planning, the Swedish Environmental Protection Agency and the Central Board of National Antiquities co-ordinate the impact assessment process and ensure that the environmental impact statement is heeded. Impact assessments are routinely subjected to post-environmental impact assessment audits. Carlberg *et al.* (1984) and Gilpin (1995: 77, 107–9) have reviewed environmental impact assessment in Sweden, and Balfors (1992) compared Swedish with Dutch environmental impact assessment.

Finland

Finland has a complex decision-making process within which impact assessment has tended to be fragmented, each body assessing areas of its own concern. There is some way to go to integrate environmental impact assessment fully into Finnish legislation. Assessment is mainly undertaken for land and water development or individual projects. There was still scope for tighter legislation in the mid-1990s, as regulations allow some proposals to 'escape' assessment (Gilpin, 1995: 104).

Austria

Progress with impact assessment in Austria has been somewhat slow. Only since 1992 has there been real progress with environmental impact assessment legislation, although there had been some application to individual engineering or industrial projects (Gilpin, 1995: 96–7). The Austrian Environmental Impact Assessment Act was introduced in 1993 and became law in 1995, when the country joined the European Union. This Act provides a regulatory framework for major developments (Davy, 1995).

Norway

Norway, not a European Community member, established a Ministry of Environment in 1972. A general requirement for environmental impact assessment was proposed in 1977 and environmental impact assessment was required by law for major projects from 1985 when the Planning and Building Act was revised. A further revision in 1989 improved provisions for impact assessment. An environmental impact assessment procedure was in force by 1990, focused on projects likely to have significant impact on the environment or human community, with legislative provision for public participation (Gilpin, 1995: 106–7).

Like Sweden, Denmark and Finland, Norway had extended environmental impact assessment requirements to its foreign aid programmes by the 1990s.

Switzerland

Switzerland too has so far stayed outside the European Union. Article 9 of the 1983 Federal Law on the Protection of the Environment, which came into force in 1985, required projects likely to cause significant environmental impacts to undergo environmental impact assessment at a pre-implementation – that is, an early planning – stage (Borlin, 1984). In Switzerland, environmental impact assessment links with spatial planning, and the latter examines alternative locations or approaches (Eggenberger, 1993). Environmental impact assessment takes place after a development has been defined and the location chosen. However, spatial planning and environmental impact assessment together do manage to ask the questions: How will a development proceed? Where will it take place? Why is it needed?

As one might expect, given the canton system of government (i.e. considerable local autonomy), environmental impact assessment is managed in a decentralized way overseen by federal authorities. The cantons implement impact assessment and enforce the provisions of environmental impact statements, the central government issues environmental impact assessment guidelines. The promoter or developer carries out impact assessment, the environmental impact statement or report is submitted to a licensing authority (which also advises the promoter) for review. It is then sent to the Environmental Protection Agency to approve, comment upon or reject. According to Gilpin (1995: 100), the focus is mainly on biophysical impacts rather than socio-economic ones.

Centrally planned economies including the former Eastern Bloc countries

Impact assessment in the People's Republic of China and in Cuba is discussed in Chapter 7. For a review of environmental impact assessment legislation in centrally planned economies, *see* Starzewska (1988) and *Environmental Impact Assessment Review* 1994: **14(2–3)**, a special issue entitled 'Environmental decision-making in Central and Eastern Europe'. The former Czechoslovakia appears to have been evolving environmental protection legislation by the 1950s, although environmental impact analysis proper has been a more recent introduction. Environmental impact analysis has been within the context of socio-economic plans and long-term strategic planning, and applied at the stage when operational plans were devised (Starzewska, 1988: 216).

Russia and the former USSR (now the CIS and Affiliate States)

The USSR had broad proposals for environmental impact assessment in 1985. All development projects that had the potential to affect the environment significantly were reviewed by the USSR State Ecological Expertise Commission before a decision was taken to approve a project. In 1988 a Central Administration of Environmental Impact Assessment was established, which since 1990 has required a favourable environmental impact statement before finance is released for development. Environmental impact assessment measures were strengthened in 1988 ('perestroika in the field of environmental conservation'); – effectively, measures were initiated for development and implementation of environmental impact assessment. Local and regional monitoring supports national monitoring for impact assessment. Govorushko (1991) reviewed the situation for the period between the 1970s and the early 1990s. Since the breakup of the USSR there is more state-by-state variation in provision of environmental impact assessment and limited resources. Gilpin (1995: 98) noted a need for improved and more tightly enforced procedures.

Estonia, Latvia, Lithuania

Wood (1995a) and Gilpin (1995: 98) have briefly reviewed environmental impact assessment legislation in the Baltic states. There is large-scale exploitation of oil shale in north-east Latvia, which has led to considerable atmospheric and water pollution, spoil-heap fires and serious health impacts on human health and the waters of the Gulf of Finland and the Baltic. There is also uranium mining near the Baltic. There is clearly a need for retrospective impact studies and for better environmental impact assessment in the future. Little information has so far been published in English.

Poland

At first, environmental impact assessment in Poland was applied on a project-by-project basis (Janikowski and Starzewska, 1986). Starzewska (1988: 221–223), Rzeszot and Wood (1992), Gilpin (1995: 100) and Wood (1995a) have reviewed its further development in Poland. Poland developed an environmental impact assessment approach through pilot studies conducted in Upper Silesia and the Legnica-Głoców copper-mining region by the Institute for Environmental Development (sponsored by the World Health Organization). The results of these studies were published in 1983 and guided application of environmental impact assessment. Environmental impact assessment has been implemented mainly through the Environment Act, the Town and Country Planning Acts and legislation governing mining (Jendroska and Sommer, 1992). By 1989 environmental impact assessment was in widespread use. The Polish Commission on environmental impact assessment of the Ministry of Environment, Natural Resources and Forestry reviews assessments and use of environmental impact statements, and carries out post-environmental impact assessment audits. By 1995 there was provision for public involvement quite early in the impact assessment process.

Poland has a computerized soil information system, BIGLEB, and a procedure for modelling the impacts of power generation-related pollution. Much of the country's electricity is derived from sulphur-rich brown coal, and, although considerable efforts have been made to clean up power-station emissions, using Dutch technology, there are still some atmospheric pollution-related health problems and acid deposition damage to soils, forests and water bodies. (I have relied on English-language publications for my information on Poland, which means that more up-to-date Polish sources have not been discussed, and correspondingly for some of the other countries discussed here.)

Czech and Slovak Republics and Slovenia

Czechoslovakia ceased to exist in 1993, splitting into the Czech and Slovak Republics. The Czech Republic passed an Environmental Act in 1991 which includes requirements for environmental impact assessment and environmental planning (Branis, 1992; Branis and Koskova, 1992). There is heavy dependence on brown coal, and some impact assessment activity has focused on the impacts of its use. Wood (1995a) and Gilpin (1995: 97) have briefly reviewed environmental impact assessment legislation in the Czech Republic, Slovak Republic and Slovenia (Slovenia was formerly part of Yugoslavia).

Hungary

Hungary has severe air pollution problems, manifest as public health and aid deposition impacts, related to industrial development (Carter and Turnock, 1993).

References

ACE 1990: *Environmental impact assessment: the French experience* (English trans.). Paris (Neuilly): Atelier Central de l'Environnement (ACE).

Alder, J. 1993: Environmental impact assessment: the inadequacies of English law. *Journal of Environmental Law* **5(2)**, 203–20.

Anon. 1989: Implementing the EIA Directive: an overview. *EIA Trainers' Newsletter* **3(2–4)**. Manchester: University of Manchester, Department of Planning and Landscape, EIA Centre.

Anon. 1995: *A review of environmental impact assessments (EIAs) prepared for proposed activities in Antarctica (submitted by the United Kingdom)*. Antarctic Treaty XIXth Consultative Meeting, Seoul, 8–19 May 1995 (XIX ATCM/INF 15), mimeograph, 18 pp.

Balfors, B. 1992: Environmental impact assessment in the Netherlands: what can Sweden learn from the Dutch EIA? *Scandinavian Housing and Planning Research* **9(4)**, 237.

Barrett, B.F.D. and **Thérivel, R.** 1989: EIA in Japan: environmental protection versus economic growth. *Land Use Policy* **6(3)**, 217–31.

Barrett, B.F.D. and **Thérivel, R.** 1991: *Environmental policy and impact assessment in Japan*. London: Routledge.

Bartlett, R.V. (ed.) 1989: *Policy through impact assessment: institutionalized analysis as a policy strategy*. New York: Greenwood Press.

Bass, R. 1990: California's experience with environmental impact reports. *Project Appraisal* **5(4)**, 220.

Beanlands, G.E. and **Duinker, P.N.** 1983: *An ecological framework for environmental impact assessment in Canada*. Halifax, Nova Scotia: Dalhousie University, Institute for Resource and Environmental Studies.

Beanlands, G.E. and **Duinker, P.N.** 1984: Lessons from a decade of offshore environmental impact assessment. *Ocean Management* **9(3–4)**, 157–75.

Benninghof, W.S. and **Bonner, N.** 1985: *Man's impact on the Antarctic environment: a procedure for evaluating impacts from scientific and logistic activities*. Cambridge: SCAR and Scott Polar Institute.

Bichard, E. and **Frost, S.** 1988: EIA in the UK planning system. *Land Use Policy* **5(4)**, 362–64.

Bonner, N. 1989: Environmental assessment in the Antarctic. *Ambio* **XVIII(1)**, 83–9.

Borlin, M. 1984: Environmental impact assessment in Switzerland: experiences and present situation. *Zeitschrift für Umweltpolitik* **7(4)**, 447–62.

Bowden, M.A. and **Curtis, F.** 1988: Federal EIA in Canada: EARP as an evolving process. *Environmental Impact Assessment Review* **8(1)**, 97–106.

Branis, M. 1992: A system for certified environmental impact assessment experts in the Czech Republic. *Environmental Impact Assessment Review* **14(2–3)**, 203–8.

Branis, M. and **Koskova, E.** 1992: The Environmental Impact Assessment Act in the Czech Republic: origins, introduction and implementation issues. *Environmental Impact Assessment Review* **14(2–3)**, 196–202.

Breakell, M. and **Glasson, J.** (eds) 1981: *Environmental impact assessment: from theory to practice*. Department of Town Planning Working Paper 50. Oxford: Oxford Polytechnic, Department of Town Planning.

British Antarctic Survey undated: *Environmental protection in Antarctica*. Cambridge: British Antarctic Survey.

British Antarctic Survey 1991: *Environmental impact assessment (EIA) within the British Antarctic Survey*. BAS Environmental Information Leaflet 2. Cambridge: British Antarctic Survey.

Brorasmussen, F. 1992: Environmental impact assessment and risk analysis in Denmark. In Colombo, A.G. (ed.), *Environmental impact assessment*, vol. 1. Dordrecht: Kluwer, 105–19.

Burgess, J.S., Spate, A.P. and **Norman, F.I.** 1992: Environmental impacts of station development in the Larsemann Hills, Princess Elizabeth Land, Antarctica. *Journal of Environmental Management* **36(4)**, 287–99.

Caldwell, L.K. 1982: *Science and the National Environmental Policy Act: redirecting policy through procedural reform*. Huntsville, AL: University of Alabama Press.

Caldwell, L.K. 1989: A constitutional law for the environment: 20 years with NEPA indicates the need. *Environment* **31(10)**, 6–11, 25–8.

Canelas, L.D. 1991. Implementation of the EEC Directive in Portugal: a case study. *Impact Assessment Bulletin* **9(3)**, 75–83.

Canter, L.W. 1996: *Environmental impact assessment*. 2nd edn (1st edn. 1977). New York: McGraw-Hill.

Carlberg, E.G., Grip, K. and **Zettersten, G.** 1984: Environmental impact assessment in Sweden. *Zeitschrift für Umweltpolitik* **7(4)**, 425–46.

Carter, F.W. and **Turnock, D.** 1993: *Environmental problems in Eastern Europe*. London: Routledge.

Catlow J. and **Thirlwall, G.** 1976: *Environmental impact analysis*. DoE Research Report 11. London: Department of the Environment.

Chambers, R. (ed.) 1970: *The Volta resettlement experience*. London: Pall Mall Press.

Cheremisinoff, P.N. and **Morresi, A.O.** 1977: *Environmental assessment and impact statement handbook*. Chichester: Wiley.

Clark, B.D, Bisset, R. and **Wathern, P.** 1981: The British experience. In O'Riordan, T. and Sewell, W.R.D. (eds), *Project appraisal and policy review*. Chichester: Wiley, 125–53.

Clark, B.D., Gilad, A., Bisset, R. and **Tomlinson, P.** (eds) 1984: *Perspectives in environmental impact assessment*. Dordrecht: D. Reidel.

Clark, M.L. 1994: The Antarctic Environmental Protocol. NGOs in the protection of Antarctica. In Princen, T. and Finger, M. (eds), *Environmental NGOs in world politics: linking local and global*. London: Routledge, 160–85.

Coenen, R. 1993: NEPA's impact on environmental impact assessment in European Community member countries. In Hildebrand, S.G. and Cannon, J.B. (eds), *Environmental analysis: the NEPA experience*. Boca Raton, FL: Lewis Publishers, 703–15.

Commission of the European Communities 1985: Council directive on the assessment of the effects of certain public and private projects (85/337/EEC). *Official Journal of the European Communities* L175/40, 5 July. Brussels: Commission of the European Communities (adopted June 1985, and supposed to be effective by July 1988).

Conacher, A. 1988: Resource development and environmental stress: environmental impact assessment and beyond in Australia and Canada. *Geoforum* **19(3)**, 339–52.

Corwin, R., Heffernan, P.H. and **Johnston, R.A.** 1975: *Environmental impact assessment*. San Francisco: W.H. Freeman.

Davy, B. 1995: The Austrian Environmental Impact Assessment Act. *Environmental Impact Assessment Review* **15(4)**, 361–76.

Delicael, A. 1995: The new Canadian Environmental Assessment Act: a comparison with the Environmental Assessment Review Process. *Environmental Impact Assessment Review* **15(6)**, 497–505.

De Vuyst, D. and **Hens, L.** 1991: Environmental impact assessment in Belgium: an overview. *Environmental Professional* **13(2)**, 166–73.

De Vuyst, D., Nierynck, E., Hens, E., Ceuterick, L., De Baere, V. and **Wouters, G.** 1993: Environmental impact assessment in Flanders, Belgium: an evaluation of administrative procedures. *Environmental Management* **17(3)**, 395–408.

Ditton, R.B. and **Goodale, T.C.** (eds) 1974: *Environmental impact analysis: philosophy and methods*. Madison, WI: University of Wisconsin, Sea Grants Program.

Dixon, J. and **Fookes, T.** 1995: Environmental assessment in New Zealand: Prospects and practical realities. *Australian Journal of Environmental Management* **2(2)**, 104–11.

DoE 1981: *A manual for the assessment of major development projects* (Department of the Environment and Welsh Office). London: HMSO.

DoE 1989: *Environmental assessment: a guide to the procedures*. London: HMSO.

DoE 1990: *This common inheritance: Britain's environmental strategy*. Cm. 1200. Department of Environment. London: HMSO.

Dreyfus, D.A and **Ingram, H.M.** 1976: The National Environmental Policy Act: a review of intent and practice. *Natural Resources Forum* **16(2)**, 243–61.

Dunn, B. 1986: Environmental impact assessment in Australia. *Northwest Environmental Journal* **29(2)**, 107–13.

Eggenberger, M. 1993: Impact assessment and spatial planning: cantonal guiding plans as an instrument for environmental protection. *Impact Assessment* **11(1)**, 87–97.

Enell, M. 1995: Environmental impact of nutrients from Nordic fish farming. *Water Science and Technology* **31(10)**, 61–71.

FCO 1995: *Guide to environmental impact assessment of activities in Antarctica*. London: Foreign and Commonwealth Office, South Atlantic and Antarctic Department, Polar Regions Section.

FEARO 1985: *Environmental assessment in Canada: survey of current practice* (Federal Environmental Assessment and Review Office). Ottawa: Canadian Council of Resources and Environment Minister.

Flamm, B.R. 1973: A philosophy of environmental impact assessment: toward choice among alternatives. *Journal of Soil and Water Conservation* **28**, 201–3.

Formby, J. 1987: The Australian government's experience with environmental impact assessment. *Environmental Impact Assessment Review* **7(3)**, 207–26.

Fortlage, C.A. 1990: *Environmental impact assessment: a practical guide*. Aldershot: Gower.

Foster, B.J. 1984: Environmental impact assessment in the UK. *Zeitschrift für Umweltpolitik* **7(4)**, 438–54.

Fowler, R.J. 1982: *Environmental impact assessment, planning and pollution control measures in Australia*. Canberra: Australian Government Printing Service.

Gazidellis, V. 1989: EIA procedures in Greece. *EIA Trainers' Newsletter* **3**, 4–5.

Geraghty, P.J. 1996: Environmental impact assessment practice in Ireland following the adoption of the European Directive. *Environmental Impact Assessment Review* **16(3)**, 189–211.

Gibson, R.B. 1993: Environmental assessment design: lessons from the Canadian experience. *Environmental Professional* **15(1)**, 12–24.

Gilpin, A. 1995: *Environmental impact assessment (EIA): cutting edge of the twenty-first century*. Cambridge: Cambridge University Press.

Glasson, J., Thérivel, R. and **Chadwick, A.** 1994: *Environmental impact assessment: principles and procedures, process, practice and prospects*. London: University College London Press.

Govorushko, S.M. 1991: Environmental impact assessment in the USSR: current situation. *Impact Assessment Bulletin* **9(3)**, 83–8.

Graham Smith, L. 1993: *Impact assessment and sustainable resource management*. Harlow: Longman.

Hamanaka, H. 1981: Technical aspects of environmental impact assessment: major issues in Japan. *Journal of the Japanese Society of Air Pollution* **16(2)**, 77–87 (English abstract, paper in Japanese).

Hawke, N. 1987: *Environmental impact assessment: North American and European developments*. Leicester: Leicester Polytechnic School of Law.

Healey, T. and **Harada, K.** 1991: Modern environmental assessment procedures for enclosed seas. *Marine Pollution Bulletin* **23**, 355–61.

Heer, J.E. and **Hagerty, D.J.** 1977: *Environmental assessments and impact statements*. New York: Van Nostrand Reinhold.

Hildebrand, S.G. and **Cannon, J.B.** (eds) 1993: *Environmental analysis: the NEPA experience*. Boca Raton, FL: Lewis Publishers.

HMSO (1956) *Volta River Project, I: Report of the Preparatory Commission (Government of UK and Gold Coast); II, Appendices to reports of the Preparatory Commission (Government of UK and Gold Coast)*. London: HMSO.

Hollick, M. 1980: Environmental impact assessment in Australia: the federal experience. *Environmental Impact Assessment Review* **1(3)**, 330–36.

Hollick, M. 1986: EIA: an international evaluation. *Environmental Management* **10(2)**, 157–78.

Hood, D.W. and **Burrell, D.C.** (eds) 1976: *Assessment of the Arctic maritime environment*. Fairbanks, AK: University of Alaska, Institute of Marine Science.

Hyland, J., Hardin, D., Steinhauer, M., Coates, D., Green, R. and **Neff, J.** 1994: Environmental impact of offshore oil development on the outer continental shelf and slope off Point Arguello, California. *Marine Environmental Research* **37(2)**, 195–229.

Jammet, H. and **Madelmont, C.** 1982: French regulations on EIA of nuclear installations. *Environmental Impact Assessment Review* **3(2–3)**, 259–70.

Janikowski, R. and **Starzewska, A.** 1986: EIA project in Poland. *EIA Worldletter* May–June, 1–4.

Jendroska, J. and **Sommer, J.** 1992: Environmental impact assessment in Polish law: the concept, development and perspectives. *Impact Assessment Review* **14(2–3)**, 169–94.

Jorissen, J. and **Coenen, R.** 1992: The EEC Directive on EIA and its implementation in the EC member states. In Colombo, A.G. (ed.), *Environmental impact assessment*, vol. 1. Dordrecht: Kluwer, 1–14.

Kennedy, W.V. 1980: Environmental impact assessment in the Federal Republic of Germany. *Environmental Impact Assessment Review* **1(1)**, 72–94.

Kennedy, W.V. 1981: The West German experience. In O'Riordan, T. and Sewell, W.R.D. (eds), *Project appraisal and policy review*. Chichester: Wiley, 155–85.

Kennedy, W.V. 1982: The European Communities Directive on Environmental Impact Assessment. *Environmental Policy and Law* **8(3)**, 84–95.

Kennedy, W.V. 1984: US and Canadian experience with environmental impact assessment: relevance for the European Community? *Zeitschrift für Umweltpolitik* **7(4)**, 339–66.

Keys, J.R. 1992: Environmental impact assessment: process and concerns. In Barrett, P.J. and Davey, F.J. (eds), *Miscellaneous series: Royal Society of New Zealand, workshop report*, vol. 28. Wellington: Royal Society of New Zealand, 27–31.

Lee, N. 1978: Environmental impact assessment in EEC countries. *Journal of Environmental Management* **6(1)**, 57–76.

Lee, N. 1983: Environmental impact assessment: a review. *Applied Geography* **3(1)**, 5–27.

Lee, N. 1995: Environmental assessment in the European Union: a tenth anniversary. *Project Appraisal* **10(2)**, 77–90.

Lee, N. and **Wood, C.M.** 1978: EIA: a European perspective. *Built Environment* **4(2)**, 101–10.

Lee, N. and **Wood, C.M.** 1985: Training for environmental impact assessment within the European Economic Community. *Journal of Environmental Management* **21(3)**, 271–286.

Leewis, R.J. 1991: Environmental impact of shipwrecks in the North Sea. II. Negative aspects: hazardous substances in shipwrecks. *Water Science and Technology* **24(10)**, 299–300.

McCormick, J.F. 1993: Implementation of NEPA and environmental impact assessment in developing nations. In Hildebrand, S.G. and Cannon, J.B. (eds), *Environmental analysis: the NEPA experience*. Boca Raton, FL: Lewis Publishers, 716–27.

McHarg, I.L. 1969: *Design with nature*. Garden City, NY: Natural History Press.

McHugh, P.D. 1994: The European Community Directive: an alternative environmental impact assessment procedure. *Natural Resources Journal* **34(3)**, 589–628.

Maclaren, V.W. and **Whitney, J.B.** (eds) 1985: *New directions in environmental impact assessment in Canada*. Toronto: Methuen.

Maeda, M. 1991: The Kansai-International-Airport Project and environmental impact assessment. *Marine Pollution Bulletin,* special issue: Proceedings of an International Conference on the Management of Enclosed Coastal Seas, Kobe, 3–6 August, 1990, 349–53.

Manheim, B.S. Jr 1994: NEPA's overseas application. *Environment* **36(3)**, 43–5.

Matarrese, G. 1989: EIA in Italy: a review. *EIA Trainers' Newsletter* **3**, 6–7.

Mayda, J. 1993: Historical roots of EIA? *Impact Assessment Bulletin* **11(4)**, 411–15.

Meehan, B. 1989: Implementation of the EC EIA Directive in Ireland. *EIA Trainers' Newsletter* **3**, 5–6.

Ministry of Environment 1991: *Environmental impact assessment in Denmark*. Copenhagen: Ministry of Environment (Denmark).

Ministry of Health and Environmental Protection 1980: *Governmental standpoint on environmental impact assessment*. Leidschendam, the Netherlands: MHEP.

Mitchell, P.A., Sherrard, J.J. and **Wright, C.J.** 1992: Accounting for greenhouse effects in environmental impact assessment. *National conference on environmental engineering – the global environment: Australian implications*. Conference Proceedings of the Institution of Engineers, 12–19 June 1992, Gold Coast, Australia. St Leonards, Australia: E.A. Books.

Monbailliu, X. 1984: EIA procedures in France. In Bisset, R. and Tomlinson, P. (eds), *Perspectives on environmental impact assessment*. Papers relating to Annual Training Courses on EIA, CEMP. Aberdeen: University of Aberdeen, Centre for Environmental Planning and Management (CEMP), 51–5.

Montz, B.E. and **Dixon, J.** 1993: From law to practice: EIA in New Zealand. *Environmental Impact Assessment Review* **13(2)**, 89–108.

Morgan, R.K. 1983: The evolution of environmental impact assessment in New Zealand. *Journal of Environmental Management* **16(2)**, 139–52.

Morgan, R.K. 1986a: Environmental management and EIA in New Zealand: a period of change – Part I. *EIA Worldletter* July–August issue, pp. 1–6.

Morgan, R.K. 1986b: Environmental management and EIA in New Zealand: a period of change – Part II. *EIA Worldletter* September–October issue, pp. 1–8.

Morris, P. and **Thérivel, R.** (eds) 1995: *Methods of environmental impact assessment*. London: University College London Press.

Norton, D.W. 1979: *Some relationships between environmental assessments and Arctic marine development*. Paper to PORT – Ocean Engineering under Arctic Conditions – Conference, 1974. Bergen: Norwegian Institute of Technology.

Nyholm, N. 1992: Environmental impact assessment and control of marine industrial wastewater discharges. *Water Science and Technology* **26(12)**, 449–56.

OECD 1979: *Environmental impact assessment*. Paris: Organization for Economic Co-operation and Development.

Ortolano, L. and **Shepherd, A.** 1995: Environmental impact assessment. In Vanclay, F. and Bronstein, D.A. (eds), *Environmental and social impact assessment*. Chichester: Wiley, 3–30.

O'Sullivan, M. 1990: *Environmental impact assessment*. Cork: University College Cork, Environmental Impact Assessment, Resources and Environmental Management Unit.

PADC 1976: *The assessment of major industrial applications: a manual* (University of Aberdeen, Project and Development Control Unit). DoE Research Report 13. London: Department of the Environment.

Parker, B.C. and **Howard, R.V.** 1977: The first environmental monitoring and assessment in Antarctica: the Dry Valley Drilling Project. *Biological Conservation* **12(2)**, 163–77.

Porter, C.F. 1984: *Environmental impact assessment: a practical guide*. St Lucia, Brisbane: University of Queensland Press.

Prasad, K. 1993: Environmental impact assessment: an analysis of the Canadian experience. *International Studies* **30(3)**, 299–318.

Renwick, W.H. 1988: The eclipse of NEPA as environmental policy. *Environmental Management* **12(3)**, 267–72.

Rivas, V., Gonzalez, A., Fischer, D.W. and **Cendrero, A.** 1994: An approach to environmental assessment within the landuse planning process: northern Spanish experiences. *Journal of Environmental Planning and Management* **37(3)**, 305–22.

Robinson, N. 1992: International trends in environmental impact assessment. *Boston College Environmental Affairs Law Review* **19(3)**, 591–621.

Robinson, N.A. 1993: EIA abroad: the comparative and transnational experience. In Hildebrand, S.G. and Cannon, J.B. (eds), *Environmental analysis: the NEPA experience*. Boca Raton, FL: Lewis Publishers, 679–702.

Rocandio, I.C. 1989: The legal framework for EIA in Spain. *EIA Trainers' Newsletter* **3**, 8–9.

Roskill: Chairman 1971: *Commission on the third London airport*. London: HMSO.

Rudback, G.T., Sandkvist, J. and **Forsman, B.** 1991: The icebreaker *Oden* in polar operations: environmental impact assessment. Eleventh International Conference on Port and Ocean Engineering under Arctic Conditions (POAC '91), 24–28 September 1991, St John's, Newfoundland, vol. 2, 812–22.

Rzeszot, U. and **Wood, C.M.** 1992: Environmental impact assessment in Poland: an emergent process. *Project Appraisal* **7(2)**, 83–92.

Sage, B. 1980: Ruptures in the Trans-Alaska Pipeline: cause and effects. *Ambio* **IX(3)**, 262–3.

Sanchez, L. 1993: Environmental impact assessment in France. *Environmental Impact Assessment Review* **13(4)**, 255–66.

Scuperholme, P.L. 1984: The role of environmental impact assessment within British Petroleum. In Bisset, R. and Tomlinson, P. (eds), *Perspectives on environmental impact assessment*. Papers relating to Annual Training Courses on EIA, PADC. Aberdeen: University of Aberdeen, Project and Development Control Unit (PADC – later CEMP), 383–9.

Shears, J. 1992: Environmental management in Antarctica. *NERC News* **23** (October 1992) (National Environmental Research Council, Swindon, UK), 28–9.

Spellerberg, I.E. 1991: *Monitoring ecological change*. Cambridge: Cambridge University Press.

Starzewska, A. 1988: The legislative framework for EIA in centrally planned economies. In Wathern, P. (ed.), *Environmental impact assessment: theory and practice*. London: Unwin Hyman, 210–24.

Taplin, R. and **Braaf, R.** 1995: Climate impact assessments. In Vanclay, F. and Bronstein, D.A. (eds), *Environmental and social impact assessment*. Chichester: Wiley, 249–64.

Thérivel, R., Wilson, E., Thompson, S., Heaney, D. and **Pritchard, D.** 1992: *Strategic environmental assessment*. London: Earthscan.

Thomas, I.G. 1987: *Environmental impact assessment: Australian perspectives and practice*. Melbourne: Monash University Press.

Wallington, T.J., Schneider, W.F., Worsnop, D.R., Nielsen, O.J., Sehested, J., Debruyn, W.J. and **Shorter, J.A.** 1994: The environmental impact of CFC replacements – HFCs and HFCs. *Environmental Science and Technology* **28(7)**, 320A-326A.

Walton, D.W.H. and **Shears, J.** 1994: The need for environmental monitoring in Antarctica: baselines, environmental impact assessments, accidents and footprints. *International Journal of Environmental Analytical Chemistry* **55(1–4)**, 77–90.

Wandesforde-Smith, G. 1979: Environmental impact assessment in the European Community. *Zeitschrift für Umweltpolitik* **1(79)**, 35–76.

Wathern, P. (ed.) 1988: *Environmental impact assessment: theory and practice*. London: Unwin Hyman.

White, G.F. 1968: Organising scientific investigations to deal with environmental impacts. Paper delivered to the Careless Technology Conference, Washington DC. Later published in Farvar, M.T. and Milton, J.P. (eds) 1972: *The careless technology: ecology and international development*. New York: Natural History Press (Doubleday), 914–26. (First published in 1969.)

Whitney, J.B.R. and **Maclaren, V.W.** (eds) 1985: *Environmental impact assessment: the Canadian experience*. Toronto: Institute for Environmental Studies, University of Toronto.

Wood, C.M. 1993: Antipodean environmental assessment: a New Zealand/United Kingdom comparison. *Town Planning Review* **54(2)**, 119.

Wood, C.M. 1995a: Environmental impact assessment legislation: Czech Republic, Estonia, Hungary, Latvia, Lithuania, Poland, Slovak Republic, Slovenia. *Project Appraisal* **10(2)**, 136–7.

Wood, C.M. 1995b: *Environmental impact assessment: a comparative review*. Harlow: Longman.

Wood, C.M. and **Jones, C.** 1992: The impact of environmental assessment on local planning authorities. *Journal of Environmental Planning and Management* **35(2)**, 115.

Yost, N.C. 1990: NEPA's promise – partially fulfilled. *Environmental Law* **20(3)**, 681–702.

7

IMPACT ASSESSMENT IN DEVELOPING COUNTRIES

The current view from the South can be summed up as follows: environment must not be ignored but development must not be impeded. (A.K. Biswas, in Biswas and Agarwala, 1992: vii).

ASSESSMENT NEEDS AND PROBLEMS IN DEVELOPING COUNTRIES

Impact assessment and related procedures such as eco-auditing are intended to aid development, so it is wise to ask what 'development' is, what it requires and what side-effects may be generated by the development process. Less developed countries are a very diverse group of nations. Although many have in common a tropical or subtropical environment, there is huge variation in politics, government approach, public attitudes, availability of resources, levels of education, degree of poverty, etc. (The use of social impact assessment in developing countries is dealt with in Chapter 8.) Clearly, impact assessment and related approaches must be adapted to suit a wide range of situations. There are various potential 'players' involved at one time or another in assessment and related procedures in developing countries; the following listing is partly based on the work of Horberry (1983a: 2):

- *Developing-country governments* – wish to promote economic development and, often with less enthusiasm, protect the quality of the environment and avoid disrupting society.
- *International development agencies or those concerned with bilateral aid* – wish to ensure that their funding activities give value for money and do not cause environmental or socio-economic damage, and, in some cases, would like to promote environmental concern.

- *Planners and managers ('decision-makers')* – wish to avoid unwanted environmental or socio-economic problems.
- *Consultants* – often expatriates, often capable, but sometimes lacking experience of the locality under study, or even of the country and the techniques being used. Sometimes consultants are part of a 'tied aid' package (in which aid is granted with the requirement that goods or services are purchased from the aid-giver nation with some of the aid funds and that the recipient behaves in certain ways) and may not have enough freedom to act.
- *Non-governmental organizations* – may focus on a wide range of economic or social development issues or environmental protection.
- *Transnational or multinational companies/corporations* – are seeking profits or raw materials (and perhaps disposal of waste or sites for dangerous activities), and wish to obtain licences to operate and avoid environmental or socio-economic problems – which would be bad for sales and public relations, or threaten legal action.
- *Special-interest groups* – those with power and money who wish to consolidate their position or gain more. These may be difficult to monitor and control, but when subject to scrutiny they may welcome 'cosmetic' impact assessment to avoid difficulties.
- *The public* – seldom a simple, single homogeneous group of people. There may be tribes or other groups that benefit or suffer from a development whereas others do not. An environmental feature might be valued by one group, but not by others. Assessment often has to watch out for several groups.

Mainstream development thinking today stresses the following:

- sustainable development;
- best (non-wasteful) use of resources;
- satisfaction of basic needs and maintenance of human dignity (i.e. reduction of human misery);
- avoidance of dependency – on developed countries; multinational or transnational corporations or expatriate consultants;
- empowerment and capacity-building – i.e. help developing countries establish their own assessment expertise.

The value of impact assessment and procedures like eco-audit in the quest for sustainable development has been discussed in Chapter 2. Developing countries may differ from developed ones in that the 'trade-offs' involved in sustainable development may be more difficult to identify because of the range of socio-economic differences within societies. As Horberry (1983a: 3) noted, environmental impact assessment is not applied to a 'clean slate'; it must fit in with political and other realities. Impact assessment should greatly assist authorities to obtain full value from resources and technology and avoid damage or unwanted side-effects.

Another question that needs to be asked is, 'Do developed and developing countries have similar expectations of impact assessment?' Often in developing countries impact assessment is seen as a way of justifying a development, rather than reducing problems and maximizing benefits. What is needed is *objective* review of environmental impact assessment, in developing countries (Biswas and Agarwala, 1992: viii). Sammy (1985) felt that all the developing countries had much in common with respect to environmental impact assessment, although each faces a unique mix of challenges and has a different natural and human endowment, Table 7.1 attempts to give a summary of the problems faced when attempting impact assessment in developing countries.

Some of these problems reflect shortage of resources, skilled manpower and funds; others are a result of socio-economic conditions. For example:

TABLE 7.1 Some developing country problems which impact assessment must cope with

Problem	Category of activity	Comments on selection
Unfulfilled basic human needs	SPA	Adjust methodologies to reflect importance of socio-economic factors
Scarcity of trained personnel	II, DAE, IPA, SPA	Use methodologies with minimal requirements for trained personnel. Consider expert systems
Lack of baseline data	DAE, IPA, SPA	Use methodologies that demand minimum baseline data inputs
Problems with public participation	SPA	Use methods that display results well or use PIAM, PRA, etc.
Cultural diversity	II, DAE, IPA, SPA	Use methods that focus on each groups needs, wishes and potential to give data
Emphasis on sectoral planning	II, DAE, IPA, SPA	Ensure that physical and social factors are considered
Assessment promoted from outside	II, DAE, SPA	Ensure local situation is understood/ train local expertise for greater future role/empowerment
Cost	II, IPA, DAE, SPA	Use low-cost methods whenever possible
Short planning horizon	II, IPA, DAE, SPA	Use rapid appraisal methods/try to get longer-term planning

Key: II = impact identification; DAE = description of affected environment/people; IPA = impact prediction and assessment; PIAM = participatory impact assessment and monitoring; PRA = participatory rural appraisal; SPA = selection of proposed action

Source: based (considerably modified) on Sammy and Canter (1982: 37, table 3) and other sources

- There may be institutional weaknesses that hinder assessment (Horberry, 1984; 1985: 219).
- A public that may have high levels of illiteracy and unfulfilled basic human needs can be difficult to consult.
- Corrupt or inefficient authorities lack the will to practise pre-emptive planning.
- There are often frequent changes of government.
- In a number of developing countries there has been accelerating urban growth, slums have grown, services have been stressed and may have broken down, or were never adequately developed. Some cities in developing countries are huge, and their poorer people face considerable physical and social impacts as a result of city expansion. The cities and peri-urban environments (together with rural marginal environments that have increasing populations) will suffer considerable development impacts and deserve more attention from environmental assessors and impact assessors. For a review of rapid urban environmental assessment, *see* Leitmann (1993).

The skills and funding problems are easier to resolve by inputs of aid and training. The socio-economic problems are likely to be more difficult to solve. Attitudes that commonly hinder assessment include the view that assessment is a 'luxury' (the poor need to earn a living and environmental damage is a price that can be paid) or that it is a 'conspiracy' by developed countries or international business to hold back less developed ones; or it is hypocrisy – insistence that developing countries care for and spend on caring for the environment when developing countries in the past did not and perhaps, as a consequence, developed. Sammy (1985: 132) noted that projects in developing countries are often on a larger scale than those in developed ones; in proportion to national budgets, with so much at stake, impact assessment assumes an important role.

In common with much modern technology and know-how, assessment methods and procedures have mainly originated in developed countries. There is now over 25 years of hindsight (Sudara, 1984; McCormick, 1993) but there is often a long way to go to adapt them to fit all situations in both developing and developed countries. Nevertheless, it should be possible to synthesize approaches for each set of circumstances (for reviews, *see* Clark *et al.*, 1984: 81–90; Ahmad and Sammy, 1985; Lim, 1985; *Environmental Impact Assessment Review* 1985: **5(3)**, special issue entitled 'EIA for developing countries: progress and prospects'; Biswas and Qu Geping, 1987; Lim, 1988; Adams, 1990: 149–60; Biswas *et al.*, 1990; Finsterbusch *et al.*, 1990; Wickramasinghe, 1990; Biswas and Agarwala, 1992; Lee and Wood, 1993; Duke *et al.*, 1994; MacDonald, 1994). There is a reasonably well-developed 'pool of know-how' for impact assessment in less developed countries in certain sectors, notably those relating to large dams, road-building, tourism development, irrigation projects, thermal power-stations, industrial facilities and petrochemical plants. Sammy (1985: 143) cautioned that assessment should not be promoted in developing countries as a pre-packaged 'import', and that the posture adopted by users in such countries should not be one of 'pupil' to developed countries' 'teacher', as the task is a common one and developed countries can learn from assessment's adoption and adaptation in developing countries.

Two contrasting generic approaches to assessment were recognized by Horberry (1983a: 24): first, the 'conventional', 1970s-US-style, one-off 'snapshot' approach with little concern for how the assessment will affect the policy, programme or project, or consideration of solutions; and second, the Holling-type adaptive, 'conceptual' approach, which uses simulation models and ongoing assessments as a response to uncertainty and instability and which seeks to integrate assessment with management throughout the project, programme, policy or plan (Holling, 1978).

Carrying out an impact assessment may not lead to its becoming an effective part of a planning or environmental management strategy. Ebisemiju (1993) noted that only nine out of 121 developing countries had established frameworks for its implementation, and even in those nine, performance had been poor. To cure the poor performance, Ebisemiju advocated that each developing country should seek a legislative rather than an administrative approach to establishing an environmental impact assessment agency; seek to establish an independent environmental agency with political influence over sectoral agencies; decentralize its environmental impact assessment activities; aim to ensure mandatory involvement of local people and concerned agencies in scoping, reviewing, monitoring and auditing; require mandatory scoping and terms of reference; require the registration of consultants; and seek to internalize environmental impact assessment in the planning process. It must, however, be stressed that it is better to carry out some impact assessment or related practices, no matter how poor, than none at all.

Learning from past experience can be difficult if government departments are jealous or secretive and refuse to release or publish information. Also, information tends to flow from developing countries to developed ones but not among developing countries. To work well, assessment needs to be integrated with an effective political, social and planning situation. Integration may be difficult, and probably requires public involvement and use of local knowledge (*see* Yap, 1990). There may be situations where authorities use approaches such as impact assessment to side-step inadequate administrative situations (typically where planning and decision-making is 'compartmentalized' by discipline or by sector), or to try to modify behaviour of existing agencies without too much fuss (where things may have become 'hidebound'; Engelmann, 1981). The role of assessment to identify and communicate impacts may be paralleled, even eclipsed, by its use as a strategy by 'actors' in development (Wandesforde-Smith *et al.*, 1985: 204). In spite of the difficulties, assessment procedures have been evolving in a wide range of free-enterprise and centrally planned economies.

Reviews of environmental impact assessment application in developing countries often indicate poor public involvement and weaknesses at the review stage. Environmental impact analysis may be required by law in many countries, but few seem actually committed to it. Sammy and Canter (1982: 39) felt that public participation in developing countries' decision-making was a new and quite radical idea – and were not surprised that citizens were happy to leave planning to planners. The solution, they felt, was actively to seek public participation. Some funding agencies encourage publication of environmental impact statements.

A recurrent problem seems to be lack of political will for effective assessment. Probably because assessment is often initiated at a later stage than in developed countries, the emphasis is more on avoidance and mitigation of development impacts than consideration of alternatives (Clark *et al.*, 1984: 82; Wood, 1995: 302). Weak legal foundations for environmental impact assessment often mean poor screening; developments 'escape' assessment (Mayda, 1985). Environment departments in developing countries are mainly quite recent creations, and can have relatively low status and lack clout compared with existing departments of agriculture, forestry, etc., making it difficult to resource, promote and enforce impact assessment and environmental management.

A problem much commented on is the difference between the 'temperate' environments typical of developed countries and the 'tropical' ones found in many developing countries. Accordingly, assessment methods, standards and data collection techniques may need 'tropicalization'. Tropical ecology can be less well studied than that of other environments. This is one reason why impact assessment faces greater challenges than it does in developed countries; others are the complex societies often found in developing countries and the pressing problems of poverty (Biswas and Agarwala, 1992: 4). Lemons and Porter (1992), drawing on a number of case-studies and the first author's work with the Georgia Institute of Technology in the USA, noted much in common between environmental impact assessment, social impact assessments and related approaches (including technology assessment) in developed and developing countries, but called for support for data-gathering, especially at local scale, in the latter and suggested that wherever possible more than one technique should be used to compensate for weaknesses. A number of researchers concerned with data collection in developing countries, notably Chambers (1984), have stressed the weaknesses and bias that often blight field studies. For instance, research may be carried out only in dry seasons; there may be a concentration on more accessible areas; interpretation is usually made by affluent urban-dwellers of the circumstances of poor, rural people; and the meetings that take place may be with unrepresentative samples of society.

Data are often a problem in developing countries. There may be little ecological, meteorological or social monitoring. Collecting information can be a challenge where roads are poor, money scarce and local people perhaps uncooperative or even hostile. Simulation modelling can sometimes help in data-poor situations (*see* Chapter 5). Sammy (1985) suggested that there is a need for an international databank to which all countries would have access. (Various bodies do offer some information on a regional or national scale.)

In developing countries there is a need to improve administrators' and the public's understanding of and support for impact assessment. People in developed countries tend to assume that in developing ones environmental concern is given as high a priority by voters and authorities as it would be in their own countries, but often that is not the case. This is not only because people in developing countries may be too poor to worry about environmental quality; it also reflects the fact that environmental impact assessment grew up in an open system of government with a relatively affluent, well-informed citizenry and wide disclosure of information, whereas many developing countries have a very different tradition. Developed countries see a chance for better quality of life through conservation and environmental care; many less developed countries see an improved quality of life as coming from the exploitation of resources. Thus environmental impact assessment evolved where an ethic of environmental concern had been espoused by many voters and was enforced by law, but in developing countries there is a struggle to obtain basic needs that is hardly supportive. Fuggle (1989: 38) recognized these difficulties and argued that environmental impact assessment in developing countries should:

- encourage the formulation of realistic options early in the planning process;
- determine the relative effects of options on different interest groups;
- foster implementation of environmental management and rehabilitation plans;
- identify developments likely to have significant consequences.

QUALITIES OF IMPACT ASSESSMENT APPROACHES SUITABLE FOR DEVELOPING COUNTRIES

Broadly, desirable qualities are:

- simplicity;
- inexpensiveness;
- speed;
- flexibility;
- incorruptibility;
- promotion of a political will to incorporate impact assessment into planning and decision-making.

Biswas and Qu Geping (1987) argued along similar lines, that environmental impact assessment methods for developing countries should be low-cost, fairly rapid, and applicable with limited available expertise (for discussion of rapid assessment and related approaches such as rapid rural appraisal, *see* Chapter 2). A decade later, these recommendations are still relevant, and many would add 'applicability to programmes, policies and planning' to the list. Sammy (1985: 133) felt that the fall in price of computers and software plus increasing data-sharing would mean that simplicity of techniques would become less important and the trend would be towards more sophisticated modelling methods.

Horberry (1983a: 47) felt that the main barriers to effective assessment in developing countries were not technical but procedural. Among the observations and recommendations he made were:

- that the authority to establish and administer environmental impact assessment should be strong enough to ensure that it has some influence on decisions and that other agencies comply with the process;
- that a rigid and elaborate environmental impact assessment method was not advisable;
- that environmental impact assessment is of limited value unless it results in improved project design and incorporation of measures for monitoring and managing environmental problems as they occur during the life of a project (one is tempted to add 'programme, policy and plan' to 'project' in this recommendation).

ADOPTION OF IMPACT ASSESSMENT BY INTERNATIONAL ORGANIZATIONS

Impact assessment and associated procedures, together with environmental management, have been promoted in both developed and developing countries by a number of international organizations: UN agencies; aid agencies; international non-governmental organizations; funding bodies; and lending banks. Gilpin (1995: 72–90) gives a recent review of the use of environmental impact assessment by international organizations.

Aid may cause unwanted impacts, fail to encourage the delivery of promising opportunities, or be upset by unexpected events (Horberry, 1983b). Some development aid to Bangladesh in the late 1960s helped attract people into areas hit by floods in 1970, when there were over 250 000 flood-related deaths. Boreholes sunk in some Sahelian countries in the 1960s helped encourage over-grazing and increased vulnerability to the drought of 1968–75 (Whyte and Burton, 1980: 136). Impact assessment can help reduce the chance of such problems and so is of great importance to aid organizations.

Most of the developed countries have a bilateral aid agency which distributes development funding to developing countries. For example, there are the US Agency for International Development (USAID), the UK's Overseas Development Administration (ODA), Canada's International Development Research Centre (IDRC) and Canadian International Development Agency (CIDA), Japan's Japan International Cooperation Agency, the Nordic countries' NORDIC Aid, the Norwegian Agency for International Development (NORAD), the Swedish International Development Agency (SIDA), the Danish International Development Agency (DANIDA) and Finland's FINAID. There are also multilateral funding agencies or aid donors, such as the World Bank, the United Nations Development Programme (UNDP), the European Development Fund (EDF), the Organization of Petroleum Exporting Countries (OPEC), and a number of Islamic development assistance organizations. Increasingly, too, non-governmental organizations, such as Oxfam, are supporting developing countries. Multilateral and bilateral agencies have been active in promoting impact assessment of their development activities (for an examination of the environmental impact assessment of bilateral aid, *see* Abel *et al.*, 1979). In some cases they have applied impact assessment to the determination of the form future development research should take; that is, assessment has been applied at a very early stage of policy-making (Wramner, 1992; Kennett and Perl, 1995).

USAID was one of the first bodies to use environmental impact assessment. Since 1974 it has required an initial environmental evaluation for all the projects it funds and an environmental impact assessment for all the activities it funds where impacts are expected to be significant (USAID, 1974). In 1975 a US public-interest group fought a court action against USAID to try to improve impact assessments relating to its activities in developing countries (Gilpin, 1995: 118–19). In 1978 the Council on Environmental Quality drafted regulations (adopted in 1979 into the Foreign Assistance Act) which more or less extended National Environmental Policy Act-like assessment to US foreign aid activities (Manheim, 1994).

Assessments of the need for and effectiveness of aid started to appear by the early 1980s. Most of them were retrospective (e.g. Appasamy, 1982; Kennedy, 1988; Finsterbusch and Van Wicklin, 1988; Berlage and Stokke, 1992; Bertlin, 1994). The United Nations Environment Programme supported a handbook on environmental impact assessment for developing countries in the early 1980s (Davies and Muller, 1983). The UN Asian and Pacific Development Corporation (UNAPDC) examined the application of impact assessment to development projects (UNAPDC, 1993), and Maddock (1993) has tried to assess how effective project-related impact assessment has been. Kennedy (1988) and Jiggins (1995) have reviewed the impacts of aid in non-Western countries, and Liu (1988) has examined the international development of environmental impact assessment. Fewer social impact assessments have been applied to aid, but one that has is Atampugre's (1993) study of the social impacts of an Oxfam soil and water conservation project in the Sahel. Aid-related social impact assessment, is dealt with in more detail in Chapter 8.

The use of impact assessment to improve aid policies, programmes, plans and projects in a proactive fashion is increasing (Kennett and Perl, 1995). A number of non-governmental organizations promote or use participatory rural appraisal approaches for environmental impact assessment and social impact assessments (e.g. Oxfam, 1995: 99–102). The adoption of environmental impact assessment by a range of international organizations is outlined in what follows.

World Bank

The World Bank established an Environmental Adviser in 1970 and, through its Office of Environmental and Health Affairs, it has issued detailed guidelines for environmental assessments. These are required for major projects before funding approval is given (USAID, 1974; World Bank, 1974, 1975). Because projects are initially prepared by potential borrowers, it was at first difficult for the World Bank to insist on assessment early in the planning process. However, by the early 1980s borrowers were bringing in consultants to do assessment during the planning process under the Bank's supervision (Horberry, 1983a: 39).

The Bank's Office of Environmental and Health Affairs has continued to issue guidelines, manuals and other documents supporting and encouraging impact assessment (e.g. World Bank, 1991a, 1991b). The Bank made a public commitment in 1986 to include environmental impact assessment in all its project appraisals. An Environment Department was created in 1987, controlling four Environmental Divisions, and in 1989 an Operational Directive (4.00) was issued which made environmental assessment mandatory for all projects likely to have significant environmental effects (World Bank, 1989). A further Operational Directive (4.01 of 1992) strengthened assessment provisions to seek to ensure that all World Bank-supported developments that might have significant negative impact were subject to environmental impact assessment to ensure that they were environmentally sound and sustainable, and that local non-governmental organizations were kept informed (World Bank, 1991a; Heuber, 1992; Goodland and Edmundson, 1994).

Since 1991 the World Bank has sorted development proposals into four categories: those requiring limited assessment; those requiring detailed environmental assessment; those not requiring environmental assessment; and activity concerned with reducing or countering environmental problems (*see* Gilpin, 1995: 84).

World Bank environmental impact statements have not always been widely distributed although the bank tries to ensure that borrowers do circulate them (the application of impact assessment to structural adjustment is discussed in Chapter 3). The Bank has made use of strategic environmental assessment (*see* Chapter 3) since the early 1990s, applying it (sectorally), for example, to a salinity control and drainage programme in Pakistan and to an African locust control programme.

USAID

As already mentioned, the US Agency for International Development (USAID) issued impact assessment guidelines as early as the mid 1970s (USAID, 1974). The agency conformed to a legally binding set of environmental assessment procedures as a consequence of revised (1978) regulations deriving from the National Environmental Policy Act and the Foreign Assistance Act 1979. USAID operates a two-tier screening, followed if need be by full impact assessment.

Asian Development Bank

The Asian Development Bank of which 30 Asian and Pacific countries were members in 1995, established an Environmental Unit in the early 1980s that was responsible for ensuring the environmental soundness of its funding activities (Biswas and Agarwala, 1992: 178–83). Horberry (1983a: 40) noted that the approach was somewhat informal. The Bank has subsequently produced a range of publications on environmental assessment which are used widely (Lohani, 1988; ADB, 1990).

United Nations Environment Programme

The United Nations Environment Programme (UNEP) was established following the 1972 Stockholm Conference on the Human Environment. In 1978 it established a Division of Environmental Assessment and adopted goals and principles for environmental impact assessment in 1987. UNEP, as the UN's lead agency on environmental issues, has made serious efforts to extend environmental impact assessment to developing countries since the 1970s (Waller, 1982; Horberry, 1985: 210; Gilpin, 1995: 82–3); in the early 1980s it published a guide to assessing industrial impacts (UNEP, 1980), and in the late 1980s it made recommendations to member states on adopting environmental impact assessment procedures, established goals and principles, and published a guidebook (UNEP, 1987, 1988).

Overseas Development Administration

The UK's overseas aid body, the Overseas Development Administration published a manual of environmental appraisal for its staff in 1989, updating it in 1992 (ODA, 1992). The ODA applies environmental assessment, social assessment and environmental impact assessment to its projects and programmes.

Scandinavian aid agencies,

The Swedish International Development Agency (SIDA) has applied environmental impact assessment, initially on an *ad hoc* basis, to its aid schemes since 1992. The Danish International Development Agency (DANIDA) and the Norwegian Agency for International Development (NORAD) apply environmental impact assessment to their aid activities (Biswas and Agarwala, 1992: 168–77).

Organisation for Economic Co-operation and Development

The Organisation for Economic Co-operation and Development (OECD) was one of the first bodies to look at the application of environmental impact assessment to aid projects, publishing helpful guidelines (Evers, 1986; OECD, 1992). In 1974 the OECD recommended that member governments adopt environmental impact assessment procedures and methods; five years later it issued guidelines (OECD, 1979), and in 1986 a further manual to strengthen efforts (OECD, 1986; Evers, 1986). Since 1992 the OECD has used environmental impact assessment when granting aid to developing countries (OECD, 1992; Wood, 1995: 4). The OECD, together with the World Bank and USAID, has supported research, development and world-wide promotion of environmental impact assessment.

Food and Agriculture Organization

The Food and Agriculture Organization of the United Nations (FAO) has developed environmental impact assessment procedures (*see* Wramner, 1987).

SPREAD OF IMPACT ASSESSMENT IN DEVELOPING COUNTRIES

In 1972 only nine governments of developing countries had environmental impact assessment departments. By 1982 over 100 had. Today, virtually all countries have them. There are a number of ways in which assessment has spread and been adopted (*see* Chapter 3). The main ones are first, by government legislation (as was the case in the USA) second, introduction by the government but with implementation left to various agencies, such as federal or professional bodies (as was the case in West Germany and Brazil); and third, by strengthening of the assessment elements of, and adding to, existing practices and legislation.

It is worth stressing that in the USA, the various components that support environmental impact assessment under the National Environmental Policy Act had in large part been separately derived. The USA had freedom of information laws, and encouraged citizen participation; its non-governmental organizations were able to fight cases in court; and the citizenry were relatively well-informed and educated. Few developing or indeed developed countries have similar conditions, particularly citizen involvement and freedom of information. France, for example, in spite of a history of citizen involvement in decision-making, has tended to have centralized impact assessment dominated by technocrats and Ministers.

As mentioned earlier, around the mid-1970s there began to be demands that environmental impact assessment be applied to American involvement abroad. The World Bank and USAID responded, issuing guidelines and manuals and making environmental impact assessment requirements (USAID, 1974; World Bank, 1974, 1975). Another significant development took place on 5 January 1979 when President Carter issued an Executive Order (Executive Office of the President, 1978: Executive Order No. 12114 on 'The effects abroad of major federal actions'), requiring all US agencies to carry out an assessment of foreign activities that involved world commons (i.e. resources that are not owned or administered by any single nation, such as the atmosphere and oceans), toxic or radioactive substances in the USA, impacts on countries uninvolved in an activity, plus special cases designated by the President.

Brown (1990: 137) felt that there was a danger of impact assessment being seen as a 'stand-alone tool' – as many administrators doubtless regard it. To see it in that way could mean that it failed to link to other things, which would be a waste, as it has much more to offer than just identification and evaluation of impacts. Others have also argued the need for developing countries to develop their own integrated assessment systems, stressing that requirement of environmental impact assessment by donor agencies is only an interim solution (Hartje, 1984, 1985).

COUNTRY EXPERIENCE

Some of the countries discussed in the following section, such as Israel, Malaysia or South Korea, should perhaps now be classed as developed countries and have been included in Chapter 6. However, the literature tends to class them with developing countries.

It is difficult to make reliable generalizations, but in many developing countries the poor are likely to be ignored, and special interest groups and foreign businesses often control development. In a number of developing countries, impact assessment has mainly been promoted by aid agencies and funding bodies (Brown *et al.*, 1991). There is considerable variation in mode of adoption, style and effectiveness of impact assessment from country to country in the developing world, although a number have adopted US-type procedures. Some developing countries adopted environmental impact assessment before European ones and have made good use of it; examples are Malaysia, the Philippines, Colombia. In the early 1980s the OECD (1986) concluded that among developing countries, the most progress with assessment had been made in South-East Asia, especially in Thailand and the Philippines. Ebisemiju (1994) came to similar conclusions.

Asia

The UN Economic and Social Council for Asia and the Pacific (ESCAP), a body which actively promotes impact assessment in Asia, noted in 1992 that there was a need for coherent sustainable development policies and less reckless exploitation of the environment (Gilpin, 1995: 81–3, 132–3). To help promote impact assessment, ESCAP has been sponsoring environmental impact assessment training workshops and has issued guidelines since the late 1980s.

A directory of impact assessment studies in Asia has been published by ENSICNET (1994) (*see also* Werner, 1992; Welles, 1995). Asian and Pacific region agricultural impact assessments have been reported on by the Asian Productivity Organization (1992). Biswas and Agarwala (1992: viii) estimated that there had been over 11,500 environmental impact assessments conducted in Asia, so there is considerable practical experience.

Pakistan

A 1983 Ordinance (No. XXXVII) requires every project to file a detailed environmental impact statement with the Pakistan Environmental Protection and Improvement Agency at the time it is planned if it is likely to affect the environment adversely (Nay Htun, 1988: 227).

India

Valappil *et al.* (1994) review the development of environmental impact assessment in India between 1972 and 1994. The 1977 Constitution (Forty-second Amendment) Act, Article 48A, obliged the Indian government to protect and improve the environment. Between 1980 and 1985 various measures were introduced to support that intent. By the early 1980s environmental impact assessment was applied to large projects on an *ad hoc* basis by the Department of the Environment, Impact Assessment Division, which was responsible for administering and promoting environmental impact assessment (Horberry, 1983a: 38). The 1986 Environmental (Protection) Act (Act No. 29) provided the Indian government with powers to take whatever action it feels fit to ensure environmental protection, including environmental appraisal. By 1988 guidelines for environmental impact assessment had been issued covering various types of development. More recently, India has established a Technology Information Forecasting and Assessment Council responsible for technology impact statements. Progress with impact assessment in India has been less than might have been expected (Khanna, 1988; Maudgal, 1988; Gopalan *et al.*, 1992; Gilpin, 1995: 131–2), and there have been calls for better procedures, more rigid enforcement and increased public involvement (Valappil *et al.*, 1994). There have been moves to modify India's Companies Act to include requirement for eco-auditing (Biswas and Agarwala, 1992: 204–13.

Nepal

There is a growing literature on environmental impact assessment in the Himalayan region, much of it on the use of impact assessment to 'stock-take' damage to ecosystems, rather than prevent problems and improve planning (Ahmad, 1993). In 1982 the Environmental Impact Study Project of the Nepalese Department of Soil Conservation and Watershed Management was established to explore environmental impact assessment, the focus being mainly on the development of renewable energy resources and the impacts of tourism development.

Bangladesh

Environmental impact assessment is carried out by the Bangladesh Planning Commission. So far it has been mainly related to major development projects. Donor agencies have promoted environmental impact assessments on flood control, salinity control and other projects. According to Schroll (1995) there had been limited progress toward mandatory national procedures, proposals having been made in 1991 by the Ministry of Environment and Forestry (they were subsequently redrafted).

Millions of Bangladesh's citizens live close to sea-level and are vulnerable to or depend upon river flows. Assessment of the impacts of global environmental change will be especially valuable to Bangladesh.

Sri Lanka

The multi-donor Mahaweli Development Programme was subjected to an environmental assessment (prompted by USAID) in the early 1980s (Horberry, 1983a: 47). In 1980 the National Environmental Act enabled the authorities to require an environmental impact assessment for all public and private development projects, and in 1984 such an assessment became mandatory. In 1986 the Act was strengthened. By 1995 the National Environmental Authority required an environmental impact assessment for all major manufacturing projects and specified a high degree of public involvement (Gilpin, 1995: 132).

People's Republic of China

The 1979 Environmental Protection Law (Articles 6 and 7) was the foundation for China's environmental impact assessment legislation (Wathern, 1988: 229). By the mid-1980s environmental impact assessment was applied to large construction projects and efforts were being made to develop methods and procedures suitable for Chinese needs (*see* Ziyun, 1986; Ning *et al.*, 1988; Wang and Ware, 1990; Wenger *et al.*, 1990; Gilpin, 1995: 129). Ortolano and Shepherd (1995: 6) felt that the People's Republic of China had developed a well-adapted programme of impact assessment that suited its needs. Public involvement has been limited.

Hong Kong

In 1996 the Environmental Protection Department administered environmental impact assessment. The developer commissions assessment and submits the environmental impact statement to the Department, which may then require a more detailed impact assessment (which it oversees). Environmental impact statements have been made available to the public since 1991. It seems likely that environmental impact assessment procedures will come more into line with those of China after 1997.

Taiwan

Environmental impact analysis has been used on most major government projects since 1979, in many cases after implementation had begun. From 1983 environmental impact assessment was extended to major private-sector as well as government projects. In 1985 the Environmental Protection Administration formulated a programme for environmental impact assessment, but in 1995 laws governing it were still not fully established (Lin Lewis, 1989; Gilpin and Lin, 1990; Gilpin, 1995: 145–6). An Environmental Impact Assessment Department has been established by the Taiwan Power and Light Company.

South Korea

Korea's experience with environmental impact assessment up to the early 1980s has been briefly outlined by Kim (1981), Clark *et al.* (1984: 89), Lim (1984: 21–8; 1985: 143–7) and Gilpin (1995: 136–9). Assessment was first mandated in 1980 by the 1979 Environmental Conservation Law, but enforcement only really took place after 1983. Environmental impact assessment has followed a centralized model, the impact assessment process having been introduced as part of the 1981 Environmental Preservation Act. Initially environmental impact assessment was restricted to developments by government agencies and government-funded projects. Starting about 1986 there were attempts to extend it to non-government projects (Wathern, 1988: 227). The 1990 Basic Environmental Policy Act required the public to be given notice of impact assessments. Kim and Murabayashi (1993) reviewed the strengths and weaknesses of the Korean environmental impact assessment system.

There have been some applications of SIA: Won Woo Hyun (1993) tried to assess the social and cultural impacts of satellite broadcasting on the people of Korea.

South-East Asia

Within South-East Asia and the 'Pacific Rim' there has been a fair degree of co-operation between the Association of South East Asian Nations (ASEAN) countries in developing impact assessment procedures (ASEAN *et al.*, 1985; ESCAP, 1985; Nay Htun, 1988; Brown *et al.*, 1991). The Asian Development Bank has been active in promoting environmental impact assessment (*see* the discussion of the Bank earlier in this chapter). By 1985 the United Nations Environment Programme regional office in Bangkok was making proposals for a regional model of environmental impact assessment to serve the ASEAN nations drawing on Holling-type adaptive environmental assessment and management (Roque, 1985).

Thailand

Thailand has had an environmental agency since 1975. In that year the Improvement and Conservation of National Environmental Quality Act was passed, and, after it had been amended in 1978, it provided a strong legislative base for environmental impact assessment (Horberry, 1983a: 36; Snidvongs, 1985). By the early 1980s legislation enabled the authorities to call for environmental impact assessment before a proposed development commenced. The types of development subject to environmental impact assessment were listed by Wathern (1988: 228). Thailand has made quite wide use of environmental impact assessment for large dam projects, and some of this experience has proved useful for other countries (James, 1994). Lohani and Halim (1987) reviewed the environmental impact assessment methods then in use in Thailand and made estimates of the costs of assessment (*see also* Werner, 1992: 17). The impact assessment process is now administered by the Office for the National Environmental Board (ONEB): the developer carries out an environmental impact assessment (normally being required to use consultants registered with the ONEB, and the ONEB defines the terms of reference/scoping). Completed environmental impact statements for public projects go to the Cabinet for a final decision to be made (ONEB, 1988; Tongcumpou and Harvey, 1992; Gilpin, 1995: 146–8).

Malaysia

Malaysia has been quicker than many developed countries to adopt and adapt environmental impact assessment. As a result of the Environmental Quality Act 1974 and the Third Malaysia Plan, the country established a Division of the Environment and an Environmental Quality Council. Before this, environmental matters had been dealt with in a fairly uncoordinated manner by a number of different bodies (Barrow, 1980). The need for environmental impact assessment was recognized in the Third Malaysia Plan of 1976. Seminars in the late 1970s initiated discussion and evaluation that led to environmental impact assessment legislation (Suhaimi, 1980; Goh, 1977; Jaafar, 1979; National Electricity Board of the States of Malaysia, 1979; Clark *et al.*, 1984: 88). Section 34A of the Environmental Quality (Amendment) Act 1984, extended environmental impact assessment legislation in 1986 to require assessment for all public or private projects likely to have major environmental effects. In 1988 full implementation with environmental impact assessment was made mandatory for certain industrial developments. Requirements and guidelines were published by the Malaysian Department of the Environment (Department of Environment, 1990a, 1990b). The focus was at first mainly on projects, and in a number of cases after an alternative had been selected (Nor, 1991). Malaysian environmental impact assessment procedures call for an initial environmental evaluation (a 'preliminary assessment'); if a full environmental impact assessment is required, the report it produces is viewed by an Environmental Impact Assessment Technical Committee composed of representatives from government departments, industry, the public, academia and the judiciary, and by an independent review panel (Gilpin, 1995: 139).

Singapore

A densely populated city-state, Singapore must, as a matter of importance, anticipate any development problems. Since independence in 1965 Singapore has made huge progress with economic development, health care and environmental management. By 1995 all development project proposals were subject to impact assessment, conducted by the Ministry of Environment, Planning and Development working closely with involved government authorities. Planning permission is issued only when all concerned are satisfied (Wathern, 1988: 239; Gilpin, 1995: 144).

Indonesia

Some environmental impact assessment procedures were in force in Indonesia by 1978 (Gilpin, 1995: 133). An Environmental Management Law was passed in 1982 (Act No. 4 of 1982, Article 16) which made some environmental impact assessment provisions and called for assessments, but up to 1983 those carried out were *ad hoc* and often insisted on by aid donors (Horberry, 1983a: 37–8; Clark *et al.*, 1984: 89; MacAndrews, 1994). In 1986 procedures were tightened by Regulation PP No. 29/1986, which required developers to conduct an initial environmental evaluation, which is then assessed to see whether a full environmental impact assessment is necessary. Guidelines prepared by the Ministry of Population and Environment are used to control the process (Wathern, 1988: 227; Ministry of Population and Environment, 1991). By the 1990s a system had evolved whereby the Environmental Impact Management Agency co-ordinated environmental impact assessment activities; this body issued environmental impact assessment guidelines in 1991.

Indonesia has been ahead of most nations, both developed and developing ones, in requiring by law from 1995 that all companies carry out eco-audits every 5 years.

Papua New Guinea

Papua New Guinea passed an Environmental Planning Act in 1978 which enabled the Minister of Environment to require an environmental plan if it seemed necessary (Wathern, 1988: 227). There have been a number of impact assessments of mining and hydroelectric developments (e.g. activities affecting the Fly River), many being 'grey literature', meaning that they are not widely circulated. Australian practices have had some influence and Australian consultants are often involved in assessments. A review of Papua New Guinea impact assessment experience, intended to judge how assessment could be better applied in the Pacific region, has been produced by Hughes and Sullivan (1989).

The Philippines

There was interest in environmental impact assessment in the Philippines perhaps before 1970, earlier than in many developed countries. The adoption of environmental impact assessment procedures dates from 1977 (NEPC, 1977, 1979), and the Philippines National Environmental Policy is modelled on the US National Environmental Policy Act (Clark *et al.*, 1984: 86–7; Lim, 1984: 14–21; 1985: 139–43; Wathern, 1988: 227; Ross, 1992). In 1977 a National Environmental Protection Council was founded by presidential decrees Nos. 1586, 1121 and 1151 of 1977–8. This formally established a requirement for environmental impact assessment and a system for preparing environmental impact statements. From 1982 impact assessment has been applied to all developments likely to affect the environment significantly. A draft environmental impact statement/initial environmental evaluation (termed a 'project description' and part of the screening process) is circulated to concerned parties and receives a public hearing. The developer then prepares a final environmental impact statement which is checked by the National Environmental Protection Council. If that body approves, a Certificate of

Environmental Impact Statement Compliance is issued and development proceeds (James, 1994: 222–6; Gilpin, 1995: 143).

Administered by the National Environmental Protection Council between 1977 and 1985, environmental impact assessment procedures were similar to those in the USA, although some developments were exempted by the President and there have been criticisms of inadequate budget and poor enforcement (Horberry, 1983a: 37; Abracosa and Ortolano, 1987). Since 1987 the Environmental Management Bureau has administered environmental impact assessment. According to Ortolano and Shepherd (1995: 11), during the Marcos regime environmental impact assessment upset the traditional power balance between national ministries, the consequence of which was an effort to hobble it. Most environmental impact assessment seems to be initiated after feasibility decisions, and some agencies and companies at least partially avoid being constrained by impact assessment.

The Pacific islands

Onorio and Morgan (1995) have reviewed Pacific environmental impact assessment activities and training, indicating that a number of nations are trying to tune it to local needs and integrate it with their national planning. Hughes and Sullivan (1989) have examined Papua New Guinea's experience with a view to its application in the Pacific.

The Mediterranean and Middle East

Cyprus

Environmental impact analysis has been applied to various development projects in Cyprus. For example, it was used in 1988 by the United Nations Environment Programme on the Larnaca sewerage system (Gilpin, 1995: 87).

Egypt

Egypt has made more progress with impact assessment than most African states. Most assessments have been carried out by development agencies involved in large projects. Since the 1980s the Egyptian government has been training impact assessors, and an Egyptian Environmental Affairs Agency has been established. Egypt has provisions for compelling developers to deposit an indemnifying bond to pay for any adverse impacts their activities cause at or for some years after completion; however, the requirement to pay can be waived on some projects.

Egypt is heavily dependent on Nile flows originating from other countries. Global environmental change and increasing demand for water might lead to problems and conflict. Egypt would be well advised to assess the impacts to develop appropriate strategies. Also, with large populations living very close to sea-level, for example in Alexandria, Egypt should be assessing the impacts of global climate change.

Israel

Israel has required environmental impact assessment on certain development projects since 1982 and has published a list of types of project for which it is mandatory (e.g. chemical plants). Assessments are prepared by the Israel Environmental Protection Service, which issues guidelines (Biswas and Agarwala, 1992: 25–35).

Kuwait

Much of the literature on impact assessment in Kuwait relates to assessments of Gulf War damage. However, there have been studies of environmental impact assessment methodology in Kuwait (Al-Sultan and Al-Bakri, 1989; Alghadban and Alajmi, 1993).

Africa

Although the use of impact assessment is spreading, Africa generally lags behind in the adoption of environmental impact analysis. South Africa, Egypt (considered earlier, with Mediterranean states), Nigeria and Ghana have made most progress with environmental impact assessment (Ofori, 1991; Gilpin, 1995: 159). When used, impact assessment has often started after project implementation and has tended to be treated as a 'stand-alone tool'; it has thus often been of limited value (Kamukala, 1992). Kakonge and Imevbore (1993) noted difficulties rooted in poor institutional frameworks and legislative foundations, lack of funding, and weak baseline data availability. To get better enforcement and more uniformity of standards and approach, it has been suggested that Africa should adopt something like the European Community 'EIA Directive', perhaps promoted by the Organization of African States (Hocutt *et al.*, 1992; Kakonge, 1994).

Adger and Chigume (1992) review environmental assessment in Zimbabwe. In Kenya environmental impact assessment is reported to have upset inter-ministry power balances, resulting in efforts to constrain it (Hirji, 1990; Ortolano and Shepherd, 1995). Ayanda (1988) and Olokesusi (1992) reviewed the use of environmental impact assessment in Nigeria, Kamukala (1992) used impact assessment in Tanzania, and Gilpin (1995: 90) has also reviewed progress in Tanzania. Benaceur (1991) looked at Algerian environmental impact assessment experience.

There has been interest in environmental impact assessment in South Africa since the 1970s with mandatory requirements since the early 1980s. Sowman *et al.* (1995) reviewed the development of environmental evaluation in that country and reported on its attempts to establish integrated environmental management. Quinlan (1993) discussed the application of environmental impact assessment to projects in South Africa, calling for a reconceptualization of procedures to better address socio-economic issues. One of the pressing needs in Africa is to divert households from wasteful use of fuelwood. Ellegard (1993) applied environmental assessment to an appropriate stove project that was intended to do that.

Latin America

For reviews of how Latin American countries had dealt with environmental problems and made use of environmental impact assessment up to the late 1980s, *see* Smith (1971) and Moreira (1988).

Argentina

Moreira (1988: 252) observed that Argentina had one of the most technically accomplished staffs for the application of environmental impact assessment in Latin America. As early as the late 1970s, Argentina had been actively reviewing environmental impact assessment approaches and had been exploring the integration of impact assessment into a regional planning (river basin development) approach (OAS, 1978: 43–58, appendices B and D). However, in the late 1980s environmental impact assessment was not enforced in all states, and there was no mandatory requirement for impact assessment because the decision had been made to back voluntary compliance. Government guidelines had been issued by the late 1980s, but Moreira (1988) felt there was a need for a better administrative framework.

Uruguay

Although some environmental impact assessments had been conducted by aid agencies or voluntarily by developers, Moreira (1988: 243) felt there was inadequate legal provision.

Brazil

Interest in environmental impact assessment began around 1974. Before that, Brazil voiced worries that environmental concern could hinder economic development in developing countries. Early experience with environmental impact assessment was mainly in the mining and electricity generation (hydroelectricity) sectors, promoted mainly by international funding bodies and electricity supply consortia. North-eastern Brazilian and Amazonian Brazilian hydroelectric and large-scale mining and ore-processing developments have generated environmental and social impact assessments (e.g. Goodland, 1978). Brazilian environmental non-governmental organizations have been active in establishing impact assessment (Horberry, 1983a: 36; Fowler and Dias de Aguiar, 1993; *see also* Chapter 4 on the Linha Vermelha case, the Chapter 8 discussion of Amazonian aboriginal peoples, and the Chapter 9 discussion of roads in Amazonia).

In 1981 the National Environmental Policy Law (Law 6931) stipulated that environmental impact assessment was required for potentially polluting activities, but there was little enforcement before the late 1980s. The Federal Secretariat for the Environment (SEMA) and more recently the National Council of the Environment (CONAMA) have been involved in overseeing environmental impact assessment. In 1983 SEMA issued regulations for its application to development projects of a certain category and scale (Lim, 1984: 28–33; 1985: 147–50).

For some years, effectiveness of application varied from state to state. From the mid-1980s state environmental agencies have had the primary role in environmental impact assessment, overseen by a National System of the Environment (SISNAMA) and CONAMA. For example, the state government of Rio de Janeiro delegated much of the development and implementation of environmental impact assessment to the State Foundation for Environmental Engineering (FEEMA), a quasi-private corporation. Wandesforde-Smith and Moreira (1985) examined how efforts were made to strengthen environmental impact assessment measures in Rio during local government reorganizations, presenting two case-studies illustrating the political difficulties and pressures from policy and élites that were faced.

Resolutions before Brazil's Senate in 1986 and 1987 legally defined responsibilities, set criteria and laid guidelines. Social and economic impacts are strongly emphasized, in addition to assessment of biogeophysical impacts. Moreira (1988: 250–2), Fowler and Dias de Aguiar (1993) and Capelli (1994) provide reviews of environmental impact assessment in Brazil.

Colombia

Colombia was the first country in Latin America to institute a formal environmental impact assessment process. In 1973 the National Congress gave the President powers to control pollution and protect the environment. In 1974 the National Code of Renewable Resources and Protection of the Environment was passed, one of its provisions was for environmental impact assessment (Moreira, 1988: 245–7).

Mexico

Mexico started applying environmental impact assessment in the mid-1970s. A Directorate of Environmental Impact Assessment was established within the Ministry of Agriculture and Water Resources. In 1980 a Law of Public Works laid down regulations for the implementation of environmental impact assessment, followed in 1982 by a federal Law on Environmental Protection which established the need for it. In 1988 the General Law on Ecological Balance and Environmental Protection specified to what developments it was to be applied. Efforts were made in 1985 to improve environmental impact assessment practice, including the issue of better guidelines and strengthened review procedures (Moreira, 1988: 248–50; Pisanty-Levy, 1993).

Venezuela

Laws passed since 1976 have established guiding principles for environmental planning and management, including legislation on environmental impact assessment. In spite of government support for environmental impact assessment, up to the late 1980s much of it took place after developments had got under-way (Moreira, 1988: 247).

Peru

In a brief review, Moreira (1988: 244) suggested that Peruvian legislation was ineffective. Environmental impact assessment had mainly been required by foreign aid agencies, and there were inadequate guidelines and procedures.

The Caribbean

As is the case in most developing countries, there is 'grey literature' coverage of environmental impact assessment (especially applied to tourism development). Jamaica introduced environmental impact analysis under the co-ordination of its National Resources Conservation Authority (Miller, undated). There have been studies of impacts of peat extraction from Jamaican swamplands but limited material in academic journals. Gamman and McCreary (1988) looked at the integrating of impact assessment into development planning. Bisset (1992) examined the development of an environmental impact assessment system for the Turks and Caicos Islands.

The 1992 Rio 'Earth Summit'

The 'Earth Summit' deserves mention because it endorsed the widespread application of impact assessment. However, it made little or no specific financial provision to assist.

Implementation of impact assessment in developing countries

Environmental and social impact assessment are now widely applied in developing countries, mainly at the project level (Lim, 1984). Certain fields seem to be attracting attention: the spread and effects of HIV/AIDS (*see* Chapter 8); the impacts of structural adjustment (Kone, 1993; FAO, 1995) and free-trade agreements, impacts associated with industrial development and pollution control (Turner, 1984), agricultural innovation impacts (Campbell, 1989) and dam-building, land development and irrigation schemes (*see* Chapter 9).

One difficulty has been the dissemination of impact assessment information among developing countries, or even within some countries. This is largely a function of 'grey literature': the production of a limited number of reports or environmental impact statements by an assessor which end up locked away in government offices and company headquarters. There are also problems generated by the tendency for information to remain in the hands of agencies and companies in developed countries and not to circulate among developing ones, perhaps because of language difficulties or lack of funds to maintain libraries and purchase academic journals. There are also situations where an agency in a developing country commissions an impact assessment and then fails to release it to other bodies, perhaps because of inter-departmental rivalry, poor communications or excessive secrecy.

Impact assessment is unlikely to be much heeded if there is not a satisfactory review process; that is, legislation and procedures (and a will) to ensure that an environmental impact statement has been properly prepared and receives adequate consideration. The review process may take the form of a board or a requirement that an environmental impact statement be sent to all concerned parties with a set time for them to report back. An increasing number of developing countries have established effective review legislation and procedures (e.g. Malaysia, Thailand, the Philippines).

Yap (1990) explored the difficulties of incorporating impact assessment into planning and decision-making in developing countries. To combat the tendency for it to become just a planning tool and to cope with deficiencies in data availability, she proposed participatory impact assessment and monitoring. This approach draws upon local people to provide and collect data, ensures that assessment adapts to local needs and facilitates a multidisciplinary consideration. Participatory impact assessment and monitoring draws upon experience gained from rapid rural appraisal and participatory rural appraisal (*see also* Chapter 2).

References

Abel, N., Pain, A. and **Stocking, M.** 1979: *Study of the environmental impact of bilateral aid programmes.* Report of the Overseas Development Group. Norwich: University of East Anglia, School of Development Studies.

Abracosa, R. and **Ortolano, L.** 1987: Environmental impact assessment in the Philippines: 1977–1985. *Environmental Impact Assessment Review* **7(4)**, 293–310.

Adams, W.M. 1990: *Green development: environment and sustainability in the third world*. London: Routledge.

ADB 1990: Environmental risk assessment: dealing with uncertainty in environmental impact assessment. ADB Environmental Paper 7. Manila: Asian Development Bank.

Adger, W.N. and **Chigume, S.** 1992: Methodologies and institutions in Zimbabwe's evolving environmental assessment framework. *Third World Planning Review* **14(3)**, 283–95.

Ahmad, A. 1993: Environmental impact assessment in the Himalayas: an ecosystem approach. *Ambio* **XXII(1)**, 4–9.

Ahmad, Y.J. and **Sammy, G.K.** 1985: *Guidelines to environmental impact assessment in developing countries*. London: Hodder & Stoughton.

Alghadban, A.N. and **Alajmi, D.** 1993: Environmental impact assessment – integrated methodology: a case-study of Kuwait, Arabian Gulf. *Coastal Management* **21(4)**, 271–98.

Appasamy, P.P. 1982: Impact assessment of international development projects. *Impact Assessment Bulletin* **2(2)**, 173–86.

ASEAN, UNEP and **CDG** 1985: *Report of a workshop on the evaluation of environmental impact assessment applications in ASEAN countries.* Bandung, Indonesia: ASEAN.

Asian Productivity Organization 1992: *Environmental impact assessment in agricultural projects in Asia and the Pacific: report of an APO Seminar, 6–16 March 1991.* Tokyo: Asian Productivity Organization.

Atampugre, N. 1993: *Behind the lines of stone: the social impact of a soil and water conservation project in the Sahel*. Oxford: Oxfam.

Ayanda, J.O. 1988: Incorporating impact assessment in the Nigerian planning process: needs and procedure. *Third World Planning Review* **10(1)**, 51–64.

Barrow, C.J. 1980: Development in Peninsular Malaysia: environmental problems and conservation measures. *Third World Planning Review* **2(1)**, 7–25.

Barrow, C.J. 1987: *Water resources and agricultural development in the tropics*. Harlow: Longman.

Benaceur, Y. 1991: Les études d'impact sur l'environnement en droit positif algérien. *Revue Algérienne des Sciences Juridiques, Economiques et Politiques* **XXIX(3)**, 443.

Berlage, L. and **Stokke, O.** 1992: *Evaluating development assistance: approaches and methods*. London: Frank Cass.

Bertlin, J. 1994: Assessing the economic and environmental impact of UK overseas aid for water development projects: sustainability lessons from the NAO review process. *Project Appraisal* **9(4)**, 263–73.

Bisset, R. 1992: Devising an effective environmental impact assessment system for a developing country: the case of the Turks and Caicos Islands. In Biswas, A.K. and Agarwala, S.B.C. (eds), *Environmental impact assessment for developing countries*. Oxford: Butterworth Heinemann, 214–34

Biswas, A.K and **Agarwala, S.B.C.** (eds) 1992: *Environmental impact assessment for developing countries*. Oxford: Butterworth-Heinemann.

Biswas, A.K., Khoshoo, T.N. and **Khosla, A.** 1990: *Environmental modelling for developing countries*. London: Tycooly.

Biswas, A.K. and **Qu Geping** (eds) 1987: *Environmental impact assessment for developing countries*. London: Tycooly International.

Brown, A.L. 1990: Environmental impact assessment in a development context. *Environmental Impact Assessment Review* **10(1–2)**, 135–43.

Brown, A.L., Hindmarsh, R.A. and **McDonald, G.T.** 1991: Environmental assessment procedures and issues in the Pacific Basin – Southeast Asia region. *Environmental Impact Assessment Review* **11(2)**, 143–56.

Campbell, M.J. (ed.) 1989: *New technology and rural development: the social impact*. London: Routledge.

Capelli, S. 1994: O estudo de impacto ambiental na realidade brasileira. *Estudos Jurídicos* **27(70)**, 49–64.

Chambers, R. 1983: *Rural development: putting the last first*. Harlow: Longman.

Clark, B.D., Gilad, A., Bisset, R. and **Tomlinson, P.** (eds) 1984: *Perspectives on environmental impact assessment*. Dordrecht: D. Reidel.

Davies, G.S. and **Muller, F.G.** 1983: *A handbook on environmental impact assessment for use in developing countries*. A report for UNEP. Nairobi: United Nations Environment Programme.

Department of Environment 1990a: *Environmental impact assessment (EIA): procedure and requirements in Malaysia*. Kuala Lumpur: Ministry of Science, Technology and Environment, Department of Environment.

Department of Environment 1990b: *A handbook of environmental impact assessment guidelines*. Kuala Lumpur: Ministry of Science, Technology and Environment, Department of Environment.

Duke, K.M., Cornaby, B.W. and **Velagaleti, R.R.** 1994: Technology-transfer of environmental impact assessment methodologies to developed and developing countries. *Journal of Scientific and Industrial Research* **53(8)**, 609–18.

Ebisemiju, F.S. 1994: Environmental impact assessment: making it work in developing countries. *Journal of Environmental Management*, **38(4)**, 247–73.

Ellegard, A. 1993: The Maputo Coal Stove Project: environmental assessment of a new household fuel. *Energy Policy* **21(5)**, 518–614.

Engelmann, R.J. 1981: Incorporating environmental considerations into planning. *Environmental Conservation* **8(1)**, 23–30.

ENSICNET 1994: *Directory of environmental impact assessment studies in some Asian countries*. Bangkok: Environmental Systems Information Centre (Asian Institute of Technology).

ESCAP 1985: *Environmental impact assessment: guidelines for planners and decision makers*. Bangkok: UN Economic and Social Commission for Asia and the Pacific.

Evers, F.W.R. 1986: Environmental assessment and development assistance: the work of the OECD. *Impact Assessment Bulletin* **4(3–4)**, 307–20.

FAO 1995: *The measurement of the impact of environmental regulations on trade*. Rome: UN Food and Agriculture Organization, Committee on Commodity Problems.

Finsterbusch, K., Ingersol, J. and **Llewellyn, L.** (eds) 1990: *Methods for social analysis in developing countries*. Boulder, CO: Westview.

Finsterbusch, K. and **Van Wicklin, W.A.** III 1988: Unanticipated consequences of AID projects: lessons from impact assessment for project planning. *Impact Assessment Bulletin* **6(3–4)**, 126–36.

Fowler, H.G. and **Dias de Aguiar, A.M.** 1993: Environmental impact assessment in Brazil. *Environmental Impact Assessment Review* **13(3)**, 169–76.

Fuggle, R.F. 1989: Integrated environmental management: an appropriate approach to environmental concerns in developing countries. *Impact Assessment Bulletin* **8(1–2)**, 31–45.

Gamman, J.K. and **McCreary, S.T.** 1988: Suggestions for integrating environmental impact assessment and economic development in the Caribbean region. *Environmental Impact Assessment Review* **8(1)**, 43–60.

Gilpin, A. 1995: *Environmental impact assessment: cutting edge for the twenty-first century*. Cambridge: Cambridge University Press.

Gilpin, A. and **Lin, S.** 1990: *Environmental impact assessment in Taiwan and Australia: a comparative study*. Hawaii: East–West Center.

Goh, Kim Seng 1977: Proposed environmental impact assessment procedure for Malaysia. Paper presented to National Seminar on Environmental Impact Assessment, Kuala Lumpur, 26–28 September 1977.

Goodland, R.J. 1978: *Environmental assessment of the Tucuruí Hydroproject (Rio Tocantins, Amazonia, Brazil)*. Brasília: Eletronorte SA (168 pp.).

Goodland, R.J. and **Edmundson, V.** (eds) 1994: *Environmental assessment and development*. Washington DC: World Bank.

Gopalan, R., Sekaran, K. and **Banerjee, M.** 1992: Proposed EIA methodology for India. In Biswas, A.K. and Agarwala, S.B.C. (eds), *Environmental impact assessment for developing countries*. Oxford: Butterworth-Heinemann, 204–13.

Hartje, V.J. 1984: Environmental impact assessments for development projects: institutional constraints in international development cooperation. *Zeitschrift für Umweltpolitik* **4**, 485–503.

Hartje, V.J. 1985: Environmental impact assessment for development projects: institutional constraints in international development cooperation. *Journal of Public and International Affairs* **5(1)**, 49–57.

Heuber, R. 1992: The World Bank and environmental assessment: the role of non-governmental organizations. *Environmental Impact Assessment Review* **12(1)**, 331–47.

Hirji, R.F. 1990: Institutionalizing EIA in Kenya. PhD dissertation: Stanford University, California.

Hocutt, C.H., Bally, R. and **Stauffer, J.R. Jr.** 1992: An environmental assessment primer for developing countries with an emphasis on Africa. *Advances in Modern Environmental Toxicology* **20(1)**, 39–61.

Holling, C.S. 1978: *Adaptive environmental assessment and management*. Chichester: Wiley.

Horberry, J. 1983a: *Status and application of environmental impact assessment for development*. Gland: Conservation for Development Centre, International Union for the Conservation of Nature and Natural Resources.

Horberry, J. 1983b: Establishing environmental guidelines for development aid projects. *Environmental Impact Assessment Review* **5(2)**, 133–53.

Horberry, J. 1984 Establishing environmental guidelines for development aid projects: the institutional factor. *Environmental Impact Assessment Review* **4(1)**, 98–102.

Horberry, J. 1985: International organizations and EIA in developing countries. *Environmental Impact Assessment Review* **5(3)**, 207–22.

Hughes, P. and **Sullivan, M.** 1989: Environmental impact assessment in Papua New Guinea: lessons for the wider Pacific region. *Pacific Viewpoint* **30(1)**, 34–55.

Jaafar, Abu Bakar Ir 1979: A review of the proposed EIA procedure for Malaysia. Paper to Seminar Kebangsaan: Pengurusan Alam Sekitar dan Penilaian Kesan Kesan Alam Sekitar, Kuala Lumpur, 16–20 October, 1978. *Asian Environment* (Kuala Lumpur) **III(3)**, 20–9.

James, D. (ed.) 1994: *The application of economic techniques to environmental impact assessment*. Dordrecht: Kluwer.

Jiggins, J. 1995: Development impact assessment: impact assessment of aid projects in non-western countries. *Impact Assessment* **13(1)**, 47–70.

Kakonge, J.O. 1994: Monitoring of environmental impact assessments in Africa. *Environmental Impact Assessment Review* **14(4)**, 295–304.

Kakonge, J.O. and **Imevbore, A.M.** 1993: Constraints on implementing environmental impact assessments in Africa. *Environmental Impact Assessment Review* **13(4)**, 299–308.

Kamukala, G.L. 1992: Application of environmental impact assessment in the appraisal of major development projects in Tanzania. In Biswas, A.K. and Agarwala, S.B.C. (eds), *Environmental impact assessment for developing countries*. Oxford: Butterworth-Heinemann, 184–90.

Kennedy, W.V. 1988: Environmental impact assessment and bilateral development aid: an overview. In Wathern, P. (ed.), *Environmental impact assessment: theory and practice*. London: Unwin Hyman, 272–85.

Kennett, S.A. and **Perl, A.** 1995: Environmental impact assessment of development-orientated research. *Environmental Impact Assessment Review* **15(4)**, 341–60.

Khanna, P. 1988: *Environmental impact assessment and land use planning and urban settlement projects in India*. Manila: Asian Development Bank.

Kim, H.C. 1981: EIA in Korea. Paper presented to Seminar on Environmental Impact assessment, Seoul. Mimeograph.

Kim, K. and **Murabayashi, D.H.L.** 1993: Recent developments in the use of environmental impact statements in Korea. *Environmental Impact Assessment Review* **12(3)**, 295–314.

Kone, T. 1993: Ajustement structurel et politique agricole en Côte d'Ivoire: l'impact environnemental. *Labor, Capital and Society* **26(1)**, 86–101.

Lee, N. and **Wood, C.M.** 1993: *EIA in developing countries*. Manchester EIA Leaflet Series 15. Manchester: Manchester University EIA Centre.

Leitmann, J. 1993: Rapid urban environmental assessment: toward environmental management in cities of the developing world. *Impact Assessment Review* **11(3)**, 225–60.

Lemons, K.E. and **Porter, A.L.** 1992: A comparative study of impact assessment methods in developed and developing countries. *Impact Assessment Bulletin* **10(3)**, 57–66.

Lim, G.C. 1984: *Implementation of environmental impact assessment in developing countries*. Development Studies Program, Woodrow Wilson School of Public and International Affairs Discussion Paper 115. Princeton, NJ: Princeton University.

Lim, G.C. 1985: Theory and practice of EIA implementation: a comparative study of three developing countries. *Environmental Impact Assessment Review* **5(2)**, 133–53.

Lim, G.C. 1988: From negligence to prevention: environmental impact assessment in developing countries. *Urban Law and Policy* **9(5)**, 421–37.

Lin Lewis, S.J. 1989: EIA in Taiwan: current status and future trends. *Environmental Impact Assessment Review* **9(1)**, 67–74.

Liu, F. 1988: The international development of environmental impact assessment. *Environmentalist* **8(2)**, 143–5.

Lohani, B.N. 1988: *Environmental assessment and management in the Bank's developing member countries*. Manila: Asian Development Bank.

Lohani, B.N. and **Halim, N.** 1987: Recommended methodologies for rapid environmental impact assessment in developing countries: experiences derived from case studies in Thailand. In Biswas, A.K. and Qu Geping (eds), *Environmental impact assessment in developing countries*. London: Tycooly International, 65–111.

MacAndrews, C. 1994: The Indonesian Environmental Impact Management Agency (BAPEDAL): its role, development and future. *Bulletin of Indonesian Economic Studies* **30(1)**, 85–104.

McCormick, J.F. 1993: Implementation of NEPA and environmental impact assessment in developing nations. In Hildebrand, S.G. and Cannon, J.B. (eds), *Environmental analysis: the NEPA experience*. Boca Raton, FL: Lewis Publishers, 716–27.

MacDonald, M. 1994: What's the difference?: a comparison of EIA in industrial and developing countries. In Goodland, R.J. and Edmunds, V. (eds), *Environmental assessment and development*. Washington DC, World Bank, 29–34.

Maddock, N. 1993: Has project monitoring and evaluation worked? *Project Appraisal* **8(3)**, 188–92.

Manheim, B.S. Jr. 1994: NEPA's overseas application. *Environment* **36(3)**, 43–5.

Maudgal, S. 1988: *Environmental impact assessment in India: an overview*. Manila: Asian Development Bank.

Mayda, J. 1985: Environmental legislation in developing countries: some parameters and constraints. *Ecology Law Quarterly* **12**, 988.

Miller, B.A. undated: *EIA: the Jamaican experience*. Kingston: Director of the Natural Resources Conservation Department. Mimeograph (7 pp.).

Ministry of Population and Environment 1991: *Documents relating to the environmental impact analysis process in Indonesia* (revised). Jakarta: Ministry of Population and Environment.

Moreira, V. 1988: EIA in Latin America. In Wathern, P. (ed.), *Environmental impact assessment: theory and practice*. London: Unwin Hyman, 239–53.

National Electricity Board of the States of Malaysia (eds), 1979: *Environmental impact study: Port Klang*. Kuala Lumpur: National Electricity Board of the States of Malaysia (LLN).

Nay Htun 1988: The EIA process in Asia and the Pacific region. In Wathern, P. (ed.), *Environmental impact assessment: theory and practice*. London: Unwin Hyman, 225–38.

NEPC 1977: *The environmental impact statement system*. Quezon City, the Philippines: National Environmental Protection Council.

NEPC 1979: Primer on the EIS system in the Philippines. Manila: National Environmental Protection Council, EIA Office. Mimeograph.

Ning, D., Wang, H. and **Witney, J.** 1988: Environmental impact assessment in China: present practice and future developments. *Environmental Impact Assessment Review* **8(1)**, 85–95.

Nor, Y.M. 1991: Problems and perspectives in Malaysia. *Environmental Impact Assessment Review* **11(2)**, 129–42.

OAS 1978: *Environmental quality and river basin development: a model for integrated analysis and planning*. Washington DC: Organization of American States, Secretary-General.

ODA 1992: *Manual of environmental appraisal* (revised – 1st edn 1989). London: Overseas Development Administration.

OECD 1979: *The assessment of projects with significant impacts on the environment* (C[79]116–8 May 1979). Paris: Organisation for Economic Co-operation and Development.

OECD 1986: *Environmental assessment and development assistance*. Environment Monograph 4. Paris: Organisation for Economic Co-operation and Development.

OECD 1992: *Good practice for environmental impact assessment of development projects*. Guidelines in Environment and Aid 1. Paris: Organisation for Economic Co operation and Development.

Ofori, S.C. 1991: Environmental impact assessment in Ghana: current administration and procedures – towards an appropriate methodology. *Environmentalist* **11(1)**, 45–54.

Olokesusi, F. 1992: Environmental impact assessment in Nigeria: current situation and directions for the future. *Journal of Environmental Management* **35(3)**, 163–71.

ONEB 1988: *Environmental impact assessment in Thailand*. Bangkok: Office of the National Environment Board (OENB).

Onorio, K. and **Morgan, R.K.** 1995: In country training in the South Pacific: an interim review and evaluation of the South Pacific Regional Environment Programme's EIA Program. *Impact Assessment* **13(1)**, 87–99.

Ortolano, L. and **Shepherd, A.** 1995: Environmental impact assessment: challenges and opportunities. *Impact Assessment* **13(1)**, 3–30.

Oxfam 1995: *The Oxfam handbook of development and relief* (2 vols). Oxford: Oxfam.

Pisanty-Levy, J. 1993: Mexico's environmental assessment experience. *Environmental Impact Assessment Review* **13(4)**, 267–72.

Quinlan, T. 1993: Environmental impact assessment in South Africa: good in principle, poor in practice. *South African Journal of Science* **89(3)**, 106–10.

Roque, C. 1985: Environmental impact assessment in the Association of Southeast Asian Nations. *Environmental Impact Assessment Review* **5(3)**, 257–63

Ross, W.A. 1992: Environmental impact assessment in the Philippines: progress, problems, and directions for the future. *Impact Assessment Review* **14(4)**, 217–32.

Sammy, G.K. 1985: Toward the internalization of environmental impact assessment in developing countries. *Impact Assessment Bulletin* **5(3)**, 129–44.

Sammy, G.K. and **Canter, L.W.** 1982: Environmental impact assessment in developing countries: what are the problems? *Impact Assessment Bulletin* **2(1)**, 29–43.

Schroll, H. 1995: Bangladesh and the environment. *Impact Assessment* **13(3)**, 317–25.

Smith, B.B. 1971: Impact assessment in S. America. *Journal of the Society of Scientists* **22**, 177–9.

Snidvongs, K. 1985: Problems and prospects of environmental impact assessment in Thailand. In Klennert, K. (ed.), *Environmental impact assessment (EIA) for development*. Proceedings of a joint DSE/UNEP international seminar. Fedalfing: Deutsche Stiftung für internationale Entwicklung, 152–70.

Sowman, M., Fuggle, R. and **Preston, G.** 1995: A review of the evolution of environmental evaluation procedures in South Africa. *Environmental Impact Assessment Review* **15(1)**, 45–68.

Sudara, S. 1984: EIA procedures in developing countries. In Clark, B.D., Gilad, A. and Tomlinson, P. (eds), *Perspectives on environmental impact assessment*. Dordrecht: D. Reidel, 81–90.

Suhaimi, A. 1980: Environmental control by legislation: Malaysia as a case study. Paper presented to the Asian–American Conference on Environmental Protection, Jakarta. Mimeograph.

Al-Sultan, Y.Y. and **Al-Bakri, D.** 1989: The development and experience of Kuwait in environmental protection and environmental impact assessment. *Impact Assessment Bulletin* **7(4)**, 57–68.

Tongcumpou, C. and **Harvey, N.** 1992: Implications of recent EIA changes in Thailand. *Environmental Impact Assessment Review* **14(2–3)**, 271–94.

Turner, E. 1984: Industrial development without environmental pollution: role of environmental impact assessment (EIA). *Asian Environment* **6(4)**, 27–36.

UNAPDC 1983: *Environmental assessment of development projects*. Kuala Lumpur: UN Asian and Pacific Development Centre.

UNEP 1980: *Guidelines for assessing industrial environmental impacts and environmental criteria for the siting of industry*. Industry and Environment Guidelines Series vol. 1. Paris and Moscow: UN Environment Programme.

UNEP 1987: *Goals and principles for EIA*. Decision 14/25 of Governing Council, UNEP. Nairobi: UN Environment Programme.

UNEP 1988: *Environmental impact assessment: basic procedures for developing countries*. Bangkok: UN Environment Programme Regional Office for Asia and the Pacific.

USAID 1974: *Environmental assessment guidelines manual* (US Agency for International Development). Washington DC: US Government Printing Office.

Valappil, M., De Vuyst, D. and **Hens, L.** 1994: Evaluation of the environmental impact assessment procedures in India. *Impact Assessment* **12(1)**, 75–88.

Waller, R.A. 1982: EIA guidelines for the United Nations Environment Programme. *Impact Assessment Bulletin* **2(1)**, 44–52.

Wandesforde-Smith, G., Carpenter, R.A. and **Horberry, J.** 1985: EIA in developing countries. *Environmental Impact Assessment Review* **5(3)**, 201–6.

Wandesforde-Smith, G. and **Moreira, I.V.D.** 1985: Subnational government and EIA in the developing world: bureaucratic strategy and political change in Rio de Janeiro, Brazil. *Environmental Impact Assessment Review* **5(3)**, 223–38.

Wang, H. and **Ware, J.** 1990: The development of environmental quality evaluation and environmental impact assessment in China. *Impact Assessment Bulletin* **8(1–2)**, 145–60.

Wathern, P. (ed.) 1988: *Environmental impact assessment: theory and practice*. London: Unwin Hyman.

Welles, H. 1995: EIA capacity-strengthening in Asia: the USAID/WRI model. *Environmental Professional* **17(2)**, 103–16.

Wenger, R.B., Wang, H.D. and **Ma, X.Y.** 1990: Environmental impact assessment in the People's Republic of China. *Environmental Management* **14(4)**, 429–39.

Werner, G. 1992: Environmental impact assessment in Asia. In Biswas, A.K. and Agarwala, S.B.C. (eds), *Environmental impact assessment for developing countries*. Oxford: Butterworth-Heinemann, 16–21.

Whyte, A.V. and **Burton, I.** (eds) 1980: *Environmental risk assessment*. SCOPE Report 15. Chichester: Wiley.

Wickramasinghe, R.H. 1990: Environmental impact assessment and developing countries. *Marga* **11(1)**, 32–47

Won Woo Hyun 1993: The social and cultural impacts of satellite broadcasting in Korea. *Media Asia* **20(1)**, 15–20.

Wood, C.M. 1995: *Environmental impact assessment: a comparative review*. Harlow: Longman.

World Bank 1974: *Environmental, health and human ecologic conservation in economic development projects*. Washington DC: World Bank.

World Bank 1975: *Environment and development*. Washington DC: World Bank.

World Bank 1989: *Operational directive 4.00: environmental assessment*. Washington DC: World Bank.

World Bank 1991a: *Operational directive 4.01: environmental assessment*. Washington DC: World Bank.

World Bank 1991b: *Environmental assessment sourcebook:* vol. 1, *Policies, procedures and cross-sectoral issues*. Environmental Department Technical Paper 139. Washington DC: World Bank. (There are two further volumes: Environmental Department Technical Papers 140 and 154.)

Wramner, P. 1987: *Procedures for EIA of FAO's field projects: a preliminary study*. Report to AGRE, FAO. Rome: UN Food and Agriculture Organization.

Wramner, P. 1992: Environmental impact assessment of development projects: experiences from Nordic Aid. In Biswas, A.K. and Agarwala, S.B.C. (eds), *Environmental impact assessment for developing countries*. Oxford: Butterworth-Heinemann, 168–77.

Yap, N.T. 1990: Round the peg or square the hole? Populists, technocrats and environmental assessment in third world countries. *Impact Assessment Bulletin* **8(1–2)**, 69–84.

Ziyun, F. 1986: Environmental impact assessment in the Yangtze Valley: synopsis. *Ambio* **XV(6)**, 347–9.

8

SOCIAL IMPACT ASSESSMENT

INTRODUCTION

This chapter considers social impact assessment and related approaches. There are differences between social impact assessment and related approaches in terms of their evolution, their exact goals, and the background of those who practise them. However, there is often a less clear division in the literature, especially between *social impact analysis*, *socio-economic impact assessment* and *social soundness analysis* (Carley and Bustelo, 1984; Freudenburg, 1986). Assessment of economic impacts and cost-benefit analysis are reasonably distinct fields (although there is overlap between social impact assessment and *social-cost–benefit analysis*), as is *cultural impact assessment* (Canter, 1977a; Pearce, 1978; Cochrane, 1979; Canter, 1996: 435–66). According to Burdge and Vanclay (1996: 59), *social impacts* include all social and cultural consequences to human populations that alter the ways in which people live, work, play, relate to one another, organize to meet their needs and generally cope as members of a society. *Cultural impacts*, Burdge and Vanclay (1996) noted, involve changes to the norms, values and beliefs of individuals that guide and rationalize their cognition of themselves and their society. Cultural impact assessment is concerned with effects on archaeological remains, holy places, cultures, etc.

I use 'social impact assessment' in this book, as does much of the literature, to refer to both social and socio-economic impact assessment. The exact definition of social impact assessment is imprecise; indeed, it means different things to different people (*see* Box 8.1). Some see it as a political means of decision-making, others as a socio-political process that facilitates negotiation among interest groups (Dale and Lane, 1994).

Graham Smith (1993: 4) argued that social and economic and physical and biological aspects of the environment are so interconnected that impact assessment should not treat them separately, but should link them. In practice, such 'total impact assessment' is more of a goal than reality, although

BOX 8.1 SOME SELECTED DEFINITIONS OF SOCIAL IMPACT ASSESSMENT AND SOCIAL ASSESSMENT

Social impact assessment

- '[E]fforts to assess or estimate, in advance, the social consequences that are likely to follow from specific policy actions (including programs, and the adoption of new policies), and specific government actions' (Interorganizational Committee on Guidelines and Principles for Social Impact Assessment, 1995);
- 'the study of the potential effects of natural physical phenomena, activities of government and business, or of any succession of events on specific groups of people' (Wolf, 1983);
- 'the systematic advance appraisal of the impacts on the day-to-day quality of life of persons and communities when the environment is affected by development or a policy change' (Burdge, 1994a: 41);
- 'the process of assessing or estimating, in advance, the social consequences that are likely to follow from specific policy actions or project development' (Burdge and Vanclay, 1996: 59);
- 'effort to identify and assess the social impacts of a proposed project, programme or policy on individual and social groups within or an entire community or communities in advance of any decision to proceed';
- 'a process examining proposed projects, programmes and policies for their possible effects on individuals, groups and communities' (Buchan and Rivers, 1990: 97);
- 'to predict and evaluate the social effects of a policy, programme or project while in the planning stage – before the effects have occurred' (Wolf, 1980: 27);
- 'assessment of the effects of physical changes and/or socio-economic changes on peoples, institutions and communities';
- 'a process which seeks to assess in advance the impacts of legislative changes, technological innovation, development projects, hazards and risks, policy or programme implementation, aid donation or social movements';
- 'application of social science methodology to assist social planning';
- 'a sociopolitical process (like planning) which facilitates bargaining and negotiation among interest groups'.

Social assessment

- 'the process of gathering data and developing the social well-being account (the latter – a formal document on the beneficial and adverse social effects of development . . .)' (Fitzsimmons *et al.*, 1978: 5).

support for it is growing. Environmental and social impact assessment share:

- a proactive approach (at least in theory);
- an attempt to conduct structured assessment;
- efforts to be as objective as possible;
- efforts to be as comprehensive as possible;
- consideration of development alternatives;

- production of a clear, concise, unbiased impact statement;
- whenever possible, involvement of the public in the planning and decision-making process;
- a growing concern for the goal of sustainable development – both environmental and social impact assessment have great potential for those seeking that goal.

Given the commonality of the broad goals and approach between environmental impact assessment and social impact assessment, and the frequent lack of a sharp, distinct division between these, and other assessment fields (plus an increasingly stated – though not universal – desire for the two to be integrated), it makes some sense to view them as opposite ends of the same spectrum (*see* Fig. 8.1), although actual integration may not be straightforward (Azqueta, 1992). It is not uncommon to find situations where biophysical impacts cause socio-economic impacts and these in turn (perhaps several steps down the chain of causation) cause biophysical impacts. For social impact assessment the chain may be socio-economic to biophysical (and perhaps back to socio-economic) or vice versa. So, environmental and social impact assessment are distinct but overlap.

Environmental impact assessment	Social impact assessment
Biogeophysical	Socioeconomic/cultural
'Hard/objective'	'Soft/subjective'

FIGURE 8.1 Environmental and social impact assessment as opposite ends of a spectrum

Social scientists and social historians were studying social impacts long before environmental impact assessment appeared (Becker, 1995: 143; Burdge and Vanclay, 1996: 63); there was certainly interest in the impacts generated by the UK and European transport and industrial revolution by the late eighteenth century. However, the focus before the early 1970s was almost always on retrospective analysis of what had happened, rather than prediction of impacts. It is the focus on prediction and control of planning and decision-making that separates social impact assessment from other fields of social research, which tend to concentrate on causal analysis (Soderstrom, 1981: 17; Pellizzoni, 1992). In 1970 Section 102(c) of the US National Environmental Policy Act required integration of natural and social sciences for impact assessment, in effect the assessment of social and cultural impacts caused by development. Burdge (1994a: 4) suggested that the first use of the term social impact assessment was in 1973 in connection with assessments of the impact of the Trans-Alaska Pipeline on the Inuit people (*see* the discussion of the Berger Inquiry later in this chapter). It took time to develop social impact assessment, and in general it has remained underfunded and neglected compared with environmental impact assessment (Freisema and Culhane, 1976; Canter, 1977b: 164; USDA, 1977; Wilkie and Cain, 1977; Glasson and Heaney, 1993). Attention in the USA increased following the Council on Environmental Quality's 1978 requirement that the National Environmental Policy Act direct more attention to assessing socio-economic as well as physical impacts. It is probably fair to say that up to the late 1980s there had been less interest in social impact assessment in Europe than in the USA or Canada. Large projects like the Alaska oil and gas pipelines and various disasters or near-disasters around the world were by the late 1970s prompting demand for social impact assessment. The Three Mile Island incident (a near-meltdown at a US nuclear facility that necessitated evacuation of householders in 1979) is seen by many as a landmark event because it used social impact assessment to assess threats and public fears before restarting (Moss and Stills, 1981; Freudenburg, 1986: 454; Llewellyn and Freudenburg, 1989).

The US Federal Highways Administration and the US Army Corps of Engineers have been active in developing social impact assessment (mainly in relation to road developments). There has also been considerable activity in New Zealand from the early 1970s, prompted by the Environmental Protection and Enhancement Procedures 1973, the Town and Country Planning Act 1977 and the Resource Management Bill 1989 (Buchan and Rivers, 1990). New Zealand had a Social Impact Assessment Working Group, established to develop and promote social impact assessment, by 1984, and in 1990 a Social Impact Assessment Association was formed. Practitioners, methods and techniques used by social impact assessment originate in a wide range of disciplines, including social welfare (Griffith, 1978), sociology, behavioural geography, social psychology, social anthropology, etc. This diversity, the complexity of social impact assessment, and its relative lack of funding has resulted in its becoming less standardized than environmental impact assessment.

A number of individual researchers have been instrumental in establishing social impact assessment. C.P. Wolf can be singled out as a particularly active promoter; in 1976 Wolf founded the New York-based newsletter *Social Impact Assessment*. Another American, R.J. Burdge, has also been at the forefront of the development of social impact assessment (for recent examples of his writings, *see* Burdge, 1994a, 1994b; Burdge and Vanclay, 1995, 1996). Yet another active American promoter of social impact assessment has been K. Finsterbusch, who, with Wolf, is widely acknowledged to have produced some of the more influential early guides and handbooks (Wolf, 1974; Finsterbusch and Wolf, 1977). (A selection of guidebooks, handbooks and bibliographies on social impact assessment are listed at the end of this chapter.) To summarize, there has been some cross-fertilization between environmental impact assessment and social impact assessment (Gale, 1984), but in most countries the latter spread more slowly and is today less widely applied, and there is less uniformity of approach (Beckwith, 1994). Finsterbusch (1995), in an excellent review of social impact assessment, felt that there had been a decline in numbers of social impact assessment since the early 1980s, a 'boom' having occurred in the USA in the 1970s. A number of social impact assessment activists feel that, relative to environmental impact assessment, social impact assessment has been neglected and deserves more resources, research and development (Glasson and Heaney, 1993). For example, the physical effects of the Chernobyl disaster received attention, but, apart from health impacts, the socio-economic effects within Russia and impacts on the green movement and green politics beyond had much less. Burdge and Vanclay (1996), in a review, felt optimistic about social impact assessment, being of the opinion that its definition and process had been clarified and that much progress had been made, although it needed to be much better integrated into the 'development process'.

Critics of social impact assessment have claimed that it is imprecise; too theoretical; too descriptive, rather than analytical and explanatory; weak at prediction; *ad hoc*; mainly applied at the local scale; likely to delay projects programmes or policies to which it is applied (like environmental impact assessment, causing 'paralysis by analysis'); and a waste of development resources (Muth and Lee, 1986). Social impact analysis has been dismissed by some as blighted by social scientists' efforts to establish their legitimacy. Another criticism levelled at it is that few of the theories that it uses are tightly defined and guarantee comparable results for successive studies.

However, critics must face the fact that social and economic impacts often greatly overshadow hoped-for development benefits. Development that is socio-culturally sound is likely to be economically and environmentally sound; if social impact assessment can identify projects, programmes and policies that generate few or no socio-economic problems, it has great value. Some of the world's most socially, politically and economically costly and embarrassing problems might well have been predicted and avoided or more effectively mitigated with social impact assessment.

Unequal control of resources generates social and economic problems and perpetuates much of the world's poverty. A vicious spiral of poverty-related environmental damage leading to increasing poverty and more damage is widely seen as a significant hindrance to achieving the goal of sustainable development. Social impact assessment is a means of investigating and countering these difficulties and should be welcomed in spite of its weaknesses. It can assist researchers in understanding the processes and can help guide the management of social change in advance of implementation of proposed developments. It also has the potential to bring together various disciplines and types of decision-makers (Soderstrom, 1981: v).

Burdge (1990: 132) tried to assess the value of social impact assessment to less developed countries and concluded that its benefits outweigh its costs. Many developing countries seem to have accepted its value and made provisions for it. A number of agencies have issued guidelines on social impact assessment or social assessment (e.g. ADB, 1991).

Just like (biophysical) environmental impact assessment, social impact assessment is, or should be, predicated on the understanding that decision-makers should be aware of the consequences of their actions before they commit themselves, and that the people involved or affected will be appraised of the impacts, and will have an opportunity to participate in the designing of their future (*Social Impact Assessment* 1984: **90–92**, special issue on 'public participation in social impact assessment'; Interorganizational Committee on Guidelines and Principles for Social Impact Assessment, 1994; 1995: 40 – this committee was founded to outline social impact assessment guidelines for the US National Environmental Policy Act). Clearly, for that to happen there must be the political will and the ability to act on social impact assessment recommendations. When social impact assessment seeks to promote public involvement it is vital that the assessors have the people's trust (Rydant, 1984), but often this is lacking and communication with local citizens is unsatisfactory. Effective public participation may require education or enhancement of awareness before there can be meaningful communication and judgement (Burdge and Robertson, 1990). Some authorities dislike public participation and prefer to use a mediator as an intermediary. However, people often suffer stress because of uncertainty. By directly involving the public in the social impact assessment process, it is possible to reduce uncertainty and therefore stress (Burdge and Vanclay, 1996: 60).

The social component of the environment differs from the biophysical in that it can react in anticipation of change. It can also adapt in reasoned ways if an adequate planning process is in place (Freudenburg, 1986; Interorganizational Committee on Guidelines and Principles for Social Impact Assessment, 1995: 40). It is also different in that reactions can be more varied and fickle, because individuals or groups in a population differ in response and their responses are more often than not inconstant. There may also be difference in timing as well as degree of impact on various sections of society. For example, property owners will probably react differently from non-property owners. As with environmental impact assessment, different socio-economic or socio-cultural impacts may be generated at various stages in a policy, programme or project cycle: for example, during construction; when the development is functioning; and after the plant has been closed down. Too narrow a temporal focus, and social impact assessment may miss impacts (Gramling and Freudenburg, 1992). Spatially it is also important to adopt a wide enough view; like environmental impact assessment, social impact assessment may have to adopt a tiered approach ('strategic social assessment').

SOCIAL IMPACT ASSESSMENT AND EFFORTS TO ATTAIN SUSTAINABLE DEVELOPMENT

Sustainable development, as discussed in Chapter 1, is not a costless approach to development. Reaching sustainable development will probably involve trade-offs that have adverse social and economic impacts, at least in the short term. It is also vital to assess whether there are any social institutions or movements that could support or hinder a sustainable development strategy before initiating it, and to monitor for changes during implementation (Ruivenkamp, 1987; Hindmarsh, 1990). Social impact assessment can be linked with public participation to ensure socio-cultural sustainability; not to make such a link is to separate physical development from related social issues, and that is unwise (Howitt, 1995). Without suitable supportive social institutions, sustainable development will probably fail. Social impact assessment should be used to help establish such institutions. A few studies (*see* Gagnon, 1995) have already applied social impact assessment to sustainable community development (seeking sustainable local development). These roles for social impact assessment are important and are likely to increase considerably. Social impact assessment may help increase public awareness of issues and promote public debate (Rakowski, 1995).

SOCIAL IMPACT ASSESSMENT AND THE PLANNING PROCESS

Social impact assessment is, among other things, a planning tool, and can monitor and assess project or programme or plan or policy changes that have taken place or that are ongoing. Much of what has already been said regarding environmental applies to social impact assessment: it needs to be better integrated into planning and decision-making (Burdge, 1994a: 7), must be undertaken early in planning, and should function as a monitoring process, not as a 'snapshot' study. There is a growing literature on social impact assessment and planning (Burdge, 1987, 1989, 1994a: 229; Boggs, 1994). Where the two are separate, rather than integrated, there appear to be broadly the same stages in the environmental and social impact assessment processes.

There have been calls for the establishment of a comprehensive framework for social impact assessment (Gramling and Freudenburg, 1992). It has been argued that both planning and social impact assessment have tended to move away from apolitical decision-making towards enhanced participation of the parties affected by development (Dale and Lane, 1994). New Zealand has made considerable progress, with the Town and Country Planning Directorate of the Ministry of Works and Development a key agency in establishing social impact assessment procedures to help manage social change and development.

Some of the basic assumptions of social (and environmental) impact assessment contradict established socio-cultural and political traditions. An example is their encouragement of public involvement in decision-making, which can be rather a 'culture shock', particularly in those countries without Western social and liberal-democratic traditions (Burdge and Robertson, 1990; Burdge, 1994a: 229). As with environmental impact assessment, lack of trained personnel is a problem. Rickson and Burdge (in Burdge, 1994a: 231) recognized resistance to social impact assessment from planning staff trained in disciplines such as economics, and a tendency for some decision-makers to seek technological, rather than social, studies because they saw it as politically 'safer'.

Social impact assessment in practice

Freudenburg (1986: 452), in a very useful review of the field, saw mainstream social impact assessment as part social science, part policy-making, part environmental sociology. A readable review of social impact assessment in the USA has been provided by Finsterbusch (1995). Evaluation research has tended to focus on expected results of programmes and policies, whereas social impact assessment tends to concentrate on the unexpected consequences of projects, population growth and technical change (and more recently global environmental change, and policies and programmes), and on the impacts of natural disasters and more 'human' developments. Examples include attempts to assess future social and health impacts of the newly introduced UK National Lottery (McKee and Sassi, 1995), or the impact of ageing populations (Restrepo and Rozenthal, 1994), or the privatization of state utilities (Ernst, 1994).

How closely environmental and social impact assessment are integrated generally depends partly on a country's or agency's definition of environment. Sometimes a multidisciplinary team deals with both environmental and social impact assessment or there may be separate specialists. Usually social impact assessment is a modest component of a wider environmental impact assessment or environmental auditing. There have been occasions where an environmental impact assessment team carried out a social impact assessment without adequate social science input; there are frequent cautions in the literature that social impact assessment should be conducted by competent, professional social scientists.

Social impact assessment often deals with a very broad range of things, and to do so comprehensively and accurately is not easy. Impacts may be felt at the individual level, family level, community level, regional level, national level or even international level. Sometimes impacts may be felt at more than one level, not necessarily at the same time; impacts can be virtually instantaneous or delayed well into the future. Family level, community level and regional level are the most studied. The crucial thing is that social impact assessment identifies undesirable and irreversible impacts. Consumers plead that it should focus on *significant* impacts and present its findings in a concise, non-jargonized form understandable by those other than social scientists. It is also important to assess social equity: the distribution of impacts among groups in the affected population(s). Social impact assessment should also consider impacts on vulnerable groups, such as the poor, the elderly and minority groups.

There are very many variables that may be of interest to social impact assessment. Some of the more important are:

- assessment of who benefits and who suffers: locals, the region, the developer, urban élites, a multinational company's shareholders, etc;
- assessment of the consequences of development actions on community structure, institutions, infrastructure, etc;

- prediction of changes in behaviour of the various groups in a society or societies to be affected;
- prediction of changes in established social control mechanisms;
- prediction of alterations in behaviour, attitude, local norms and values, equity, psychological environment, social processes, activities.;
- assessment of demographic impacts (Becker, 1995);
- assessment of whether there will be reduced or enhanced employment and other opportunities;
- prediction of alterations in mutual support patterns (coping strategies, etc.);
- assessment of mental and physical health impacts (more extensive listing can be found in Murdoch *et al.*, 1986; Burdge, 1989: 87; Little and Krannich, 1978);
- Gender impact assessment has been developed – a process that seeks to establish what effect development will have on gender relations in society (*see* Jiggins, 1995: 270; Verloo and Roggeband, 1996).

To recap, social impact assessment seeks to assess whether a proposed development alters quality of life and sense of well-being, and how well communities adapt to change(s) caused by development (Wolf, 1983). To do that, suitable indicators must be identified, monitored and assessed. They may be single or complex, composite indicators – like the United Nations Development Programme Human Development Index (UNDP, 1991) or a quality-of-life index or a social well-being account (Fitzsimmons *et al.*, 1978). Kurian (1995) argued that impact assessment is generally insufficiently gender-sensitive.

As with environmental impact assessment, it is possible for social impact assessment to focus on known vulnerable components of the social or socio-economic environment; for example, the poor, the elderly, the young, unemployed people, women, ethnic minorities, social 'underclasses'. Communities are a unit that can be monitored for changes using demographic, employment and human well-being data (for a bibliography, *see* Bowles, 1981), and so a community development model is probably the most frequently adopted approach (Gramling, 1992; Burdge, 1994b; Gagnon, 1995). Sometimes the focus, especially with aid donors, is 'target groups', typically the people(s) investment is supposed to help. It is also possible to focus on social indicators. There has also been interest in applying social impact assessment to regions (e.g. Cramer *et al.*, 1980; McDonald, 1990), or through a systems approach (Palinkas *et al.*, 1995) or via an issues-oriented approach (Finsterbusch *et al.*, 1990). When a regional approach is adopted it is possible to make use of rapid rural appraisal and participatory rural appraisal methods (Gow, 1990).

Like environmental impact assessment, social impact assessment has been applied more at project level than programme, plan or policy level, although it has been spreading to the latter (Wolf, 1980; Finsterbusch, 1984; Derman and Whiteford, 1985: 141–59; Freudenburg and Keating, 1985; Freudenburg, 1986:

471; Finsterbusch *et al.*, 1990; Freeman and Frey, 1991). It can also be used to predict the effects of changes in social norms; for example, what will happen if society shifts its position to accept increasing numbers of illegitimate births or, in the case of the USA, to control gun ownership. Social impact assessment can be used to try to predict the effects of alterations in services, and it has been so used a good deal in health care studies. Examples are the closure of mental health facilities in favour of care in the community and the ending of state monopoly in rail transport or telecommunications (Seelman, 1983).

There is a well-developed literature on social impact assessment applied to road construction, boom towns, large projects, voluntary relocatees or refugees, with the recent addition of that on 'eco-refugees' – people dislocated by global environmental change and/or their own misuse of resources (e.g. Shields, 1975; Cernea, 1985). These topics are dealt with later in this chapter and in Chapter 9. Social anthropologists have been active in social (e.g. Partridge, 1984; Green, 1986) and cultural impact assessment. Cultural impact assessment includes studies of how people change their culture in the face of immigration or contact with tourism or new tastes, and exposure to media pressures, threats, opportunities, new concepts, etc. (Cochrane, 1979).

The theory and methods of social impact assessment are still evolving (*see Environmental Impact Assessment Review* 1990:**10(1–2)**, a special issue entitled 'Social impacts of development: putting theory and methods into practice'; Burdge *et al.*, 1995). The difficulties in developing a reasonably universal approach and integrating social and environmental impact assessment have been attributed to the lack of clear definition of basic social impact assessment concepts, in spite of all the handbooks produced. Calls for conceptual and methodological development have been widely voiced since the 1980s (e.g. Finsterbusch, 1985; Murdoch *et al.*, 1986; Dietz, 1987).

Social impact assessment often uses qualitative data and may deal with more 'intangibles' than environmental impact assessment, and consequently has often attracted the criticism that it is 'soft' and imprecise (*see* Fig. 8.1). Some of the fields social impact assessment seeks to deal with are difficult to measure and quantify; they include sense of belonging, community cohesion (maintenance of functional and effective ties between a group), lifestyle, feelings of security, local pride, perception of threats and opportunities (Burningham, 1995) and psychological distress. Quantification of the last of these 'intangibles' has been examined by Egna (1995). Frequently social impact assessment relies on 'social indicators', but these are not perfect. It still needs to address the problem of how to obtain a 'credible' and 'feasible' measure of social impact (Soderstrom, 1981: 70).

TECHNIQUES AND METHODS OF SOCIAL IMPACT ASSESSMENT

It is not my intention to provide instruction in the techniques and methods of social impact assessment, but rather to give an overview to inform the 'consumer', 'participant', or 'commissioner'. Techniques and methods used in the field include social surveys (Maclaren, 1987); questionnaires; interviews (Derman, 1990); use of available statistics such as census data, nutritional status data and findings from public hearings; operations research; systems analysis; social-cost–benefit analysis; the Delphi technique (Yong *et al.*, 1989); marketing and consumer information; reports from social, health, crime prevention and welfare sources; and field research by social scientists (Wildman, 1990). Of these, census and demographic data tends to present fewest challenges and problems. Behavioural psychologists are often involved in social impact assessment to ascertain things like perceptions, likely reactions, whether stress has been or will be suffered, what constitutes a sense of well-being, etc. (Broady, 1985).

Useful reviews of social impact assessment methods have been provided by Kent (1979), Armour (1988), Finsterbusch (1995: 241–5) and Burdge (1994a: 13–24). Ideally, social impact assessment methods are:

- systematic;
- reduce bias to a minimum (i.e. are as objective as possible; Albrecht and Thompson, 1988);
- allow consistent comparison;
- allow reasoned judgement;
- convey the results of assessment to the social impact assessment user(s), who may give some degree of feedback into assessment (Rydant, 1984).

Field research techniques can be divided into *direct* and *indirect*. Direct observation of human behaviour may be open or discreet (an example of the latter is the use of street video cameras), conducted during normal times or times of stress (an example of the latter might be a study of how people react to a storm warning or drought). Indirect observation includes, among other techniques, study of changes in social indicators, patterns of trampling, telephone enquiries directed at selected members of the public, historical records, prices real estate will command and suicide rates.

Phenomena such as community cohesion and community pride are not constant, and are often related to factors that may be difficult for outsiders to assess: an annual carnival, a vague sense of place (for example, the concept of a 'cockney' in London – traditionally someone 'born within the sound of Bow Bells'), some past historic event, or the like. Even an extended period of fieldwork might overlook such factors, social impact assessment is therefore often incomplete and imprecise. The equivalent of environmental impact assessment baseline study is the preparation of a *social profile*; as in environmental impact assessment, it is necessary to establish what might be changed and what would probably happen if no development took place.

The social well-being account, basically an environmental impact statement for social impact assessment, presents positive and negative socio-economic impacts for various development options. Much energy has been expended in seeking to quantify what may be (or should be) unquantifiable. Given the complexity of identifying and assessing direct socio-economic impacts it is not surprising that much less progress has been made with cumulative impact assessment than is the case with environmental impact assessment.

A commonly used framework for social impact assessment is first, to initiate a baseline survey of social conditions; second, to predict changes associated with the proposed development; third, to establish what ongoing changes would take place in the locality or region affected if development did not take place; and fourth, to try to discover what hazards or risks are possible. Finsterbusch *et al.* (1990: 65) recognized two aspects of social impact assessment: *process evaluation* – establishing whether planned or contracted social impact assessment tasks are being carried out; – and *impact evaluation* – assessing the positive and negative impacts. Burdge (1994a: 14) outlined an approach based on the study of past developments similar to that contemplated, as a basis for predicting future impacts of the proposed development. Another approach is to seek a series of 'snapshot' views and then try to 'fill in' between them. Also, social, like environmental, impact assessment has used matrix approaches (King, 1981).

Social impact assessment often makes use of *social evaluation*, *social development evaluation*, *social analysis* and *social cost-benefit analysis* to gather information and assess how things are valued (*see* Finsterbusch *et al.*, 1990; Marsden and Oakley, 1990; ODA, 1993a, 1993b). Social development evaluation was described by Marsden *et al.* (1994: 9–34) as '"a learning process", usually retrospective and interpretive, often reliant on "indicators", yet holistic"'. This approach offers little for *ex ante* assessments, Marsden *et al.* (1994: 9–10) argued that social impact assessment and social cost-benefit analysis have been unable to 'accurately and adequately reflect the dynamics' of change involved in social development. They felt that social development evaluation could adopt a people-oriented focus to enable an understanding that would be vital if sustainable development were a goal. The stress is on an 'evaluation of evaluation perspectives' – addressing the problems of assessment (outsider bias, etc.). Social development evaluation may thus help social impact assessment to improve its approach to evaluation and, especially if social impact assessment were participatory, could help identify the strengths and weaknesses of 'developer' and 'recipient'.

APPLICATION OF SOCIAL IMPACT ASSESSMENT

Social impact assessment has been applied in a very wide variety of situations, including: development projects; investigations of impacted groups; communication and technology change; institutional and social change; structural adjustment; and community development. There is a wide range of variation in level of social impact assessment adoption from full and regulated to partial and *ad hoc*. It is impossible to give a comprehensive overview; the following sections focus on the application of social impact assessment to selected fields, mainly those that have generated an accessible literature.

Social impact assessment and large projects

Large projects can affect country people, urbanites or both, and include among others water supply dams and hydroelectric schemes; hydrocarbon exploitation; mining; agricultural development; flood control; land development; irrigation schemes; drainage schemes; resettlement and improved communications; and industrial development (Scott, 1978, 1981). Some developments impact on national or even global environment or social conditions; for example, Henry Ford's production-line innovations had worldwide socio-economic effects.

A seminal, well-documented social impact assessment application to a large project is the Huntley Monitoring Project, a study over several years, largely by Waikato University academics, of a large (1000 MW) thermal power-station and associated offshore natural-gas facilities constructed in the 1970s close by the New Zealand town of Huntley (whose population before development started in 1973 was 5300) on the Waikato River, North Island (Fookes, 1980, 1984; Buchan and Rivers, 1990: 99; Cocklin and Kelly, 1992). The Huntley social impact assessment measured over a hundred parameters, and revealed impacts that ran far beyond the area that was expected to be affected. It found that there was a need to heed local fears; that developers tend to stress benefits and understate disbenefits; that the public should be informed and involved early on; that local people suspect authority, so planners need to work through local channels; and that social attitudes and values change over time as development progresses. In connection with the last of these points, similar attitudinal variation was recorded in the UK during construction of the Humber Road Bridge. When proposed in the 1960s, this project was popular enough to be seen as a way of winning voters for the government, but after its construction by the 1980s many felt it a waste of money and even a threat to their community.

The Huntley study helped develop social impact assessment procedures and techniques but did not adequately consider impacts on the Maori peoples of the Huntley region. Social impact assessment has also been applied to power-station development in the UK (Power Station Impacts Research Team, 1979).

Social impact assessment and tourism

Tourism is an important area of development in both developed and less developed countries, and is growing rapidly. It is generating an expanding social impact assessment literature (e.g. Beekhuis, 1981; Shera and Matsuoka, 1992). There are a number of ways in which tourism can affect a region's culture, society, economy and health, as well as its physical environment. Impacts are likely to be greater where a community has been relatively isolated and when innovations are rapid. There is no shortage of post-development assessments (i.e. retrospective social impact assessment), which are valuable for establishing typical patterns of impact and for assessing the effectiveness of social impact assessment (Brougham and Butler, 1976).

Proactive tourism-focused social impact assessment is less common, but one example is a study of a hotel complex proposed for a relatively undeveloped Hawaiian island, which suggested how assessment policies and procedures could be improved (Shera and Matsuoka, 1992). There has been some recent friction between local communities and developers of large leisure complexes in France and the UK, suggesting inadequate social impact assessment. Because tourists are likely to respond to local hostility or other social problems by going elsewhere, investors might be expected to make better use of social impact assessment than seems to have been the case.

Social impact assessment of different energy scenarios

Social impact assessment has been applied to possible future energy scenarios and proposals for various national energy policies, for example, proposed 'carbon taxes' or taxes on pollution.

Social impact assessment and land development

Land development is a broad subject that can encompass many things from the development of unspoilt natural areas to redevelopment of occupied land or derelict sites. The associated impacts are thus varied. Social impact analysis has been hailed as a valuable tool for land development planning (Death, 1982), and bodies such as the Urban Institute (Washington DC) have studied the social impacts of city land development (Christensen, 1976).

Land settlement or resettlement schemes have been implemented on a large scale in a number of countries since the 1950s, notably in Indonesia, Malaysia, Brazil and Bolivia (especially their Amazonian territories), Tanzania, South Africa and other African states, and have generated a large, mainly retrospective, literature, primarily under the heading of resettlement or migration studies (e.g. Chambers, 1969).

Social impact analysis and agricultural change

Modernization of agriculture and the spread of technology, especially biotechnology, in less developed countries have been the focus of a substantial number of social impact assessment studies. The social impacts generated by agricultural change have attracted considerable attention (Cernea, 1985; Chamala, 1990; Pinhero and Pires, 1991; Campbell, 1992). Tremendous socioeconomic impacts have been associated with 'Green Revolution' innovations since the 1960s, some largely unexpected and so serious that strategies have had to be altered. The Green Revolution 'package' of innovations was not scale-neutral, and problems have often countered beneficial results: employment changes, debt, loss of land, marginalization, social changes, regional economic disparity, dependency for inputs and credit, unrest, migration, etc.

The replacement of subsistence farming with cash crop production has affected huge areas of the world in the past half-century, and has generated serious impacts. Those used to cash crop production have also had to face many developments; for example, some have turned to contract production (agreeing to supply a retailer or packaging company and thereby being bound to restrictive schedules and production options), which involves huge socio-economic impacts.

Approaches such as farming systems research, participatory rural appraisal and rapid rural appraisal (see Chapter 2) were introduced to try to reduce unwanted social and economic impacts and maximize benefits, and might offer routes for more effective social impact assessment – in particular, a structured multidisciplinary teamwork approach.

Social impact assessment and accidents

There have been many social impact assessment of accidents, mostly carried out after the event to clarify nature and cause, and assist with mitigation and future avoidance. The coverage has some overlap with hazard and risk assessment or civil defence planning, and typically considers oil spills, and nuclear disasters, such as, Three Mile Island, Windscale and Chernobyl (Impact Assessment Inc., 1990). Disasters like the Seveso and Bhopal chemical escapes, and domestic gas or petroleum explosions, oil spills and nuclear accidents since the 1960s, have worried the public, administrators and planners enough to consider socio-economic as well as physical impacts, thereby encouraging the use of social impact assessment. They have also helped stimulate environmental awareness and the green movement.

Social impact assessment and workforces

Social impact assessment has been applied to the problem of workforce impacts on host populations and local environments, and to post-employment scenarios (when construction finishes or an established employment declines or terminates). In the UK in the 1980s and 1990s, more could have been done to assess in advance the social impacts of the decline of heavy industry, especially coal-mining, possibly averting some of the serious social difficulties and costs. There are many retrospective studies of workforce impacts associated with employment decline (e.g. Putt and Buchan, 1987).

Because of the impacts that can be generated on both biophysical and socio-economic conditions in the source region and at the employment locality, labour migration deserves more attention. In a number of countries, opportunities for work in mining, tourism, etc. draw the more dynamic working-age population; the effect on the labour source region's agriculture, environmental quality and local socio-economic well-being can be marked.

Social impact assessment applied to structural adjustment and trade changes

The issues of structural adjustment and trade changes are briefly covered in Chapter 7. There have been a number of economic impact assessments of structural adjustment and world trade changes, but fewer social impact assessment studies, almost all retrospective and mostly focused on particular countries.

Social impact assessment and 'boom towns'

The application of social impact assessment to rapid settlement growth has received a good deal of attention in North America. One finding (which also applies to large construction projects) has been that problems are often associated with less obvious developments that escape monitoring and regulations. Examples include pollution from accommodation areas rather than construction sites and speculative provision of services (*see Pacific Sociological Review* 1982: **25(3)**, a special issue on 'boom towns' England and Albrecht, 1984).

Social impact assessment and developing countries

The problems, needs and potential for environmental impact assessment in developing countries were discussed in Chapter 7. A good deal of what was said there is relevant to social impact assessment, as is much of what has been said earlier in this chapter about large projects, native peoples, etc. The literature indicates a need and great potential for social impact assessment in developing countries but also serious challenges (Derman and Whiteford, 1985; Burdge, 1990; Henry, 1990; Finsterbusch *et al.*, 1990; Suprapto, 1990; *Environmental Impact Assessment Review* 1990: **10(1–2)**, a special issue entitled 'Social impacts of development: putting theory and methods into practice'). Finsterbusch, *et al.* (1990) argued that social and poverty issues are as often as not at the core of environmental impacts in developing countries, and must therefore be addressed.

The use of social impact assessment in both developed and developing countries has lagged behind biophysical environmental impact assessment (Henry, 1990). Much of what experience there is, is associated with large aid donor-funded projects where the assessment team has included social scientists. The World Bank has employed a full-time sociologist since 1974 (Burdge, 1990: 125), and by the late 1980s most funding or aid agencies had social advisers or social impact assessment specialists on their staff. In 1986 the World Bank made a public commitment to use impact assessment in all project appraisals. The US Agency for International Development has employed development soundness analysis from 1983, and by the 1990s most large donors used some form of social impact assessment (Finsterbusch and VanWicklin, 1989). The crucial development goals in most developing countries are economic and social improvements; social impact assessment is very relevant both as a means of checking in advance whether a target group will actually benefit and because it generates knowledge about the development process. However, Finsterbusch (1995: 239) suggested that, in practice, much of the 'social impact assessment' undertaken in developing countries is really social feasibility study aimed at finding out what might hinder a development.

Decision-making in many countries, including most developing countries, is not open to scrutiny or accountable to the public or the media to the extent that it is in North America or Europe. In such circumstances social impact assessment tends to lack teeth. In both the developed and the developing world, social impact assessment is often done in tense political situations with the assessors pressured or faced with efforts to discredit them (Rickson *et al.*, 1990a, 1990b). Some developing countries do not support Western-style participatory democracy, the tradition under which both environmental and social impact assessment have evolved. There may be little or no history of planners consulting the public, and the public may consist of more than one group which differ markedly in aspirations, needs and level of education. Social impact assessment may help inform and educate people, but it can also be hindered by lack of an educated public (Rakowski, 1995). Aid agencies and governments may seek the views of local people in developing countries, and sometimes those views are listened to, but often they are ignored. (It is not particularly common for people's views to be considered adequately in developed countries, either.) Public participation seldom goes as far as people having a real say in decision-making (Daneke *et al.*, 1983; Singer, 1984; Burdge and Robertson, 1990). There may be little tradition of interdisciplinary study, rivalry between ministries and agencies, unequal distribution of political power and poor resources (Fu-Keung Ip, 1990). For a review of the logistic and cultural difficulties confronting social impact assessment in developing countries, *see* Finsterbusch *et al.*, 1990: 71–87).

The Philippines has formal social impact assessment regulations, and a number of developing countries have revised their environmental impact assessment procedures to include assessment of social impacts. As social impact assessment, like its environmental counterpart, evolved first in the USA and Canada, it must be adapted to the needs and circumstances of developing countries. The differences between developed and developing countries' traditions, conditions and levels to which the public has been educated are often stressed, but there are many parallels, and social impact assessment experience gained in developed countries may be relevant or offer developing countries a substantial foundation on which to build, and vice versa. For example, Burdge (1990) noted the relevance of social impact assessment conducted in the Appalachian coalfields of the USA when applied to native peoples in the USA and Canada.

Reviewing over 100 US Agency for International Development projects (in effect, carrying out a post-social impact assessment audit), Finsterbusch and VanWicklin (1989: 76) found the following recurrerent faults:

- The original objectives of a project were seldom achieved.
- Project schedules were frequently revised as deadlines were missed.
- Project outputs are often not used as intended.
- Projects are often not adequately maintained and outputs not sustained.

- Design documents have inadequate social analyses and there are often false claims on the part of aid recipients of commitment to maintenance and sustainability.

Social impact assessment and native peoples

Rapid resource exploitation, forest clearance, the spreading railways, roads and airstrips, agricultural projects and settlement by colonists have taken place around the world, seldom following any consultation with the native population. Social impact assessment could be more widely applied to communities affected by resource development (Charest, 1995; Rickson *et al.*, 1995). Many groups have suffered impacts with little documentation of their plight; the selection following is therefore far from complete.

New Zealand's native peoples

Rights of the Maori were first established by the 1840 Treaty of Waitangi, but were often not honoured. Social impact assessment has been used to help try to ensure that Maori rights are not infringed by development. In practice there can be difficulties for social impact assessment teams if they are unfamiliar with Maori culture and beliefs. For example, the moving of dwellings or facilities is not straightforward, as the sites, rivers and other resources often have important cultural, and still more so religious, values that must be considered (Nottingham, 1990). Social impact analysis has been used to develop healthcare provisions for the Maori (Association for Social Assessment, 1994).

Australian native peoples

A large literature has grown up relating to resource development conflicts including those relating to mining, landholdings, nuclear tests, etc. (Australian Institute of Aboriginal Studies, 1984; Ross, 1990a). Howitt (1989, 1993) discussed the problems associated with mining and expressed concern that if local people were not empowered and involved there would be only 'technical' (meaning cosmetic) social impact analysis which could ignore the Aborigines' perspective and reduce the chances for them to exercise citizen veto. The question of empowerment of Aborigines through social impact assessment has also been examined by Gagnon *et al.* (1993). Ross (1990b) observed that in the past, social impact assessment relating to Aborigines has relied too much on ethnographic data and too little on actually consulting and involving them (which may not always be straightforward for cultural reasons – such as reluctance to discuss real feelings with outsiders; Chase, 1990).

Amazonian native peoples

Since the 1940s the plight of Amazonian native peoples has been better-documented, although mostly in the form of retrospective assessment. Nations with Amazonian territory have established agencies to look after the interests of native peoples, with varying success, and a number of non-governmental organizations are also active. Government agencies and consortia embarking on large developments have carried out some social studies, usually too late for their plans to be varied, and thus not strictly social impact assessment (Goodland, 1978; Aspelin and Coelho dos Santos, 1981). Some of the Amerindian groups are still little known, which makes assessment of social impacts difficult.

North American native peoples

Inuit and Eskimo rights in Alaska were improved by the 1971 Alaska Native Claims Settlement, and subsequently social impact assessment has been applied more often and with more effort to ascertain the views and needs of local peoples and threats to their lifestyles or opportunities. One of the earliest social impact assessment was conducted in 1973 to explore changes to Inuit society and culture likely to be caused by the Alaskan Pipeline from Prudhoe Bay to Prince William Sound (Burdge and Vanclay, 1996: 62). From 1970 the National Environmental Policy Act in the USA, and from 1973 the Environmental Assessment Review Process in Canada, helped stimulate the development and application of social impact assessment, because both these legislation and review processes required assessment of social impacts. The seminal Berger Inquiry of 1974–7, chaired by Mr Justice Berger, into the likely impacts of hydrocarbon exploitation plans for the Mackenzie Bay region and Mackenzie River Valley of British Columbia Province, northern Canada, led to a 10-year moratorium on the building of the proposed pipeline through the Mackenzie Valley (Gray and Gray, 1977; Gamble, 1978; Berger, 1977, 1983; Derman and Whiteford, 1985: 119–37; Burdge, 1994a: 4). It also involved some of the earliest pre-project social impact assessment, and helped stimulate social impact assessment in Canada, stressed the needs of native peoples and aided the acceptance of the concept of local public participation in resources management (Graham Smith, 1993: 108; Burdge and Vanclay, 1996: 63). In northern Canada, oil exploitation in the Beaufort Sea, involving a pipeline from the coast in Yukon Territory to Edmonton, Alberta, had socio-economic impacts on the Inuit and was subject to a social impact assessment and an environmental review from 1980 to 1984 (for an introduction to social and environmental impact assessment in the Beaufort Sea, *see Social Impact Assessment* 1983: **85–86,** 2–19). The James Bay Projects also greatly affected native peoples (Waldron, 1984).

There has been increasing interest in the application of social impact assessment to native peoples in North America. Most of this has focused on the impact of natural resources development, especially minerals and water exploitation (Geisler *et al.*, 1982; Waldron, 1984; Shapcott, 1989). In 1985 a precedent was set when the Northern Cheyenne sued the US Department of the Interior over an environmental impact statement relating to a federal coal lease sale because it had included little or no social impact assessment to determine the effects on the tribe.

So far a Mexico is concerned, the social impacts of development on native peoples have been considerable, as has been the case in most Central and

South American countries, but there have been few accessible published social impact studies and virtually no social impact assessment. Between the late 1940s and mid-1950s river basin commissions modelled on the USA's Tennessee Valley Authority brought a number of Mexico's native peoples into contact with development, and some social impact studies were carried out in advance of implementation (these were essentially social impact assessment). For example, the Papaloapan Project (initiated in 1947 by President Alemán) used 'social impact assessment' before the relocation of over 20 000 Mazatec Indians. Unfortunately the efforts were not especially successful, one reason being that administrators were not accountable to the people and so saw no need to heed the findings of social impact assessment – something that is still likely to be a widespread problem in developing countries (Derman and Whiteford, 1985: 57–60).

SOCIAL IMPACT ASSESSMENT AND CONSERVATION AREAS

The creation of national parks and reserves has generated a social impact assessment literature that has helped conservationists realize the vital need to consult and involve local people (e.g. Wolf, 1979; Rao and Geisler, 1990; Hough, 1991; Rickson, 1991; Peters, 1994).

SOCIAL IMPACT ASSESSMENT AND COMMUNICATIONS

Physical communications – roads, railways, air transport, etc., and media communications – telecommunications, news publications, television, radio and satellite broadcasting – generate social and economic impacts.

Road communications-related social impact assessment may be subdivided into that concerned with urban and that dealing with rural or inter-city links. Literature on highway impacts in the USA has been growing since the 1970s (Llewellyn, 1976). In urban areas negative impacts include noise, pollution, division of communities, altered land values and sense of identity; positive impacts include relief of congestion associated with new public transport and bypass roads, and better access to services, etc. Rural communication changes can result in altered employment and agricultural opportunities, settlement growth, health impacts, in- or out-migration, altered crime patterns, and changes in community cohesion and sense of identity. Impacts are greatest where isolated areas are opened up by communications (Snow, 1984). Air transport developments may trigger new employment, often in tourism.

Developments in telecommunications have attracted social impact assessment studies; for example, Dutton (1992) has examined the likely impacts of telephone caller identification technology, and Davis (1993) assessed the social impact of portable telephones in Hawaii (which presumably is not too dissimilar from other parts of the world). Rosenberg (1994) and others have looked at the socio-economic impacts of personal computers, and Schwab (1995) has considered the spread of the Internet 'information highway'. The impact of television has had considerable attention from social impact assessment and technology assessment attention (*see* Box 8.2).

Box 8.2 Social impacts of television

1st-order impact: People have a new source of entertainment in their homes.
2nd-order impact: People stay at home more and do not mix in the community.
3rd-order impacts:

- reduced community cohesion, less chance of uniting on issues of mutual advantage; ignorance of problems in other households or the community; reduced movement beyond house ('couch potatoes');
- damaged physical health in some cases owing to lack of exercise and late-night viewing;
- stress within households, in part because of isolation – psychological problems;
- schoolchildren underachieve owing to excessive TV-watching;
- there is a possibility that TV triggers antisocial behaviour, violence, deviant behaviour etc. (by no means proven).

There may be other impacts, part of the above chain of causation or separate:

- access to information and teaching (Open University and other distance learning, health and agricultural advice and weather warnings, especially valuable in less developed countries) and opportunity for personal advancement;
- periods of intense electricity consumption in countries with widespread TV;
- increased 'consumerism' as a consequence of exposure to advertising?;
- with satellite broadcasting, a possibility that households have access to programmes the government would rather they had not (pornography, material likely to incite unrest, etc.);
- there is a possibility that TV can be used by government or other bodies (e.g. religious broadcasting) to try to 'control minds';
- TV may entertain people and reduce crime driven by boredom.

Source: Various sources, including information in Coates (1971).

Social impact assessment relating to relocation, resettlement and migration

Resettlement, whether voluntary or enforced, has caused marked social impacts and has generated a growing social impact assessment literature (Hansen and Oliver-Smith, 1982; Cernea, 1988). Much of this relocation has resulted from large projects. In the future there is a possibility that global environmental change will become another driving force creating 'eco-refugees' (*see also* Chapter 9).

Studies should extend from before the point at which news of dislocation is released well beyond relocation and, if need be, go on for some years as impacts often continue long after resettlement.

Social impact assessment and aid

A lead was taken by US aid agencies as a consequence of post-National Environmental Policy Act legislation reinforced by group action litigation. For example, the US Agency for International Development issued guidelines for aid-related 'social soundness analysis' in 1975 (*see also* USAID, 1993). From the late 1970s most large US-funded developments have had some form of social impact assessment or social soundness assessment. In 1980 the World Bank issued guidelines for social aspects of project appraisal (World Bank, 1980; Ingersoll *et al.*, 1981; Hansen, 1985). Critics of aid programmes such as Hayter (1989) have stimulated interest in the environmental and social impact assessment of aid efforts and proposed aid. In practice, aid-related impact assessment still seems to have quite a low profile: as recently as the late 1980s, even a book on the 'greening' of aid devoted only 2 out of 302 pages to environmental and social impact assessment (Conroy and Litvinoff, 1988).

Social impact assessment and health

Social impact has been applied to a wide range of health issues in both developed and developing countries, and the literature is expanding fast, for example, in relation to HIV/AIDS (Palumb, 1993; Danziger, 1994; Keogh *et al.*, 1994), ageing and dependent populations (at present a problem in some developed countries, with the indications that it will have serious impact in a number of developing countries soon; Restrepo and Rozenthal, 1994), migraine, cancer, heart disease, tobacco consumption, illicit narcotics, medical treatment innovations, new pharmaceuticals, etc. Impact assessment in relation to health is considered in Chapter 9.

Social impact assessment and technology change

Technology innovation and change is a field where social impact assessment has been applied in both developing and developed countries (ICPE and INSTRAW, 1993). Technology affects employment opportunities, habits and quality of life, and even apparently minor changes may have far-reaching social and economic impacts. For example, few in the 1960s would have foreseen the impact of personal computers or the fax machine; the latter has had a huge impact on the use of telex.

Social impact assessment and biotechnology

Biotechnology has had and will probably have huge future impacts on a wide range of human affairs. The critical sectors where it is likely to have most effect are food and commodity production, health care and pollution control. Most people are at present relatively poorly informed about biotechnology, and this has been manifest in various scares or inertia over serious issues, rather than informed reaction. Social impact assessment can help predict public reaction and suggest how to inform people about proposals. There has been some application of social impact assessment to biotechnology innovations (e.g. Ruivenkamp, 1987; Hindmarsh, 1990), and this is likely to expand in the future.

SOCIAL IMPACT ASSESSMENT AND HAZARDS OR DISASTERS

Social impact assessment has been applied to human-induced 'accidents' and natural hazards and disasters, particularly those that are recurrent such as floods in Bangladesh, tropical storms, droughts, tornadoes, avalanches. Some hazards result from human action: industrial threats and accidents, military activity (landmines are a widespread problem), re-entry of space debris, etc. Although social impact assessment has been applied to hazards and risks, hazard and risk assessment have also begun to focus on social impacts.

ESTABLISHING WHETHER SOCIAL IMPACT ASSESSMENT WORKS AND IS WORTH THE COST

There is some overlap between post-environmental impact assessment and post-social impact assessment audits; however, the focus is more often on environmental impact assessment. Wenner (1988) looked at the effectiveness of social impact assessment applied to oil and natural gas, minerals and forestry development. Social impact assessment for a large development is likely to make up far less than 10 per cent of total costs and, given that social problems often overshadow development and might be avoided or reduced through assessment, is likely to be money well spent. Finsterbusch (1995) was firmly convinced that in the USA, social impact assessment was cost-effective and valuable. However, he stressed that there is no such thing as a 'complete' or 'accurate' social impact assessment, noting that social science has failed to predict most of the important human changes of the twentieth century! The reason for this failure of prediction, he argued, is that social change is complex, and further complicated because people interact and adapt as change occurs (Finsterbusch, 1995: 230). Also, social impact assessment is often unheeded. Rakowski (1995) felt that in addition to production of a social environmental impact statement there were more subtle social impact assessment outcomes, such as improvement in public debate. Rakowski (1995) stressed the need to consider these more subtle effects of social impact assessment even if immediate adoption of recommendations fails.

SOCIAL IMPACT ASSESSMENT: PROBLEMS AND FUTURE CHALLENGES

Burdge and Vanclay (1996: 66–74), in an excellent review of social impact assessment, recognized four major categories of problems: first, difficulties in applying the social sciences to it; second, difficulties in the process itself; third, problems with procedures for applying social impact assessment; and fourth, a prevailing 'anti-social impact assessment mentality'.

Like environmental impact assessment, social impact assessment faces the problem of cumulative impacts, and must develop better ways of dealing with these. It must also be integrated more closely with planning and management.

It has proved difficult to establish common units to allow the comparison of social impacts. One means of doing this which has been tried is to identify the impacted group(s) and to establish what percentage of utility the group(s) could obtain with the last 1 per cent of their income, that is, asking what they would be prepared to pay to avoid an impact. Finsterbusch *et al.* (1990: 3, 10) noted that the social equivalents of 'carrying capacity' or 'pollution' have not yet been adequately formulated.

In their opinion (Finsterbusch *et al.*, 1990: 10), social costs are roughly equivalent to pollution and environmental degradation, and so 'social polluters' should be made to pay those who suffer from their conduct – a social-polluter-pays principle. Gagnon (1995) critically examined the integration of social impact assessment into the regional planning process, drawing on experience in Quebec, to give local control and improve chances of sustainable development.

Disciplines such as sociology, ecology and geography, which are relevant to environmental and social impact assessment, are often marginal to planning and administration, which are likely to be in the hands of economists, engineers, agronomists, etc. Although there has been some decline of interest in the application of social impact assessment since the late 1970s, there are indications that it will be increasingly applied to hazardous waste management (Finsterbusch, 1995: 238).

There have been calls for social, like environmental, impact assessment, to adopt a 'continuous', rather than a 'snapshot', approach (Geisler, 1993). The application of computing to social impact assessment may help resolve some of the latter's shortcomings (Leistritz and Murdoch, 1981b; Leistritz *et al.*, 1994).

There may well be increased use of participatory and more politically aware social impact assessment which considers strategic perspectives (Dale and Lane, 1994). Howitt (1995) called for social impact assessment to involve the public or local community in development; that is, for participation. However, social impact assessment alone is not likely to empower local people; it must be woven into planning and management (Gagnon *et al.*, 1993). Burdge and Vanclay (1996: 66) suggested that the social sciences tend to be 'critical and discursive' rather than 'predictive and explanatory' – and this fails to assist social impact assessment. In common with environmental impact assessment, there is need for social impact assessment to improve how it deals with cumulative impacts; to achieve better public involvement; to integrate more closely with planning and management; and to adopt a more adaptive and ongoing (and less of a 'snapshot') approach.

FURTHER READING

Social impact assessment journals and journals that publish social impact assessment articles:

Social Impact Assessment (Box 587, Canal Street Station, New York 10013)
Impact Assessment Bulletin (became *Impact Assessment* in 1994)
Environmental Impact Assessment Review
Environment and Behaviour
Project Appraisal
Human Organization
Society and Natural Resources
Community Development Journal
Rural Sociology
Technological Forecasting and Social Change
American Anthropologist

Social impact assessment handbooks, guidebooks and reviews

Branch, K., Hooper, D.A., Thompson, J. and **Creighton, J.C.** 1984: *Guide to social impact assessment: a framework for assessing social change*. Boulder, CO: Westview.

Burdge, R.J. 1994: *A conceptual approach to social impact assessment: collection of writings by Rabel J. Burdge and colleagues*. Middleton, WI: Social Ecology Press.

Burkhardt, D.F. and **Ithelson, W.H.** (eds) 1978: *Environmental assessment of socioeconomic systems*. New York: Plenum.

Burtchell, R.W. and **Listokin, D.** 1972: *The environmental impact assessment handbook*. New Brunswick, NJ: Center for Urban Policy Research.

Canter, L.W., Atkinson, B. and **Leistritz, F.L.** 1985: *Impact of growth: a guide for socio-economic impact assessment and planning*. Chelsea, MI: Lewis Publishers.

Finsterbusch, K. 1980: *Understanding social impacts: assessing the effects of public projects*. Beverly Hills: Sage.

Finsterbusch, K, Ingersol, J. and **Llewellyn, L.** (eds) 1990: *Methods for social analysis in developing countries*. Boulder, CO: Westview.

Finsterbusch, K., Llewellyn, L.G. and **Wolf, C.P.** (eds) 1983: *Social impact assessment methods*. Beverly Hills: Sage.

Finsterbusch, K. and **Wolf, C.P.** (eds) 1977: *Methodology of social impact assessment*. Stroudsburg, PA: Dowden, Hutchinson & Ross. (A 2nd edition appeared in 1981.)

Fitzsimmons, S.J., Stuart, L.I. and **Wolf, P.C.** 1978: *Social assessment manual: a guide to the preparation of the social well-being account for planning water resource projects*. Boulder, CO: Westview.

Halstead, J.N., Chase, R.A., Murdoch, S.H. and **Leistritz, F.L.** 1984: *Socioeconomic impact management*. Boulder, CO: Westview.

Leistritz, F.L. and **Murdoch, S.H.** 1981: *The socio-economic impact of resource development: methods of assessment*. Boulder, CO: Westview.

McEvoy, J. III and **Dietz, T.** (eds) 1977: *Handbook for environmental planning: the social consequences of environmental change*. New York: Wiley.

Morris, P. 1977: *Guidelines for social impact assessment methodology*. Calgary: Petro-Canada.

Soderstrom, E.J. 1981: *Social impact assessment: experimental methods and approaches*. New York: Praeger.

Taylor, C., Nicholas, C., Hobson, B. and **Goodrich, C.C.** 1990: *Social assessment: theory, process and techniques*. Studies in Resource Management 7. Lincoln, New Zealand: Lincoln University, Centre for Resource Management (PO Box 56).

Tester, F.J. and **Mykes, W.** (eds) 1981: *Social impact assessment: theory, methods and practice*. Calgary: Detselig Enterprises.

Theys, J. 1978: *Environmental assessment of socioeconomic systems*. New York: Plenum.

Vanclay, F. and **Bronstein, D.A.** (eds) 1995: *Environmental and social impact assessment*. Chichester: Wiley (chapters 2, 6, 7 especially).

Vlachos, E., Buckley, W. and **Filstead, W.J.** 1975: *Social impact assessment: an overview*. Boulder, CO: Colorado State University.

Waldron, J.B. 1984: Native people and social impact assessment in Canada. *Impact Assessment Bulletin* **3(2)**, 56–62.

Wildman, P.H. and **Baxter, G.B.** 1985: *The social assessment handbook: how to assess and evaluate the social impact of resource development on local communities*. Sydney: Social Impact Publications.

Social impact assessment bibliographies

Carley, M.J. and **Bustelo, E.S.** 1984: *Social impact assessment and monitoring: cross-disciplinary guide to the literature*. Boulder, CO: Westview.

Carley, M.J. and **Derow, E.O.** 1980: *Social impact assessment: a cross-disiplinary guide to the literature*. Policy Studies Research Institute Research Paper. London: Policy Studies Institute.

Leistritz, F.L. and **Ekstrom, B.L.** 1986: *Social impact assessment and management: an annotated bibliography*. New York: Garland.

Shields, M.A. 1974: *Social impact assessment: an annotated bibliography*. Fort Belvoir, VA: US Army Engineer Institute for Water Resources.

Wolf, C.P. 1979: *Quality of life, concept and measurement: a preliminary bibliography*. Monticello, IL. PO Box 229, Illinois 61856.

REFERENCES

ADB 1991: *Guidelines for social analysis of development projects*. Manila: Asian Development Bank.

Albrecht, S.L. and **Thompson, J.G.** 1988: The place of attitudes and perceptions in social impact assessment. *Society and Natural Resources* **1(1)**, 69–80.

Armour, A. 1988: Methodological problems in social impact assessment. *Environmental Impact Assessment Review* **8(3)**, 249–65.

Aspelin, P.L. and **Coelho dos Santos, S.** 1981: *Indian areas threatened by hydro-electric projects in Brazil*. IWGIA Document 44. Copenhagen: International Work Group for Indigenous Affairs.

Association for Social Assessment 1994: Social assessment and Maori policy development. *Social Impact Assessment Newsletter* **35**, 10–11.

Australian Institute of Aboriginal Studies 1984: *Aborigines and uranium*. Canberra: Australian Government Publishing Service.

Azqueta, D. 1992: Social project appraisal and environmental impact assessment: a necessary but complicated theoretical bridge. *Development Policy Review* **10(3)**, 255–70.

Becker, H.A. 1995: Demographic impact assessment. In Vanclay, F. and Bronstein, D.A.(eds), *Environmental and social impact assessment*. Chichester: Wiley, 141–51.

Beckwith, J.A. 1994: Social impact assessment in Western Australia at a crossroads. *Impact Assessment* **12(2)**, 199–213.

Beekhuis, J.V. 1981: Tourism in the Caribbean: impacts on the economic, social and natural environment. *Ambio* **X(6)**, 325–31.

Berger, T.R. (Mr Justice) 1977: *Northern frontier, northern homeland*. Report of the MacKenzie Valley Pipeline Inquiry, vol. 1. Ottawa: Department of Supply and Services Canada.

Berger, T.R. 1983: Resources development, and human values. *Impact Assessment Bulletin* **2(2)**, 129–47.

Boggs, J.P. 1994: Planning and the law of social impact assessment. *Human Organization* **53(2)**, 167–74.

Bowles, R.T. 1981: *Social impact assessment in small communities: an integrative review of selected literature*. Scarborough, Toronto: Butterworths.

Broady, J.G. 1985: New roles for psychologists in environmental impact assessment. *American Psychologist* **40(9)**, 1057–60.

Brougham, J.E. and **Butler, R.W.** 1976: *The social and cultural impact of tourism: a case study of Sleat, Isle of Skye*. Edinburgh: Scottish Tourist Board.

Buchan, D. and **Rivers, M.J.** 1990: Social impact assessment: development and application in New Zealand. *Impact Assessment Bulletin* **8(4)**, 97–105.

Burdge, R.J. 1987: Social impact assessment and the planning process. *Environmental Impact Assessment Review* **7(2)**, 141–50.

Burdge, R.J. 1989: Utilizing social impact assessment variables in the planning model. *Impact Assessment Bulletin* **8(1–2)**, 85–99.

Burdge, R.J. 1990: The benefits of social impact assessment in third world development. *Environmental Impact Assessment Review* **10(1–2)**, 123–34.

Burdge, R.J. 1994a: *A conceptual approach to social impact assessment: collection of writings by Rabel J. Burdge and colleagues*. Middleton, WI: Social Ecology Press.

Burdge, R.J. 1994b: *A community guide to social impact assessment*. Middleton, WI: Social Ecology Press.

Burdge, R.J., Fricke, P., Finsterbusch, K., Freundenburg, W.R., Gramling, R., Holden, A., Llewellyn, L.G., Petterson, J.S., Thompson, J. and **Williams, G.** 1995: Guidelines and principles for social impact assessment: Interorganizational Committee on Guidelines and Principles for Social Impact Assessment. *Environmental Impact Assessment Review* **15(1)**, 11–43.

Burdge, R.J. and **Robertson, R.A.** 1990: Social impact assessment and the public involvement process. *Environmental Impact Assessment Review* **10(1–2)**, 81–90.

Burdge, R.J. and **Vanclay, F.** 1995: Social impact assessment. In Vanclay, F. and Bronstein, D.A. (eds), *Environmental and social impact assessment*. Chichester: Wiley, 31–65.

Burdge, R.J. and **Vanclay, F.** 1996: Social impact assessment: a contribution to the state-of-the-art series. *Impact Assessment* **14(1)**, 59–86.

Burningham, K. 1995: Attitudes, accounts and impact assessment. *Sociological Review* **43(1)**, 100–22.

Campbell, M.J. 1992: *New technology and rural development: the social impact*. London: Routledge.

Canter, L.W. 1977a: Prediction and assessment of impacts on the cultural environment. In Canter, L.W. (ed.), *Environmental impact assessment*. New York: McGraw-Hill, 153–62.

Canter, L.W. 1977b: Prediction and assessment of impacts on the socio-economic environment. In Canter, L.W. (ed.), *Environmental impact assessment*. New York: McGraw-Hill, 163–72.

Canter, L.W. 1996: *Environmental impact assessment*. 2nd edn. New York: McGraw-Hill.

Carley, M.J. and **Bustelo, E.S.** 1984: *Social impact assessment and monitoring: cross-disciplinary guide to the literature*. Boulder: Westview.

Cernea, M.M. (ed.) 1985: *Putting people first: society and variables in rural development*. New York: Oxford University Press.

Cernea, M.M. 1988: *Involuntary resettlement in development projects: policy guidelines in World Bank financial projects*. World Bank Technical Paper 80. Washington DC: World Bank.

Chamala, S. 1990: Social and environmental impacts of modernization of agriculture in developing countries. *Environmental Impact Assessment Review* **10(1–2)**, 213–19.

Chambers, R. 1969: *Settlement schemes in tropical Africa*. London: Routledge & Kegan Paul.

Charest, P. 1995: Aboriginal alternatives to mega projects and their environmental and social impacts. *Impact Assessment* **13(4)**, 371–86.

Chase, A. 1990: Anthropology and impact assessment: development processes and indigenous interests in Australia. *Environmental Impact Assessment Review* **10(1–2)**, 11–23.

Christensen, K. 1976: *Social impacts of land development: an initial approach for estimating environmental impacts*. Washington DC: Urban Institute.

Coates, J.F. 1971: Technology assessment: the benefits . . . the costs . . . the consequences. *Futurist* **5**, 225–31.

Cochrane, G. 1979: *The cultural appraisal of development projects*. New York: Praeger.

Cocklin, C. and **Kelly, B.** 1992: Large-scale energy projects in New Zealand: whither social impact assessment? *Geoforum* **23(1)**, 41–60.

Conroy, C. and **Litvinoff, M.** (eds) 1988: *The greening of aid*. London: Earthscan.

Cramer, J.C., Dietz, T. and **Johnson, R.A.** 1980: Social impact assessment of regional plans: a review of methods and issues and a recommended process. *Policy Science* **12(1)**, 61–82.

Dale, A.P. and **Lane, M.B.** 1994: Strategic perspectives analysis: a procedure for participatory and political social impact assessment. *Society and Natural Resources* **7(3)**, 253–67.

Daneke, G.A., Garcia, M.W. and **Priscoli, J.D.** (eds) 1983: *Public involvement and social impact assessment*. Boulder, CO: Westview.

Danziger, R. 1994: The social impact of HIV/AIDS in developing countries. *Social Science and Medicine* **39(7)**, 905–17.

Davis, D.M. 1993: Social impact of cellular telephone usage in Hawaii. In Savage, J.G. and Wedemeyer, D.J. (eds), *Proceedings of the Pacific Telecommunications Council 15th Annual Conference, 17–20 January 1993, Honolulu* (2 vols). Honolulu, 641–8.

Death, C. 1982: Social impact assessment: a critical tool in land-development planning. *International Social Science Journal* **34(3)**, 441–50.

Derman, W. 1990: Informant interviewing in international social impact assessment. In Finsterbusch, K., Ingersoll, J. and Llewellyn, L. (eds), *Methods for social analysis in developing countries*. Boulder, CO: Westview, 107–26.

Derman, W. and **Whiteford, S.** (eds) 1985: *Social impact analysis and development planning in the third world*. Boulder, CO: Westview.

Dietz, T. 1987: Theory and method in social impact assessment. *Sociological Inquiry* **57**(winter), 55–69.

Dutton, W.H. 1992: The social impact of emerging telephone services. *Telecommunications Policy* **16(5)**, 377–87.

Egna, H.S. 1995: Psychological distress as a factor in environmental impact assessment: some methods and ideas for quantifying this intangible intangible. *Environmental Impact Assessment Review* **15(2)**, 137–55.

England, J.L. and **Albrecht, S.L.** 1984: Boom towns and social disruption. *Rural Sociology* **49**, 230–46.

Ernst, J.S. 1994: *Whose utility?: the social impact of public utility privatization and regulation in Britain*. Milton Keynes: Open University Press.

Finsterbusch, K. 1984: Social impact assessment as a policy science methodology. *Impact Assessment Bulletin* **3(2)**, 37–43.

Finsterbusch, K. 1985: State of the art in social impact assessment. *Environment and Behaviour* **17(2)**, 193–221.

Finsterbusch, K. 1995: In praise of SIA: a personal review of the field of social impact assessment: feasibility, justification, history, methods, issues. *Impact Assessment* **13(3)**, 229–52.

Finsterbusch, K., Ingersol, L. J. and **Llewellyn, L.G.** (eds) 1990: *Methods for social impact analysis in developing countries*. Boulder, CO: Westview.

Finsterbusch, K. and **VanWicklin, W.A. III** 1989: Project problems and shortfalls: the need for social impact assessment in AID projects. In Bartlett, R.V. (ed.), *Policy through impact assessment: institutionalized analysis as a policy strategy*. New York: Greenwood Press, 73–84.

Finsterbusch, K. and **Wolf, C.P.** (eds) 1977: *Methodology of social impact assessment*. Stroudsburg, PA: Dowden, Hutchinson & Ross. (2nd edn. 1981)

Fitzsimmons, J., Stuart, L.I. and **Wolf, P.C.** 1978: *Social assessment manual: a guide to the social well-being account for planning water resource projects*. Boulder, CO: Westview.

Fookes, T.W. 1980: Monitoring social and economic impacts (a New Zealand case study). *Environmental Impact Assessment Review* **1(1)**, 72–6.

Fookes, T.W. 1984: Social monitoring: the Huntley Monitoring Project in retrospect, PhD thesis, University of Waikato (New Zealand), Hamilton.

Freeman, D. and **Frey, R.S.** 1991: A modest proposal for assessing social impacts of natural resource policies. *Journal of Environmental Systems* **20(4)**, 375–404.

Freisema, H.P. and **Culhane, P.J.** 1976: Social impacts and the environmental impact statement process. *Natural Resources Journal* **16**, 339–56.

Freudenburg, W.R. 1986: Social impact assessment. *Annual Review of Sociology* **12**, 451–78.

Freudenburg, W.R. and **Keating, K.M.** 1985: Applying sociology to policy: social science and the environmental impact statement. *Rural Sociology* **50(4)**, 578–605.

Fu-Keung Ip, D. 1990: Difficulties in implementing social impact assessment in China: methodological considerations. *Environmental Impact Assessment Review* **10(1–2)**, 113–22.

Gagnon, C. 1995: Social impact assessment in Quebec: issues and perspectives for sustainable community development. *Impact Assessment* **13(3)**, 273–88.

Gagnon, C., Hirsch, P. and **Howitt, R.** 1993: Can SIA empower communities? *Environmental Impact Assessment Bulletin* **13(4)**, 229–55.

Gale, R.P. 1984: The evolution of social impact assessment: post functionalist view. *Impact Assessment Bulletin* **3(2)**, 27–36.

Gamble, D.J. 1978: The Berger Inquiry: an impact assessment process. *Science* **199(3)**, 946–52.

Geisler, C.C. 1993: Rethinking SIA: why ex-ante research isn't enough. *Society and Natural Resources* **6(4)**, 327–38.

Geisler, C.C., *et al.* (eds) 1982: *Indian SIA: the social impact assessment of rapid resource development on native peoples.* Ann Arbor, MI: University of Michigan, Natural Resource Press.

Glasson, J. and **Heaney, D.** 1993: Socio-economic impacts: the poor relations in British environmental impact statements. *Journal of Environmental Planning and Management* **36(3)**, 335–43.

Goodland, R.J. 1978: Environmental assessment of the Tucuruí Hydroelectric Project, Rio Tocantins, Amazonia. *Survival International Review* **3(1)**, 11–14.

Gow, D.D. 1990: Rapid rural appraisal: social science as investigative journalism. In Finsterbusch, K., Ingersol, L.J. and Llewellyn, L.G. (eds), *Methods for social impact analysis in developing countries*. Boulder, CO: Westview, 142–63.

Graham Smith, L. 1993: *Impact assessment and sustainable resource management*. Harlow: Longman.

Gramling, R. 1992: Employment data and social impact assessment. *Evaluation and Program Planning* **15(3)**, 219–25.

Gramling, R. and **Freudenburg, W.R.** 1992: Opportunity–threat, development, and adaption: toward a comprehensive framework for social impact assessment. *Rural Sociology* **57(2)**, 216–34.

Gray, J.A. and **Gray, P.J.** 1977: The Berger Report: its impact on northern pipelines and decision making in northern development. *Canadian Public Policy* **3(4)**, 509–14.

Green, E.C. 1986: *Practising development anthropology*. Boulder, CO: Westview.

Griffith, C. 1978: A welfare approach to social impact assessment. *Social Impact Assessment* **34**, 9–12.

Hansen, A. and **Oliver-Smith, A.** (eds) 1982: *Involuntary migration and resettlement: the problems and responses of dislocated people*. Boulder, CO: Westview.

Hansen, D.O. 1985: Social soundness analysis: an institutionalized role for social scientists in AID's assistance program. *The Rural Sociologist* **5(1)**, 37–42.

Hayter, T. 1989: *Exploited earth: Britain's aid and the environment*. London: Earthscan.

Henry, R. 1990: Implementing social impact assessment in developing countries: a comparative approach to the structural problem. *Environmental Impact Assessment Review* **10(1–2)**, 91–101.

Hindmarsh, R. 1990: The need for effective assessment: sustainable development and the social impacts of biotechnology in the third world. *Environmental Impact Assessment Review* **10(1–2)**, 195–208.

Hough, J. (ed.) 1991: Social impact assessment: its role in protected area planning and management. In West, P.C. and Brechin, S.R. (eds), *Resident peoples and national parks: social dilemmas and strategies in international conservation*. Tucson, AZ: University of Arizona Press, 274–82.

Howitt, R. 1989: Social impact assessment: lessons from the Australian experience. *Australian Geographer* **20(2)**, 153–66.

Howitt, R. 1993: Social impact assessment as 'applied peoples' geography'. *Australian Geographical Studies* **31(2)**, 127–40.

Howitt, R. 1995: SIA, sustainability, and developmentalist narratives of resource regions. *Impact Assessment* **13(4)**, 387–402.

ICPE and **INSTRAW** 1993: Social impact assessment of investment/acquisition of technology projects in developing countries, with particular reference to the position of women. *Public Enterprise* **13(3–4)**, 239.

Impact Assessment Inc. 1990: *Economic, social, and psychological impact assessment of the 'Exxon Valdez' oil spill*. La Jolla, CA: Impact Assessment Inc.

Ingersoll, J., Sullivan, M. and **Lenkerd, B.** 1981: *Social analysis of AID projects: a review of the experience*. Agency for International Development Report. Washington, DC: Bureau for Science and Technology.

Interorganizational Committee on Guidelines and Principles for Social Impact Assessment 1994: Guidelines and principles for social impact assessment. *Impact Assessment* **12(2)**, 107–52.

Interorganizational Committee on Guidelines and Principles for Social Impact Assessment 1995: Guidelines and principles for social impact assessment. *Environmental Impact Assessment Review* **15(1)**, 11–43.

Jiggins, J. 1995: Development impact assessment: impact assessment of aid projects in non-western countries. In Vanclay, F. and Bronstein, D.A. (eds), *Environmental and social impact assessment*. Chichester: Wiley, 265–281.

Kent, J.A. 1979: *Social resource management guidelines: a 10–step process for social impact assessment*. Billings, MT: USDA Forest Service, Surface Environment and Mining Program.

Keogh, P., Alen, S., Almedal, C. and **Temahagili, B.** 1994: The social impact of HIV infected women in Kigali, Rwanda: a perspective study. *Social Science and Medicine* **38(8)**, 1047–53.

King, B.K. 1981: *What is SAM? A layman's guide to social accounting matrices*. World Bank Staff Working Paper 463. Washington DC: World Bank.

Kurian, P.A. 1995: Environmental impact assessment in practice: a gender technique. *Environmental Professional* **17(2)**, 167–78.

Leistritz, F.L., Coon, R.C. and **Hamm, R.R.** 1994: A microcomputer model for assessing socioeconomic impacts of development. *Impact Assessment* **12(4)**, 373–84.

Leistritz, F.L. and **Murdoch, S.H.** 1981a: *The socioeconomic impact of resource development: methods of assessment*. Boulder, CO; Westview.

Leistritz, F.L. and **Murdoch, S.H.** 1981b: Computerized socioeconomic impact assessment models: criteria for evaluation and comparison of selected models. *Social Impact Assessment* **63–64**, 4–16.

Little, R.L. and **Krannich, R.S.** 1978: A model for assessing social impacts of natural resource utilization on resource dependent communities. *Impact Assessment Bulletin* **6(2)**, 21–35.

Llewellyn, L.G. 1976: *Social impact assessment: a sourcebook for highway planners:* vol. 3. *The social impact of urban highways: a review of empirical studies*. Washington DC: US Department of Transportation.

Llewellyn, L.G. and **Freudenburg, W.R.** 1989: Legal requirements for social impact assessment: assessing the social-science fallout from Three Mile Island. *Society and Natural Resources* **2(3)**, 193–208.

McDonald, G.-T. 1990: Regional economic and social impact assessment. *Environmental Impact Assessment Review* **10(1–2)**, 25–36.

McKee, M. and **Sassi, F.** 1995: Gambling with the nation's health? The social impact of the National Lottery needs to be researched. *British Medical Journal* **311(7004)**, 521–2.

Maclaren, V.W. 1987: The use of social surveys in environmental impact assessment. *Environmental Impact Assessment Review* **7(4)**, 363–75.

Marsden, D.M. and **Oakley, P.** (eds) 1990: *Evaluating social development projects*. Oxfam Development Guidelines 5. Oxford: Oxfam.

Marsden, D.M., Oakley, P. and **Pratt, B.** (eds) 1994: *Measuring the process: guidelines for evaluating social development*. Oxford: Intrak.

Moss, T.H. and **Stills, D.L.** (eds) 1981: Three Mile Island nuclear accident: lessons and implications. *Annals of the New York Academy of Sciences* **365**. New York: New York Academy of Sciences.

Murdoch, S.H., Leistritz, F.L. and **Hamm, R.R.** 1986: The state of socioeconomic impact analysis in the United States: an examination of existing evidence, limitations and opportunities for alternative futures. *Impact Assessment Bulletin* **4(3–4)**, 101–32.

Muth, R.M. and **Lee, R.G.** 1986: Social impact assessment in natural resource decision making: toward a structural paradigm. *Impact Assessment Bulletin* **4(3–4)**, 170–83.

Nottingham, I. 1990: Social impact reporting: a Maori perspective – the Taharoa case. *Environmental Impact Assessment Review* **10(1–2)**, 175–84.

ODA 1993a: *Social development handbook*. London: Overseas Development Administration.

ODA 1993b: *Project evaluation: a guide for NGOs*. London: Overseas Development Administration.

Palinkas, L.A., Harris, B.M. and **Petterson, J.S.** 1995: *A systems approach to social impact assessment: two Alaskan case studies*. Boulder, CO: Westview.

Palumb, D.R. 1993: The social impact of AIDS: facing the challenges. *American Journal of Pharmacy and the Sciences Supporting Public Health* **165**, 29–34.

Partridge, W.L. (ed.) 1984: *Training manual in development anthropology*. Washington DC: American Anthropological Association.

Pearce, D.W. 1978: *The valuation of social costs*. London: Macmillan.

Pellizzoni, L. 1992: Sociological aspects of environmental impact assessment. In Colombo, A.G. (ed.), *Environmental impact assessment 1992*, vol. 1. Dordrecht: Kluwer, 313–34.

Peters, D. 1994: Social impact assessment of the Ranomafan National Park Project of Madagascar. *Impact Assessment* **12(4)**, 385–408.

Pinhero, P. and **Pires, A.R.** 1991: Social impact analysis in environmental impact assessment: a Portuguese case study. *Project Appraisal* **6(1)**, 2.

Power Station Impacts Research Team 1979: *The socio-economic effects of power stations on their locations*. Oxford: Oxford Polytechnic, Department of Town Planning.

Putt, B. and **Buchan, D.** 1987: *Project wind-down: an experience in community consultation*. Wellington: Town and Country Planning Directorate, Ministry of Works and Development.

Rakowski, C.A. 1995: Evaluating a social impact assessment: short-term and long-term outcomes in a developing country. *Society and Natural Resources* **8(6)**, 525–40.

Rao, K. and **Geisler, C.** 1990: The social consequences of protected areas development for resident populations. *Society and Natural Resources* **3(1)**, 19–32.

Restrepo, H. and **Rozenthal, M.** 1994: The social impact of aging populations: some major issues. *Social Science and Medicine* **39(9)**, 1323–38.

Rickson, R.E. 1991: Environmental impact assessment in the USSR: current situation. *Impact Assessment Bulletin* **9(3)**, 83–8.

Rickson, R.E., Hundloe, T., McDonald, G.T. and **Burdge, R.J.** (eds) 1990: Special issue: 'Social impact of development: putting theory and methods into practice'. *Environmental Impact Assessment Review* **10(1–2)**, 375.

Rickson, R.E., Lane, M., Lynch-Blosse, M., and **Western, J.S.** 1995: Community, environment, and development: impact assessment in resource-dependent communities. *Impact Assessment* **13(4)**, 347–70.

Rickson, R.E., Western, J.S. and **Burdge, R.J.** 1990b: Social impact assessment: knowledge and development. *Environmental Impact Assessment Review* **10(1–2)**, 1–10.

Rosenberg, R.S. 1994: *The social impact of computers*. 2nd edn. (1st edn 1992) Boston: Academic Press.

Ross, H. 1990a: Progress and prospects in aboriginal social impact assessment. *Australian Aboriginal Studies* **1**, 11–17.

Ross, H. 1990b: Community social impact assessment: a framework for indigenous peoples. *Environmental Impact Assessment Review* **10(1–2)**, 185–295.

Ruivenkamp, G. 1987: Social impacts of biotechnology in agriculture and food processing. *Development* **4**, 58–9.

Rydant, A.L. 1984: A methodology for successful public involvement. *Social Impact Assessment* **96–98**, 4–9.

Schwab, U. 1995: The information highway and its social impact [in German]. *Nachrichten für Dokumentation* **46(5)**, 293–301.

Scott, A. 1978: *The social impact of large scale industrial developments: a literature review*. Glasgow: Planning Exchange.

Scott, A. 1981: *The social impact of large scale industrial developments*. London: Social Science Research Council.

Seelman, K.D. 1983: Social impacts of the AT&T disinvesture: a preliminary assessment. *Social Impact Assessment* **81–82**, 2–9.

Shapcott, C. 1989: Haida case study: implications for native people of the north. *Canadian Journal of Native Studies* **9(1)**, 55–83.

Shera, W. and **Matsuoka, J.** 1992: Evaluating the impact of resort development on an Hawaiian island: implications for social impact assessment policy and procedures. *Environmental Impact Assessment Review* **12(4)**, 349–62.

Shields, M.A. 1975: Social impact assessment: an expository analysis. *Environment and Behaviour* **7(3)**, 265–84.

Singer, C. 1984: Differential public participation in SIA. *Social Impact Assessment* **90–92**, 5–7.

Snow, R.T. 1984: Pastoral nomads, aid agencies, and international highways: the Turkana and the Kenya–Sudan Road Link. *Social Impact Assessment* **93–95**, 11–15.

Soderstrom, E.J. 1981: *Social impact assessment: experimental methods and approaches*. New York: Praeger.

Suprapto, R.A. 1990: Social impact assessment, and environmental planning: the Indonesian experience. *Impact Assessment Bulletin* **8(1–2)**, 25–8.

UNDP 1991: *Human development report 1991* Oxford: Oxford University Press for United Nations Development Programme.

USAID 1993: *Handbook No. 3: project assistance (appendix 3F – social soundness analysis)*. Washington DC: US Agency for International Development.

USDA 1977: *Social impact assessment: an overview*. Washington: US Department of Agriculture, Forest Service.

Verloo, M. and **Roggeband, C.** 1996: Gender impact assessment: the development of a new instrument in the Netherlands. *Impact Assessment* **14(1)**, 3–20.

Waldron, J.B. 1984: Native people and social impact assessment in Canada. *Impact Assessment Bulletin* **3(2)**, 56–62.

Wenner, L.N. 1988: Social impact analysis in retrospect: was it useful? *Impact Assessment Bulletin* **3(4)**, 11–24.

West, P.C. and **Brechin, S.R.** 1991: *Resident peoples and national parks: social dilemmas and strategies in international conservation*. Tucson, AZ: University of Arizona Press.

Wildman, P. 1990: Methodological and social policy issues in social impact assessment. *Environmental Impact Assessment Review* **10(1–2)**, 69–79.

Wilkie, A.S. and **Cain, H.R.** 1977: Social impact assessment under NEPA: the state of the field. *Western Sociological Review* **8**, 105–8.

Wolf, C.P. (ed.) 1974: *Social impact assessment: man–environment interactions*. Stroudsburg, PA: Dowden, Hutchinson & Ross.

Wolf, C.P. 1979: *Quality of life, concept and measurement: a preliminary bibliography*. Monticello, IL: PO Box 229, Illinois 61856.

Wolf, C.P. 1980: Getting social impact assessment into the policy arena. *Environmental Impact Assessment Review* **1(1)**, 27–36.

Wolf, C.P. 1983: What is social impact assessment? *Social Impact Assessment* **83–84**, 9–18.

World Bank 1980: *Human factors in project work*. World Bank Staff Working Papers 397. Washington DC: World Bank.

Yong, Y.W., Keng, K.A. and **Leng, T.L.** 1989: A Delphi forecast for the Singapore tourism industry: future scenario and marketing implications. *International Marketing Review* **6(3)**, 35–46.

9

IMPACT ASSESSMENT IN PRACTICE

INTRODUCTION

To offer some insight into impact assessment practice, authors frequently present detailed case-studies. I have tried to follow a slightly broader approach, selecting sectors that give some indication of impact assessment performance and challenges:

- large dam construction
- health
- road construction
- pipelines
- tourism development
- resettlement, relocation and migration
- industrial development.

It has become easier to get access to environmental impact statements; for example, American initial environmental evaluations are available on the World-Wide Web (contact the Institute for Global Communications +1–415–422–0220; e-mail: support@igc.org), and other bodies are increasingly 'publishing' grey material.

APPLICATION OF IMPACT ASSESSMENT TO SELECTED SECTORS

Construction of large dams

Many large dams have been built since the late 1950s (Obeng, 1978; Barrow, 1981). Environmental and social problems have often overshadowed these dams, even though they may have been engineering, and even economic, 'successes'. Before the mid-1970s most large dam projects escaped impact assessment; nowadays, most would be subjected to an assessment. With decades of hindsight experience, there *should* have been a steady improvement in the reduction of unwanted impacts. In practice, the record has often been disappointing. The cause may be poor impact assessment performance or failure to heed its findings, or both (*see* Fig. 9.1).

During the 1970s and 1980s impact assessment applied to dams generated a critical mass of knowledge. Prior to this, too little was known of the

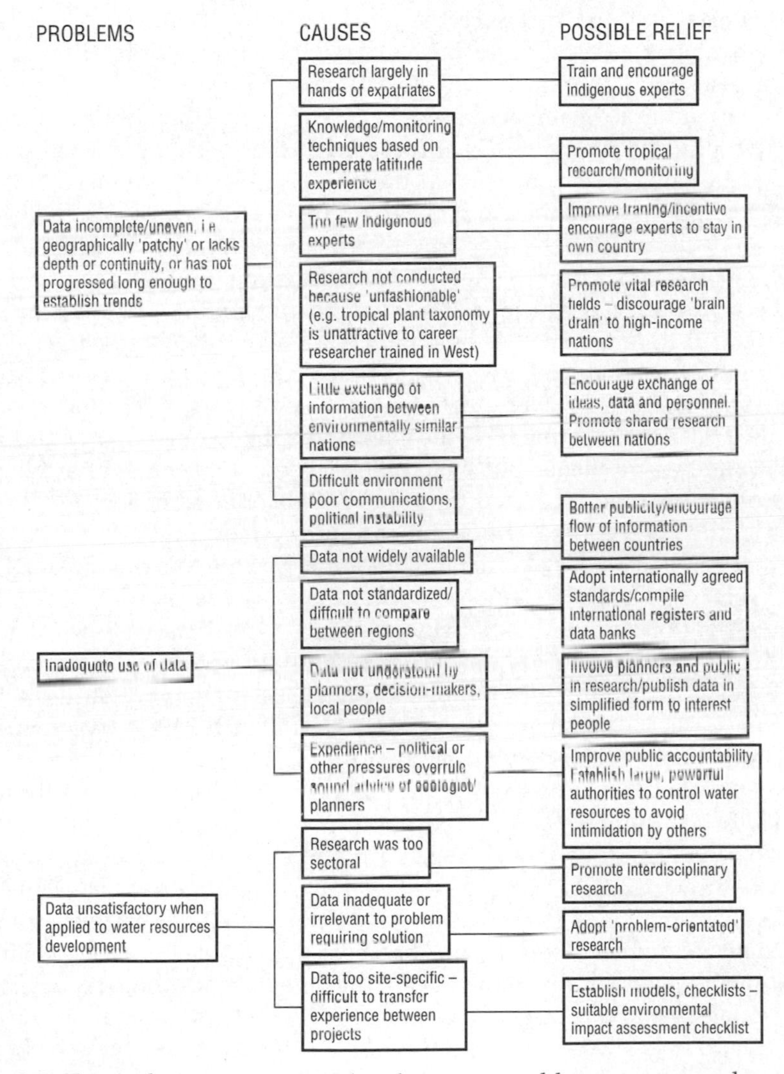

FIGURE 9.1 Tropical water resources development: problems, causes and possible relief.
Source: Redrawn from Barrow (1983: Fig. 13.1)

biological, physical and chemical characteristics of tropical rivers and reservoirs and their surroundings or of socio-economic impacts, to allow reliable environmental impact statements. Not only was little known of tropical environments, but those carrying out assessments were often of 'temperate outlook'; that is, they did not come from tropical environments and consequently were not adequately in tune with them. Environmental and social impact assessment had mainly been undertaken after project implementation, too late for satisfactory responses, even if environmental impact statements were reliable and heeded (Adams, 1992: 128–54 examined why impacts have often not been avoided in Africa). Too many assessments have been descriptive, lacking critical analysis, and impact assessment has often simply been loosely integrated into discussion of future costs and benefits.

The application of impact assessment has been slow, patchy and often poorly conducted. White (1972) observed that the construction of large dams seemed to have spread faster than the ability and willingness to anticipate problems and opportunities, and deal with them satisfactorily. There has been a tendency for dams to be seen as enlightened manipulation of the environment, so it is not surprising that 'problem-shooting' with impact assessment has been neglected. Politicians and other people holding the reins of power have tended to judge success by engineering or economic criteria; a dam is after all essentially an 'engineering approach'. A large dam was seen to meet its design criteria by generating so many megawatts, while the misery of relocated people or the threats to the environment were effectively overlooked. There might be problems, but the politician could expect a supportive press-spectacle as he (or she) opened the dam he had commissioned.

Gradually, bodies like the Organization of American States (OAS, 1978) and that responsible for planning development of the Mekong Basin of Thailand, Vietnam, Laos and Cambodia (Interim Committee for the Coordination of Investigations of the Lower Mekong Basin, 1982), as well as individual researchers, published river development and impact assessment guidelines and articles that improved funding agencies' and planners' awareness of the need for impact assessment (Rees, 1981; Herren *et al.*, 1982; Goldsmith and Hildyard, 1984, 1986a, 1986b; Marchand and Toornstra, 1986; Finney *et al.*, 1988; Atkinson, 1992; Rosenberg *et al.*, 1995). A 'quantum leap', was however, needed from 'cosmetic' or half-hearted impact assessment to adequate studies and their genuine integration with planning and decision-making, and, as part of the process, a better incorporation of social impact assessment.

Non-governmental organizations have protested in various countries over dam-related impacts, and this has helped promote more serious impact assessment. Another thrust has been provided by funding bodies, which have either seen the need or have been required by laws in their home countries to act. In a few cases those providing loans for large dams have withdrawn their support because of inadequate impact assessment, one notable case being the Narmada Project, a proposal to build dams on the Narmada River in north-west India (See Fig. 9.2). By the early 1990s bodies like the

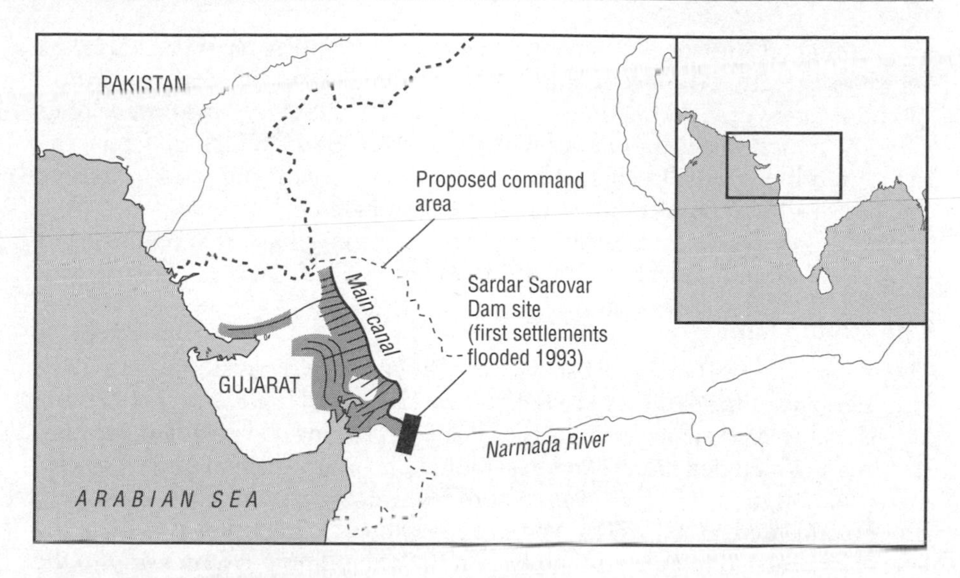

FIGURE 9.2 Narmada Project, India

World Bank were proclaiming new policies on impact assessment of dams (Goodland, 1989, 1990) – a significant step, given that as much as half of the organization's funding in the past few years has gone to large dams. Growing debts on the part of developing countries, protests from non-governmental organizations and local people, and recession in the 1980s have also prompted more caution toward large dams.

There may still be a failure to initiate sufficiently multidiscipinary studies, and assessment tends to take 'snapshot' views starting too late in the planning process. Impact assessment may still fail to consider a wide enough area. Commonly, downstream impacts are ignored, and there may also be failure to check catchments where stream flows originate (vital if silting is to be monitored and controlled).

Why do those who commission impact assessments frequently do it so poorly and why do authorities so often fail to heed or adequately use the results? It may be that one or more of the following causes operate. Development projects may have late-completion penalties or even early-completion bonus payments which encourage rushed assessment, planning and construction. It would be better if funds were withheld as an indemnity against impacts to encourage serious impact assessment early on and compensate for lack of funds to implement environmental impact statement recommendations or cure problems. Governments and those implementing projects may be ashamed to admit failure and so suppress impact assessments, or the development may have a 'juggernaut' quality (Adams, 1992), and be too unwieldy to respond to warnings or problems that appear. It is not unusual for development to be driven by government or commercial pressures or both, and to be pushed to a conclusion that ignores assessment findings. For a discussion of this 'irreversibility' and unwillingness to heed environmental impact assessment, with respect to an Amazonian dam, *see* Fearnside (1989). Another feature of many large dams is that one group (perhaps a few groups) of people benefit and others lose out; the 'losers' tend to be the poor in the vicinity or downstream of the dam and the 'winners' the rich, often living at some distance – and it is the latter who tend to initiate development and then commission and use impact assessments.

The following cases illustrate a number of strengths and weaknesses of large dam-related impact assessment practice.

The Volta Dam

The Volta Dam in Ghana, West Africa, whose reservoir was filled in 1965, although one of the earliest to be built in the 'decade of big dams' (*c.* 1958–68), is reasonably representative of large dams built during the past few decades. Its planning included studies that resembled an environmental impact assessment, although conducted well before environmental impact assessment proper appeared in 1970. The Volta Preparatory Commission planning the Volta Dam carried out a broad study into the likely impacts quite early in the planning process (the Preparatory Commission's report, in some ways similar to an environmental impact statement, was published in the mid-1950s; HMSO, 1956a, 1956b). Although not known as an impact assessment at the time, this study was better than many later impact assessments that have had a narrow focus or have been undertaken too late. Resettlement impacts were foreseen, but not enough was done to mitigate the plight of the *c.* 84 000 relocatees. Consequently, a number of retrospective studies of relocatee impacts were produced (for a bibliography, *see* Barrow, 1981).

Post-impoundment studies of the Volta Dam gave information on relocatees and their resettlement, reservoir fisheries developments, health impacts (especially malaria and schistosomiasis) and downstream impacts which post-1970s impact assessments elsewhere have drawn upon.

The Tucuruí Dam

The Tucuruí Dam, built in the early to mid-1960s, was the first really large dam to be constructed within Brazilian Amazonia, although there have been more recent ones, such as the Balbina (Fearnside, 1989). The motives for building Tucuruí were to reduce oil imports into Brazil; to help satisfy electricity demands in eastern and central southern Brazil without damming the North-East's rivers, which might better serve irrigation; and to stimulate and support mineral exploitation and development in Amazonia. The main environmental impact assessment was conducted during development, 6 or 7 years before completion (taking 1 man-month). The environmental impact statement was rather a 'snapshot' view (Goodland, 1978a, 1978b), although a research staff was established at a national research institution (INPA-Manaus), with a field station at Tucuruí to monitor impacts as the project progressed.

Impact studies were started after the site for the dam had been selected and construction had begun. This and a poor database meant that the planners failed to predict the severity of floods during the first years of dam-building, so the project very nearly suffered a catastrophe during construction. Relevant hindsight experience from Africa, India, North America and areas of Brazil beyond Amazonia was available when the decision was made to construct Tucuruí.

It is clear today that developers should have tried to heed environmental impact assessment warnings about the need for fish-ladders (for migratory species) and the need to protect people and the environment downstream (Barrow, 1987). Heeding the warnings was difficult: fish-ladders were not a very practical option once the dam was under construction as it was so high (Goodland, 1978a).

The danger that land development in surrounding uplands would cause excessive siltation of the reservoir was identified in impact studies, but failed to receive enough attention until it became clear that parts of the Tucuruí Reservoir were in danger. Reservoir-related resettlement proved a fairly sensitive topic, probably because the authorities feared land speculation. There was incomplete data on the Tucuruí region's native peoples in the late 1970s, making social impact assessment difficult. The 1978 impact assessment warned of the localized occurrence of snails with the potential to spread schistosomiasis; the threat is that agricultural development stimulated by the dam might lead to pollution resulting in widespread alkaline conditions favourable to the snails and consequently widespread human schistosomiasis.

One interesting aspect of Tucuruí impact assessment is that it provides an example of a situation where environmental impact assessment, at least in part, triggered impacts. If Eletronorte had not been warned by environmental impact assessment of the negative effects of leaving trees in the area impounded it might not have embarked on a clearing programme which some claim resulted in the use and loss of large quantities of dangerous herbicide (Barrow, 1987).

Tucuruí increased the number of turbines it operated some years after completion of the first phase of the scheme. Clearly, this development has altered downstream flows, reservoir storage characteristics, etc. Unfortunately, the 'snapshot' impact assessment did not focus on the possible consequences of this increased generation, even though it was foreseeable. As more dams are built in the Amazon Basin there will need to be vigilance for cumulative impacts.

The Victoria and Samanalawewa Dams (Mahaweli Development Programme)

The Victoria and Samanalawewa Dams are two of the 14 major dams planned for construction during the 30-year-duration Mahaweli Programme and took 62 per cent of UK aid to Sri Lanka between 1976 and 1989, so by tying up funds had considerable indirect impact on other aspects of development (NAO, 1992: 28). The Victoria Dam was completed in 1985 (its target date), and in 1986 was subjected to an extensive *ex post* evaluation which, among other findings concluded that the planning had been rushed and had given inadequate attention to environmental and social impacts. The retrospective impact assessment seems to have suffered from a poor database and lack of a comprehensive approach (NAO, 1992: 29). Resettlement had not been a success and by 1992 there were worries about the sustainability of the dam and its failure to provide the economic returns hoped for, and about difficulties in finding recurrent funds to operate and maintain it. The dam was also failing to provide as much hydroelectricity as expected (in some years 40 per cent below original estimates) and progress with associated irrigation had been poor (Pearce, 1992a), although what had been achieved was welcome and showed great promise for improved crops and rural employment.

The decision was made to proceed with the Samanalawewa Dam in 1982, and work on construction began in 1986, leading to completion in 1991. Impacts similar to those of the Victoria Dam resulted. Both these dams were funded by foreign aid agencies, including the UK's Overseas Development Administration, and Japan's Japan International Cooperation Agency. UK aid was provided on a 'trade-for-aid' basis; that is, it was tied aid (*see* the Glossary). Funding bodies ensured that there had been environmental impact assessments *ex ante* and more than once during construction, but these either missed problems or were not fully acted upon. These projects have clearly run into a wide range of planning problems: impact assessment is not the only aspect of planning to give less than optimum results.

The Three Gorges

There are huge potential benefits from the China's Three Gorges Project (which involves damming the Yangtze River), such as the flood control, hydropower and navigation improvements associated with one of the component developments, the Shanxia Reservoir (See Fig. 9.3). There are also likely to be considerable resettlement problems, perhaps affecting more than 1.4 million people, and ecological impact problems (La Bounty, 1982; Long Li, 1990; Barber and Ryder, 1993). Some researchers also fear the risk of a dam failure, given the dense human population downstream.

Fearnside (1994) was critical of the pre-construction impact studies, made between 1986 and 1988 by Western specialists brought in by the Chinese government. The assessment has mainly been of value in stock-taking and warning of problems and opportunities, rather than playing an adequate part in the development decision-making process to select optimum approach and inform local people.

Impact assessment and health

A preliminary word of warning: in the medical literature the abbreviation 'EIA' has appeared comparatively recently and has a different meaning to 'environmental impact assessment'. (It means enzyme immunoassay; and unfortunately 'SIA' too has a medical meaning – strip immunoblot assay – as well as its use to mean social impact assessment.) This is unfortunate, and might make database maintenance and literature searching difficult.

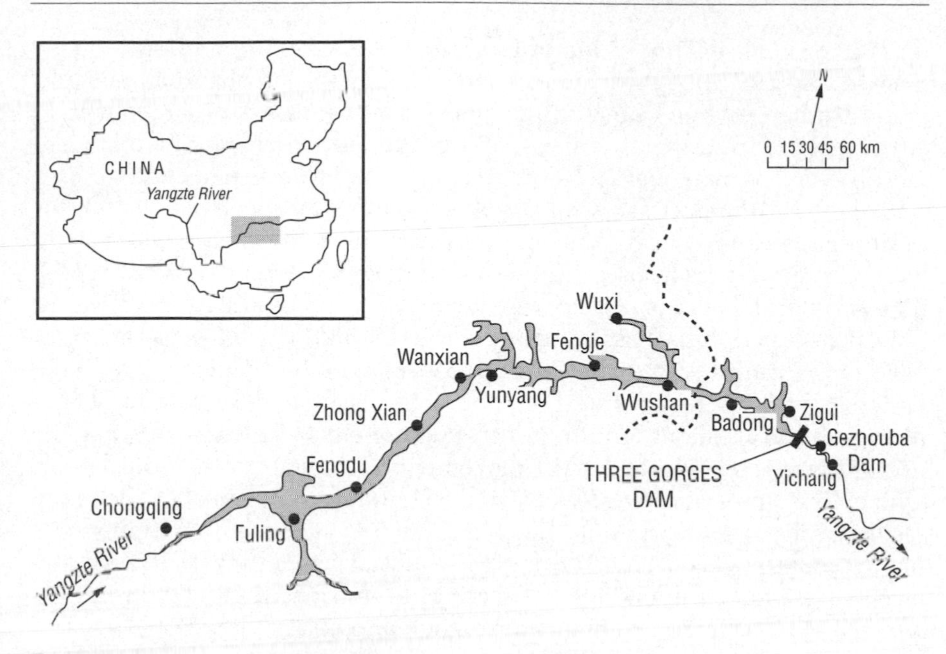

FIGURE 9.3 The Three Gorges Project

There is a large, complex and expanding literature, for example on impact assessment relating to diseases and illness (this aspect can be sub-divided into communicable and non-communicable diseases), use of drugs (legal and illegal), stress and behavioural changes, aged populations, new treatment practices, sensitivity to low levels of pollutants, and many other health fields. Studies may seek to predict positive or negative health impacts triggered by some proposed development including policy or programme changes, using environmental impact assessment, social impact assessment or economic impact assessment, or, alternatively, might review and take stock of the effects of some development that has occurred or a health change already apparent.

Retrospective (*ex post*) impact assessments have been used to assess likely health effects of environmental changes due to natural and human causes (McDowall, 1986; Harvey, 1990; *Environmental Impact Assessment Review* 1994: **14 (5–6)**, a special issue entitled 'Human health and the environment: unanswered questions, unquestioned answers'; *Environmental Impact Assessment Review* 1990: 'Environmental change and public health'), notably the impacts of global climate change on diseases such as malaria (Haines, 1990; Martens *et al.*, 1995); the impacts of stratospheric ozone thinning; the impacts of transboundary pollution; the impacts of radioactive fall out; the impact of electromagnetic pollution associated with electricity transmission and use; the impact of proposed and newly adopted changes in health regulations and treatment practices; and so on.

Health changes can lead to social, economic and biophysical impacts: heart disease, cancer, migraine and back disorders have considerable impact on earnings and quality of life of families, not just of affected individuals; disease may cause people to avoid an area leading to land abandonment and less disturbance of wildlife and the land (as is the case over large areas of Africa because of trypanosomiasis, and in the past when marshlands were left unsettled in many parts of Europe for fear of the ague – malaria).

Some health impact assessment fields are especially well-developed, for example safety and work, disease transmission or illness associated with water contact (e.g. as a consequence of irrigation development; WHO, 1983); illness or disease transmission associated with air quality, especially air pollution-related impact assessment (Simpson, 1990). The latter has an input from long-established military research concerned with predicting and measuring the effects of air transport and impacts of nuclear, chemical and biological and weapons, and from hay fever research. This research may be useful to those seeking to predict movements of fungal spores, aphids, etc., which affect crops and livestock as well as humans (movements which may result from global warming). There is growing interest in environmental impact assessment of trace substances – substances that even at very low levels may affect health. There has recently been concern in the UK, Europe and the USA at falling human sperm counts, possibly as a result of trace substances in the environment. Assessing the impacts of such pollutants will probably require the adoption of a transboundary, ideally global, stance. Stress-related diseases have attracted impact assessment study seeking to predict the effects of cultural disruption, enforced movements, civil unrest, overcrowding, etc. Most diseases have specific vectors, and are affected by human habits and environmental factors (dengue is an example) on which impact assessors and epidemiologists can focus to determine future incidence.

Another very active area of health impact assessment is that focused on agricultural chemicals and activities that affect health, such as the use of hormones on livestock for milk or meat production, or the increased incidence of malaria as a consequence of agrochemical-related pesticide resistance and land use changes. Increased pig production might lead, because of transmission via mosquitoes, to an increase in viral encephalitis. At the time of writing this, there was widespread concern in Europe over the possibility of runaway transmission of BSE ('mad cow disease') to humans; were it to happen, the social and economic costs would be huge. The possibility of the risk of BSE was noted by some before 1989 but there was no impact assessment; even in the absence of firm or complete data, impact assessment could have urged caution and triggered research more quickly. This might well prove to be one of greatest missed opportunities for the beneficial application of environmental impact assessment. Biotechnology, especially genetic engineering is attracting impact assessment studies, some of it *ex ante* risk assessment.

Much of the health impact coverage focuses at the project level (ADB, 1992; Turnbull, 1992; Birley and Peralta, 1995) or deals with a specific disease or illness or causative factor. However, there are studies of the impact of development policies on health (for a review of the literature, *see* Cooper Weil *et al.*, 1990). Since 1982, the World Health Organization has assessed environmental impact assessment and recommended environmental health impact assessments for all large developments (WHO, 1987, 1989).

A number of impact assessment professionals have observed deficiencies in the health coverage of many environmental impact assessments (Giroult, 1988) and fear that there might be duplication of efforts by environmental impact assessment and environmental health impact assessment, so they have called for better integration of health impact assessment into environmental impact assessment (*see* Martin, 1986; Westmore, 1991; Arquiaga *et al.*, 1994). For discussion of human health assessment within environmental impact assessment, *see* Gilad (1984), and a review of health risk assessment methods was provided by Carpenter (1995: 164–6), including discussion of the assessment of risk of cancer as a consequence of human activities, such as pollution from a factory. Health impacts are often indirect and may be the result of cumulative impacts, making it vital that any attempt at impact assessment consider such impacts.

Giroult (1988: 271) noted a problem with health impact assessment: that public statements releasing morbidity or mortality forecasts may be too sensitive, and in some situations likely to cause public unrest.

Social and economic impact of ageing populations

The social and economic impact of ageing populations has become an important issue in some developed countries, where demographic trends (increased survival of the elderly and falling birth rates) have begun to present or promise to cause severe (one author suggested 'stunning') social and economic problems. Some developing countries will soon face this problem, one being the People's Republic of China, where provision for the elderly is causing concern. Housing stock will have to be adapted, public transport, health care, social services and many other aspects of life will need to adapt, and the problem, given its scale, is likely to cause severe stress on taxation and welfare systems. The impacts of ageing populations on other age sectors need to be assessed, and also the impact caused when the ageing population is predominantly female. There may be situations where ageing acts to stabilize or destabilize society, or where the economy is stimulated or held back (Restrepo and Rozenthal, 1994). Countries are likely to react to the challenge of ageing populations in different ways and might be greatly helped in their preparations if social impact assessment could be focused on the impacts of ageing.

Impact assessment and disease transmission

Rapid modern transport between crowded cities and improved local communications within those cities mean the potential for transmission of disease via infected people, pests and goods in vehicles, baggage or freight. The challenges to epidemiologists and public health officials concerned with human, livestock and crop disease control have increased. Impact assessment is needed to predict threats, transmission routes and degrees of resistance to infection, and to develop contingency plans, mitigation or prevention measures. Without impact assessment it is probably unlikely the that Channel Tunnel (between the UK and France) would have been equipped with devices to deter wildlife and stray pets from making the crossing.

An area where impact assessment can be of value is in predicting and taking stock of disease problems associated with specific development activities. There are regions where land development, especially forest clearance, has brought people into contact with unknown or rare diseases that then threaten to spread to more settled areas; examples include Lassa fever, Ebola virus and Marburg virus. Some researchers have speculated that AIDS may originally have 'jumped' from monkeys that came into contact with people settling newly cleared land. In Amazonia, outbreaks of yellow-fever clearly follow a pattern whereby people are exposed to mosquitoes that are infected from a wildlife 'pool' of animals – usually when land is newly settled or cleared. A similar environment development–human contact relationship has developed in parts of North America, such as New York State, where deer act as a 'pool' for Lyme disease, which, via ticks, infects humans moving about in areas once little settled and hunted clear of deer. Chagas' disease, which is caused by the protozoan *Trypanosomiasis cruzi*, is a Latin American form of sleeping sickness that debilitates humans and reduces their life expectancy. There were at least 17 million sufferers in 1990. The disease is spread by species of bedbug associated with certain land uses, communications and housing conditions, and impact assessment could be used to predict risk localities (Butcher and Schofield, 1981; *The Economist*, 5 May 1990). Malaria, schistosomiasis and African sleeping sickness (trypanosomiasis) outbreaks are very much affected by development activities: clearance of new areas, removal of game species, presence of people in areas where they were previously absent or at times of day when they are more vulnerable (for example, in parts of South-East Asia, rubber-tappers often work at night when there is better sap flow). There is a large literature on the relationship between environment and illness or diseases (World Bank, 1974; Stanley and Alpers, 1975; WHO, 1979; Howe and Loraine, 1980). If patterns of disease transmission are understood, impact assessment should be capable of predicting health risks in advance.

Impact assessment of changes in health care

Impact assessment is increasingly being applied to proposed health care changes, such as altered procedures, changes in entitlement, or new drugs and techniques (Chu, 1990; Sutcliffe, 1995). Mitchell and Mitchell (1979) stressed the need for social impact assessment before new health laws or policies were adopted. Edmunds (1989) studied environmental impact assessment's application to Australian health care policy, and Davies (1991) explored its application to health issues in Canada, identifying a number of difficulties and potential applications.

Impact assessment of HIV/AIDS

There is a rapidly growing literature on the impacts of HIV/AIDS on communities. Most work has taken the form of simple retrospective or 'stock-taking' studies rather than being true proactive impact assessments attempting to assess what the health situation will be in the future. Impact assessment could be invaluable, because the problem has spread fast and the normal gathering of statistics may not yield useful insight fast enough. Very high prevalence of HIV is found in some urban areas: Keogh *et al.* (1994) suggest that in Rwanda it affects more than 30 per cent of urban adults, and clearly in the future services will be needed to support the infected and their offspring. There has begun to be some focus on future demographic, agricultural, labour, social and environmental impacts, and attempts are being made to identify priority areas, especially in developing countries (Palumb, 1993; Danziger, 1994).

Health risk assessment

Health-related impact assessment and health risk assessment have undergone a good deal of parallel development and overlap. Health risk assessment concentrates on establishing the probability and nature of human morbidity or mortality as a consequence of development and change with hazard and risk assessment.

Road impacts

Impact assessment relating to road development can be split into assessment prior to route selection, assessment following route selection and assessment after construction. Further subdivision is possible: urban roads, rural roads, inter-urban and strategic highways. Impact assessment experience gained from studies of power transmission lines, pipelines, canals and railways (e.g. Carpenter, 1994) can be relevant to road impact assessment, and vice versa.

Road-related impacts can be divided into:

- those resulting during or relating to construction;
- those resulting from improved communication;
- those due to road use (dust, noise, vibration damage to buildings, etc., air pollution, de-icing salt, road deaths and injuries);
- those caused by a highway's obstruction of access to or division of a residential area, farmland or natural vegetation;
- Those resulting from road construction and constant maintenance using funds that might have been better used for other things – opportunity costs.

Seldom, it seems, are the true overall costs of road communications assessed. There are recurrent costs, opportunity costs and indirect costs, such as pollution and traffic accidents, and perhaps government reluctance to invest in rail or river transport.

Impact assessment and route selection: assessing road construction impacts

Assessing road impacts and weighing communication improvements against ecological and social impacts has generated an expanding literature in both developing and developed countries (Hopkinson *et al.*, 1992). There is a well-established application of overlay techniques and air photographs to road impact assessment, recently reinforced by the use of geographical information systems and computer techniques (Dooley and Newkirk, 1976; *see also* Chapter 5). Checklists have been widely used to assess road impacts (Adkins and Burke, 1971). For a brief introduction to route planning, *see* Doornkamp (1982, 1985), and for road impact assessment; *see* Dooley (1977), Kiravanich *et al.* (1980), Watkins (1981), Kennedy (1983), Wilson and Stonehouse (1983), Dickey and Miller (1984), Standing Advisory Committee on Trunk Road Assessment (1992) and Treweek *et al.* (1993). Routes may be selected on the basis of being the safest, the least disruptive the most direct or a strategically ideal link; or there may be a need to consider availability of construction materials (laterite, crushed rock, etc.).

The application of environmental impact assessment to roads has generated a large literature, much of it retrospective; examples include UNECE's (1987) case-studies of the Banff Highway, Canada (which went through a national park and had social impacts); of Highway 5, Finland (*see also* Gilpin, 1995: 110); of Highway 69, the Netherlands; of Franconnia Notch Highway, USA; of Highway E-18, Norway; and of the Wiesbaden bypass, Germany.

The Darien Gap Highway is a link between the USA and South America lying between Panama and Colombia. Initiated in 1971, it is about 600 km long and joins northern and southern sections of the Pan-American Highway. A crucial 200 km section was the subject of an environmental impact assessment before construction reached that point and this led to a halt in 1975. The environmental impact assessment was prompted by, and explored a concern about, the possible northward movement of plant and animal diseases and pests, especially the cattle disease aftosa (foot-and-mouth disease). There was also a possibility the link would lead to criminal activities. Without the highway, there were natural barriers to the spread of the diseases and pests, and difficulty for criminals, especially drug smugglers, to travel north by land. Construction was delayed until a satisfactory control programme for aftosa was under way in Colombia. The environmental impact assessment also examined threats to forest and to the indigenous Choco and Cuna Indians.

Impacts during or relating to road construction

Road construction generally means noise, dust, traffic from heavy construction vehicles, the presence of a temporary population of construction workers and the possibility of the use of explosives. Where roads pass through remote regions, there may be adverse impacts when contact is made with local peoples. Excavations may benefit science by revealing material of archaeological or palaeontological and geological interest, but if there is no monitoring or advanced warning such evidence may be lost. In developed countries 'rescue digs' have become a well-established activity. There may be cultural impacts where roads encroach on burial sites, temples or other sacred or culturally significant sites. Impact assessment can forewarn of these situations.

Impacts resulting from improved communication

Impacts resulting from improved communication can be very difficult to predict. Negative impacts may include

- contact with isolated people who have little resistance to diseases and who may have their culture damaged;
- movements of people and wildlife via roads or roadside vegetation and ditches into new areas, where they may cause havoc;
- economic changes that may trigger unwanted township development or urban sprawl;
- movement of migrant workers, leaving rural areas or small towns short of intelligent, able-bodied people, consequently damaging farmland and placing established industries under stress;
- drift of young people to cities from rural areas;
- traffic congestion in built-up areas served by the roads;
- attraction of traffic and funds away from rail or water communication routes (for some transport, like the carriage of toxic or explosive materials, roads can be less safe; there may be more unpredictable delays due to traffic congestion; and for heavy goods road transport is more expensive);
- a change in crime patterns as a result of the improved access;
- a change in shopping patterns, with some people abandoning small towns and the centres of larger cities to travel to out-of-town malls or shopping centres.

In respect of this last point, in some parts of the UK people's preferrence for out-of-town malls is now a cause of town centre degeneration, and can disadvantage those without access to a car.

A final point is that road access may make it easier for poachers to to reach wildlife and for settlers to clear land.

Positive impacts include the sometimes profound changes to regional or even national economic development. Indeed, improvements in communications are usually seen as a key part of development strategies. Improvements in communications may have a double impact, in that the areas served enjoy better access and probably economic growth, while those missed find it increasingly difficult to compete (*see* Chung, 1988 for a review of roads in developing countries). The role of local road networks in alleviating poverty has been examined by Howe and Richards (1984), and Brokensha (1980) and Beenhakker (1979) have explored rural road appraisal and planning. The impact of a rural road in Fiji has been examined by Chung (1988), and that of roads in Liberia by AID (1980). The provision of rural roads can have considerable impact on markets: where local roads are good, there may be as much as five times the market activity as in areas with poor communications however, Porter (1995) found negative impacts on communities away from roads, especially upon women. There has been considerable land speculation in Brazilian Amazonia by ranchers who buy property along the route of a planned highway and later enjoy much-improved land values (often returns of over 15 per cent) (*see* Box 9.1).

Roads have often been built for strategic reasons, or to get to and exploit natural resources, as in parts of Amazonia, or to reinforce claims to or control over territory, as in the construction of the Alaskan Highway in the 1940s to aid US troop movements in event of an invasion, the 1970s–1990s building of the Calha Norte Highway in northern Brazilian Amazonia, Amazonian highway construction in general, roads in Kashmir (Kreutzmann, 1991), Nepal (Blaikie *et al.*, 1977), Tibet and Burma, and the West Bank (Palestine) (Grossman and Derman, 1989).

When development is driven by strategic motives – territorial claims or access to important mineral resources – there is a strong likelihood that impact assessments will be neglected, 'controlled' or ignored. Transport development in Amazonia might have been better advised to concentrate on river communications, which would cost less in construction and fuel use, and would allow access to more people with less environmental damage and with less seasonal disruption – but would have less strategic 'presence'.

Box 9.1 Amazonian highways

Since the late 1960s over 14 000 km of highways have been built in Brazilian Amazonia alone. These include the infamous 'Transamazonica', started in 1970, and the 6368 km long BR230 from the Atlantic coast west to the Peruvian border (thanks to its human impacts, often known as the 'Transmiserianica'), started in 1970 and completed within a few years. Also in Brazilian Amazonia are the 2465 km-long (1747 km completed) Perimitral Norte – the BR210 – and the 1747 km long Cuiabá – Santarém highway (*see* Fig. 9.4).

The main impact has been an influx of spontaneous would-be settlers and ranchers. Before the mid-1970s there were efforts to support large-scale roadside colonization, complete with model settlements at regular intervals. Settlers and ranchers have cleared forest and killed or driven away game. Forests have been divided up by roads and huge sums of money have been spent on a road system that is still in large part unsurfaced and thus seasonally difficult to use. A number of 'boom towns' have grown up thanks to the road building. They include Maraba, Jatobal and Itaituba, which are experiencing rapid population growth, something that demands a supply of food and electricity and pollutes the rivers.

To some extent the Brazilian road network has provided a 'safety valve' for the poor and overcrowded North-East, but might equally be said to have helped reduce pressures for land reform in that region.

Ecuador has constructed an extensive road network in Western Amazonia, mainly to service oil exploration and exploitation. Peru has constructed its Carretera Marginal de la Selva ('Peripheral Jungle Highway'), initiating settlement and logging pressures in Western Amazonia.

Most impact assessment has been retrospective: Moran (1988) examined why settlers failed to establish a sustained livelihood; Goodland and Irwin (1974) and Moran (1988) review impacts.

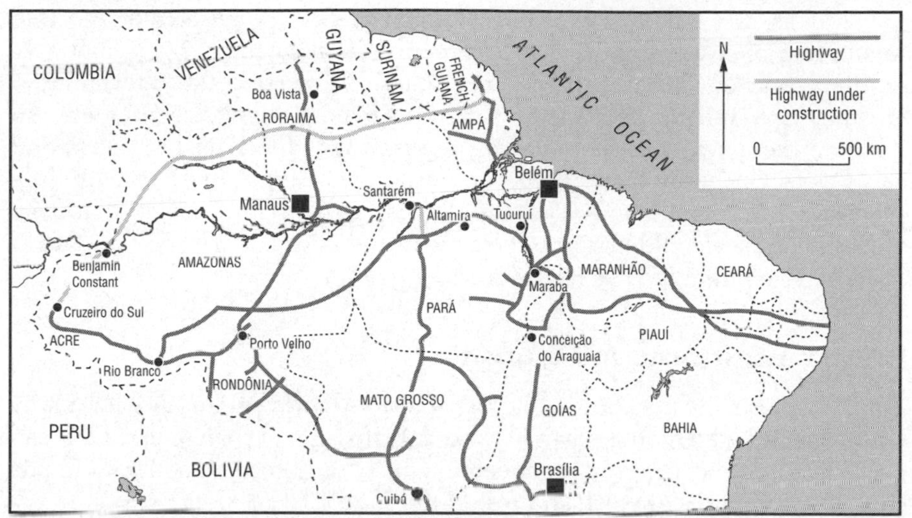

Some of the highways are unsurfaced; sections marked in a lighter tint were not completed in the 1990's

FIGURE 9.4 Road network constructed in Brazilian Amazonia since the mid-1960s (*see* Box. 9.1)

There may be unexpected benefits; for example, runoff from surfaced roads that farmers may be able to use for irrigation, or oppourunities for wildlife conservation on roadside verges.

Where road construction is likely to trigger rapid urban development, the planning authorities should prepare contingency plans or initiate measures to try to restrict or control it.

Impacts due to road use

Road use is likely to generate dust, noise, vibration (which damages buildings), air pollution, in some countries pollution from de-icing, human road deaths and injuries, wildlife 'road kill' and hindrance to movements. Positive impacts (Watkins, 1981) include access to increased opportunities and services, possibly better regional–national cohesion and opportunities for some wildlife (in roadside habitats).

Impacts caused by highways obstructing access or dividing a residential area, farmland or natural vegetation

Care needs to be taken to ensure that construction does not hinder land drainage or movement of wild animals and local people (bridges or subways may be required, and are difficult to add later). A common sight near Amazonian roads is areas of dead forest where trees have been drowned because local drainage has been disrupted by construction.

Logging roads

Road impacts are not confined to regular communication routes: 'skid trails' and logging roads, mainly temporary structures used to extract timber from forest areas, cause serious problems. These trails can suffer erosion, leading to siltation of streams and reservoirs or irrigation systems. They may give wildlife poachers, illegal loggers and squatter farmers access to terrain that is otherwise difficult to penetrate. Some governments have problems in policing the network of abandoned logging roads for drug producers and smugglers. Generally these temporary structures are put in place with little planning and with an eye to the lowest cost. With proactive impact assessment, they could be placed so as to minimize erosion or perhaps act as fire-breaks or scenic routes for tourism.

Impact assessment and tourism

Tourism is one of the largest items of trade world-wide, and for many countries is a major, often virtually the only, foreign exchange earner. It can be defined as the temporary movement of people to destinations outside their normal places of work and residence.

A number of tourism managers have expressed a wish to modify their country's or resort's approach to enhance positive and reduce negative effects of tourism. Impact assessment should enable optimum tourism planning, and there has been considerable application of environmental impact assessment and social impact assessment to try to achieve this (OECD, 1980; McTagert, 1980; Beekhuis, 1981; Jefferies, 1982; Mathieson, 1982; Clark *et al.*, 1984: 479–516; Edington and Edington, 1986). Impact assessment can also help ensure that tourism development is sustainable development.

It should be stressed that tourism is an amalgam of many different activities and attractions, and, in any given situation, is likely to change its character over time, possibly 'unpredictably'. As with much of the environmental impact assessment literature, there tends to be almost exclusive concern with negative impacts. Tourism can, however, have positive impacts (*see* Mieczkowski, 1995). There is a tendency to divide tourism into 'mass' tourism, often giving a low return for the host country, and 'high-quality' tourism, giving better rewards for the hosts. In practice, there is a wide range of types of tourism within such divisions, some of it spontaneous and difficult to control.

The impacts depend on whether the tourists use local services and contaminate the environment with waste or culturally damaging contacts, or are very short-stay (for example, ship-based). Short-stay tourists usually make much smaller demands on local services, have limited cultural impact and cause little pollution. Numbers of visitors and length of stay are clearly important, but so too are the nature of the demands made by the tourists and the vulnerability of the locality or attraction. (The impact of tourism on cities in developing countries is explored by McLaren, 1993.)

Small islands and delicate environments and cultures are easily damaged. A small number of visitors to environments such as the Himalayas, the English Lake District, Australia's coral reefs, the Arctic or Antarctica can cause ecological havoc, for example by eroding paths and dropping litter. Prehistoric cave paintings in France have suffered because of the moisture exhaled and the mould spores brought in by tourists. Where large numbers of people visit areas that develop a dry vegetation cover, recurrent bushfires are likely to increase in number. Forewarned of the threat, developers could establish fire-breaks, mount publicity programmes and adopt other fire-prevention or fire-fighting measures.

Some activities indulged in by tourists can be especially damaging if not tightly controlled. Trekkers can trample vegetation and soil, and cause land degradation; ski resorts can cause serious environmental damage if they plan and manage their facilities badly and fail to control their skiers. Visitors to conservation areas such as game reserves can disturb flora and fauna; power-boat users and marinas can injure coral and marine mammals and a much wider variety of wildlife if they are careless about speeding, sewage disposal or use of anti-foul paints (which poison many organisms). In the UK's Norfolk Broads (a scenic wetland), increasing numbers of pleasure boats since the 1950s have caused severe bank erosion and pollution, leading to serious environmental damage. Another impact is the increased dependency on tourism of a group of people or the host country as a whole (DeKadt, 1979).

Much can be done to control hotel building and sewage disposal, and to ensure that the supply of foodstuffs and services benefits local communities by providing employment. These are things that, with the use of impact assessment, can be identified early in tourism development. Cultural and social impacts are less easy to assess in advance, but given the serious problems that have developed in many countries it is well worth the effort to try to assess them.

A growing field is 'eco-tourism', which broadly means either tourism that is based on some attractive environmental attribute or tourism that is 'environmentally friendly'. Sometimes the two are combined. In practice, eco-tourism ranges from the minimizing of impacts to the coupling of tourism and conservation so that the former pays for and supports the latter. Tourism can give an incentive for conservation: something for local people to be proud of and to become involved with, especially if it provides a steady income in some way. Tourism may go beyond employing local people to providing funds for environmental management, restoration of land and buildings, and so on. Many forms of wildlife survive today mainly because of tourist interest in hunting, fishing and photography.

In some developed countries, various problems such as nitrate pollution of groundwater have prompted governments to adopt set-aside measures restricting agricultural use. Tourism and recreation are likely to play a vital role in providing alternative, more environmentally friendly, land uses (golf-courses, recreational woodlands, etc.). If eco-tourism is to be successful, there has to be careful pre-development assessment to ensure that the right facilities are provided and the right policy adapted. Where tourism is seasonal, it may provide useful off-season employment and income for those involved in agriculture or fishing; sometimes, however, labour migration occurs and the source areas are deprived of farmhands and other vital labourers. When this happens, there may be serious socio-economic difficulties and land degradation because land management is neglected, and these effects may appear in areas quite distant from the tourism development. Tourism impact assessment therefore has to deal with indirect impacts and cumulative impacts.

Concepts such as carrying capacity have been applied to tourism planning and management (for example, in relation to an island's water supplies, the trampling of vegetation or pathway erosion), but should be used with caution. Tourism impact assessment should include some allowance for changes in attitudes and fashions; investment in tourism may suddenly be irrelevant if people decide not to visit or in some other way change their habits. To some extent, marketing and advertising can be used to control attitudes and perhaps fashion, provided there is enough warning of an unwanted trend developing. Again impact assessment and monitoring may play a vital role.

The OECD (1980) reviewed the impacts of tourism on the environment. Farrington and Ord (1988) carried out an environmental impact assessment on a tourist railway.

Resettlement, relocation and migration

Millions of people are displaced by civil unrest, natural disasters and planned developments each year, mostly in the developing countries (Wilks and Hildyard, 1994). For example, between 1.5 and 2.0 million people fled Rwanda to refugee camps in Zaire between July 1994 and early 1996, causing serious environmental damage (deforestation and pollution, poaching from game reserves and health problems) to their host country (Biswas and Quiroz, 1996) and diverting funds that might have been used for environmental management or development.

Impacts associated with relocation of people are an important field of study and one likely to expand in the future, given the possibility of global environmental change and demographic trends (*see* Black, 1994a, 1994b; Newland, 1994). Some universities and international agencies have established migrant or refugee studies departments, and there are journals that focus on the subject, including *Journal of Refugee Studies* and *International Migration Review*. Retrospective studies and studies of ongoing movements help establish hindsight knowledge and allow the development of suitable guidelines for predictive impact assessment. There have been some forward-looking (so far, perhaps, too speculative) assessments of likely future movements of people; for example, movements generated by global environmental change. However, use of impact assessment could be improved. Thus apparently fewer than 55 per cent of World Bank-assisted projects have been subjected to proper resettlement appraisal before development was approved (Wilks and Hildyard, 1994; World Bank, 1994). A major problem with resettlement-related impact assessment is that it has not been given very high priority.

Movements of people can be crudely divided into combinations of *planned* and *unplanned*, *forced* and *voluntary* movement. It is also possible to consider either movements caused by *human* (political, economic and insecurity/warfare) actions or movements resulting from *environmental* causes, although the latter may sometimes be indirectly caused by the former and vice versa (McGregor, 1993). Impact assessment applied to unplanned movements tends to be retrospective or, at best, is undertaken once the movement is under way. Assessment can focus on the impacts on relocatees, on the receiving environment and peoples or on the environment and peoples left in the relocatees' 'source area'.

Knowledge of the impact that refugees and relocatees have on the environment and the people with whom they come into contact and on how it relates to policy is steadily growing (e.g. Pichon, 1992; Thapa, 1993; Unruh, 1993; Black, 1994a, 1994b; Prothero, 1994; Biswas and Quiroz, 1996). Research suggests that impact assessment will need to consider social, economic and environmental issues and the policies that affect them.

Dam- and reservoir-related resettlement

Dam-related resettlement has been subject to impact assessment studies – both environmental and social impact assessment – since the 1970s, although much of it has been initiated after developments were under way. These studies have built up considerable hindsight knowledge, especially from experience in Africa, Thailand, India, Brazil and, more recently, China (*see* Chambers, 1970; World Bank, 1980; Hansen and Oliver-Smith, 1982; Scudder, 1993; Fearnside, 1993; Gutman, 1994).

There have been attempts to provide guidelines for resettlement policy and practice based on what has been learned about impacts (Butcher, 1971; Cernea, 1988; Cernea and Guggenheim, 1993), but relatively few published pre-relocation impact assessments or studies of suitable impact assessment methods. Gutman (1994: 203) stressed that the best resettlement performance followed social impact assessment made early in the project preparations by independent specialists. There may be advantages in a participatory ('grass-roots') approach to impact assessment which ties in with such trends in resettlement planning (Hall, 1994).

Resettlement and land development schemes

Large resettlement and land development schemes have been undertaken in many countries, including (Federal Land Development Authority), Malaysia, Brazil (Projeto Integrado de Colonização – 'Integrated Colonization Project'), Indonesia (Transmigration), Tanzania (Ujamaa), South Africa (various relocation schemes), Sudan (notably the Gezeira Scheme) and others (Chambers, 1969; Palmer, 1974; Scholtz, 1992; Shami, 1993). The reasons for resettlement, the mode of relocation and the environments selected vary but there is a rich hindsight knowledge on which to draw for developing impact assessment approaches. An unforeseen problem in a number of these countries has been the failure of relocatees to adapt to their new environment and settle. For example, the failure of settlers to relocate successfully in Brazilian Amazonia has been examined by Moran (1988). Sometimes it is the settlers who move on, sometimes their grown-up children (movement of offspring has become a problem with Federal Land Development Authority relocation schemes in Malaysia).

Refugee camp impacts

Large concentrations of people, often coming about with little reference to the suitability of the local environment, cause considerable environmental, social and health impacts. Simply bringing together numbers of refugees from many different settlements, especially if they had been isolated, is likely to encourage transmission of disease and lead to conflicts. Poor water and fuelwood supplies, insanitary conditions, crowding, bad diet, stress and feelings of insecurity, inadequate housing and heat contribute to the risks of illness. Host populations may suffer real or perceived hardships as a consequence of refugee camps nearby, and can become jealous enough to come into conflict.

Although hindsight knowledge is accumulating (*see* Allan, 1987; Martin, 1992; Black, 1994b), in practice refugee camps often spring up suddenly with little chance for adequate impact studies or planning. Consequently, impact assessment is less likely to improve siting; however, it may aid management and monitoring, and improve speed of response by facilitating contingency planning. Clearly, refugee-related impact assesment must be fast and flexible.

Smallholder settlement

Smallholders may be settled in a planned manner. However, in many countries there are serious impacts arising as a consequence of less well-organized settling, and there may be completely spontaneous movements. The 'target areas' tend to be peri-urban lands, where people congregate having been attracted by hopes of city jobs or having been driven from the countryside, or marginal lands so far unattractive to others.

To some degree, settler impact will reflect the nature of landholding. If it is formalized and there is security, there is less chance that people will over-exploit and then abandon the land or be forced to move on to escape intimidation from those seeking to acquire their land. Because the settler movements are often spontaneous it has been rare for any *ex ante* assessment to take place. A good deal of information has nevertheless been provided by retrospective studies (e.g. Collins, 1986), and this should assist future impact assessments and settler management.

Eco-refugees

People can be directly or indirectly forced to relocate by environmental change, and in recent years have often been termed 'eco-refugees'. Eco-refugees might become a major development problem if present fears about global warming and pollution prove correct (Westing, 1992; Woehlcke, 1992; Myers, 1993; McGregor, 1994). Mass movements of people could be triggered by climatic change and associated sea-level rise or by problems with agriculture. Fluctuating production in grain-exporting countries might drive up international food prices enough to cause severe economic problems and unrest in some countries. Eco-refugees could drastically affect development within a host country to which they were forced to relocate, and might cause international conflicts (Brown, 1989; Widgren, 1990). It is easy enough to recognize lands at risk and vulnerable groups, so predictive impact assessments are possible and should be undertaken soon so that contingency planning can be done.

Environmental changes may cause spectacular natural disasters or more gradual, insidious problems or local deterioration. These may lead to hurried or more measured relocation or expenditure on mitigation measures. Once a cause is identified and the likely movements and numbers of relocatees can be estimated, an assessment should be made to determine likely impacts on host populations and environment. If the assessment shows that serious problems are likely, it may be possible to find alternative relocation strategies, establish contingency plans or mitigation measures, or raise enough interest to counter the root cause.

Marginalization

People may elect to move, or be forced, into marginal environments where it may be difficult to win a comfortable, steady living in return for efforts – they have become marginalized. A group of people may become disadvantaged relative to others for a wide range of reasons, possibly without moving – they have become marginalized. People in a locality may find that some environmental or socio-economic change or innovation may render their livelihoods more difficult or precarious – they too have become marginalized.

Marginal areas may be difficult to reach, or subject to periodic threats that make production precarious. They may also be environments that are unattractive areas in which to live and work. Strategic reasons or the attraction of resource exploitation may prompt a government or company to pay a premium to those willing to settle marginal areas. Conditions of marginality may suddenly change if there is new technology, or increased or reduced investment, or better communications, environmental change or even attitudinal changes that increase or reduce the demand for some product of marginal land. In the former USSR there is less money now with which to compensate those settling extreme environments, and consequently people have been relocating from regions like Siberia. Impact assessment can be used to determine why people settle marginal areas, how well they might exploit the environment and what factors are likely to alter their fortunes.

Seasonal and employment-related migration

In most parts of the world there is a seasonal pattern of disease and illness, food abundance, employment opportunities, stress and depression. Sometimes a seasonal disadvantage can escalate into a more persistent or deteriorating condition as has been examined by Chambers *et al.* (1981) and Gill (1991).

Impact assessment and oil and natural gas pipelines

Pipelines are linear features that can both disrupt movement of people and animals and facilitate movements. Animals seeking to cross the pipeline may be discouraged by the obstacle, noise, foundation materials and likelihood of increased human activity. On the other hand, people, plants and animals may follow pipeline routes, especially service tracks across otherwise inhospitable environments. In some regions, poachers and pest species have gained access via pipeline routes to areas where they cause problems. Pipelines constructed in Canada, Alaska, the former USSR and northern USA since the 1970s have helped improve environmental and social impact assessment procedures (e.g. Rees, 1980). Relatively warm pipelines across cold temperate and sub-Arctic landscapes or the Arctic tundra in particular have almost certainly generated far fewer impacts than would have been the case had impact assessments not been conducted.

There is also a reasonable knowledge-base on pipeline construction in semi-arid and arid environments and for subtropical and tropical humid conditions (Suman, 1987), there have been retrospective studies of pipelines in Amazonia and the Far East, for example (Wellner, 1994).

Some of the experience with pipeline impacts can be applied to other linear features such as electricity transmission lines, canals and roads, and the impact assessment approaches are generally similar; for example, overlays are useful.

Impact assessment and industrial development

Industrial development is a sector in which impact assessment associates closely with hazard and risk assessment and, increasingly, eco-auditing (*see* Chapter 2). The application of impact assessment extends from the materials supply (mining, water supply, hydroelectricity generation – all of which regularly use environmental impact assessment) to industrial processing and the impacts of products and packaging. There is no shortage of guidelines for industrial environmental impact assessment, some of which appeared quite soon after the USA's National Environmental Policy Act was passed (e.g. Gunnerson, 1978; World Bank, 1978; UNEP, 1980; Clark *et al.*, 1981; 1984: 253–67), and there are hazard or risk assessment guidelines (e.g. World Bank, 1985; Otway and Peltu, 1985).

Hazard assessment, risk assessment, technology assessment and environmental impact assessment for industry have probably received more attention than those other sectors as a consequence of various disasters and accidents suffered since the 1970s, including Flixborough (UK), Love Canal (USA), Seveso (Italy) and Bhopal (India). In particular, attention has focused on chemical plants (UNEP, 1982): the US Environmental Protection Agency has released a number of handbooks and guidelines on industrial risk and impact assessment since the mid-1980s, as has the World Bank (1985).

The insurance industry, the need for a good public image to maintain sales and the threat of legal action have probably prompted adoption of assessment as much as have regulations. Nevertheless, there are many industrialists who support and lobby for better environmental management, and groupings of 'green businessmen' were active at the 1992 Rio 'Earth Summit'.

There has been growing interest in assessing industrial development impacts in developing countries, and in checking the impacts of materials or methods 'driven out' of developed countries by tougher environmental regulation, environmental taxes, profit motives or public opinion (this has been termed the 'export of hazards'; Ives, 1985; Gourlay, 1992). Assessment of industrial impacts in rural and tropical areas of developing countries has been conducted by several researchers, including Christiansson and Ashuvud (1985) and Pernea and Pernea (1986).

In some developed countries, there has been interest in the assessment of the impacts of decline of once important industrial activities, especially socio-economic impacts, for example in the former coal-mining or steel-making towns of Yorkshire, Nottinghamshire and South Wales in the UK, and in Pennsylvania in the USA.

Industrial development generally means employment opportunities, which generates competition between authorities or cities over where the plant should be built. The combination of an often powerful industrial concern, various government authorities and city administrators plus possible public worry and activity by non-governmental organizations means that impact assessment can take place in one of the most volatile and difficult environments. An examination of such conflicts was provided by Bowander and Chetri (1984) in relation to the siting of a fertilizer complex. The greatest challenges facing impact assessment's application to industrial development are its satisfactory integration with planning and environmental management and adequate enforcement. The methodologies have been under trial and development since the 1970s, and accreditation of assessors should become more widespread and help ensure better assessment standards together with stronger regulations and the introduction of eco-auditing.

Retrospective impact assessment is valuable when there is confusion or the available information is limited, in order to establish a baseline before remedial action is taken or ongoing monitoring started. In the case of the Chernobyl, Bhopal, Windscale and Three Mile Island accidents there were considerable delays before attempts were made to assess health and other impacts.

The impacts of pollution are of interest to industrial environmental and social impact assessment and health environmental impact assessment. Pollutants may have immediately obvious effects, have less obvious long-term effects, or be locked up or in some way sequestered and pose a long-term insidious threat; or pollution episodes may constantly recur. Cumulative impacts may be difficult to predict when there is local pollution superimposed on regional and global pollution and affected by differing environmental conditions and patterns of exposure. The scale of the pollution impact threat is worrying; according to Johnson (1994: 379), 'more than half the population of the USA [a few years ago] lived in residential ZIP code areas with one or more uncontrolled toxic waste sites'.

Impact assessment and forestry

A sector where predictive environmental and social impact assessment seems to have been relatively neglected, in both developed and developing countries, is forestry. There are retrospective studies, notably on the establishment of tree plantations and the problems caused by clear-felling (the cutting of large swathes of trees, leading to loss of wildlife and, often, soil erosion problems) and damaging modes of logging. Comparisons have been made of the impacts caused by hand-felling and 'gentle' removal of logs with the use of heavy equipment and even the burning of remaining vegetation. Eucalyptus plantations in particular have generated a good deal of interest (almost wholly retrospective, and often too subjective) and have a reputation for resulting in negative socio-economic and physical impacts. Many forestry research establishments and commercial foresters conduct *ex ante* trials to see whether a tree species will flourish under given conditions. It would make sense to extend this work to consider visual impact, impact on wildlife, hydrological impacts, fire risk, vulnerability to global change, etc. There have been retrospective studies to assess the impact of acid deposition and other pollution on forests, and some attempts to assess future pollution impact scenarios for various forested countries. Carbon 'lock-up' (CO_2-sink) forests have had some assessment attention seeking to identify their potential, the best sites, cost-effectiveness, etc. There have been efforts to assess the impacts of large-scale forest clearance on regional or even global climatic conditions (e.g. the effects of removing extensive swathes of Amazonian rain forest on airmass movements leading to reduction of north-eastern Brazil's rainfall).

Impact assessments applied in advance of forestry programmes and projects are not abundant (e.g. Adger and Whitby, 1991; Epp, 1995). Given the pressures faced by many of the world's forests and the time needed to establish new tree cover or develop new plantation varieties, impact assessment should be more widely used to stimulate wise developments and prompt remedial actions. Global warming may necessitate marked changes in tree varieties being planted at a given site; it may take decades to develop suitable new stock able to resist wetter or warmer conditions or more drought. Impact assessment can help warn of the need for such breeding and point to qualities likely to be required, and identify the best planting sites for long-term viability.

Impact assessment of warfare and munitions

Perhaps one of the first applications of impact assessment was in seeking to predict war scenarios ('war games') to facilitate the development of military tactics, civil defence measures and munitions supply strategies. In the 1960s and 1970s there were a number of efforts to establish the likely impacts of the partial or extensive use of nuclear weapons, some of which generated the much-debated 'nuclear winter' scenario. There have been retrospective impact assessments of conflicts such as the Gulf War (Barnaby, 1991; Small,

1991; Kordagui and Alajmi, 1993), and attempts to assess the impact and ongoing threat of discarded munitions, especially anti-personnel mines (Westing, 1984). The latter studies are important, given the huge numbers of civilians killed or injured each year, and might support international agreements, or better manufacturing practices by some or all of the makers of the weapons that could reduce long-term threats to civilians. For example, mines could be manufactured with explosives or triggers that gradually decay so that after a few years they would become harmless.

The indirect impacts of war are considerable, and have seldom been properly traced. Many economies have been blighted by destruction of infrastructure, manpower and trade. Huge sums of money that might otherwise be spent on environmental management or other forms of development are committed to armaments purchases. However, some countries profit from war if they are producers of a strategic resource. The impacts of war can be long-lived: the effects of defoliant use during the Vietnam War on vegetation and people's health are still apparent. They can also be very widespread: a significant proportion of Africa's environmental and human welfare problems are related to conflicts. On the other hand, sometimes war can have beneficial side-effects, for example the development of remote sensing for peaceful use from military surveillance satellites and air photography.

Geneva Conventions have outlawed certain types of bullet, fléchettes (steel darts) delivered by missiles or dropped from aircraft), and chemical and biological weapons. As a means of generating information for such international agreements, impact assessment has great potential. Wars have often been fought over disagreements concerning resources. In the future, natural resources and global environmental change could well trigger conflict. Impact assessment is a means of recognizing these risks and initiating agreements, treaties, resource-sharing arrangements or a search for alternatives to the resource. Already the problem of shared river basins has attracted interest from those keen to prevent future conflict (Pearce, 1992b).

REFERENCES

Adams, W.M. 1992: *Wasting the rain: rivers, people and planning in Africa*. London: Earthscan.

ADB 1992: *Guidelines for the health impact assessment of development projects*. ADB Environmental Paper 11. Manila: Asian Development Bank.

Adger, W.N. and **Whitby, M.C.** 1991: Environmental assessment in forestry: the initial experience. *Journal of Rural Studies* **7(4)**, 385–395.

Adkins, W.G. and **Burke, D.** 1971: Interim report: social, economic and environmental factors in highway decision-making. Austin: Texas A & M University, Texas Transportation Institute.

AID 1980: Impact of rural roads in Liberia. Project Impact Evaluation 6. Washington, DC: US Agency for International Development.

Allan, N.J.R. 1987: Impact of Afghan refugees on the vegetation resources of Pakistan's Hindukush–Himalaya. *Mountain Research and Development* **7(3)**, 200–4.

Arquiaga, M.C., Canter, L.W. and **Nelson, D.I.** 1994: Integration of health impact considerations in environmental impact statements. *Impact Assessment* **12(2)**, 175–197.

Atkinson, J. 1992: *Narmada Dam: environmental and social impact of a World Bank project*. Fizroy, Victoria: Community Action Abroad.

Barber, M. and **Ryder, G.** (eds) 1993: *Damming the Three Gorges: what dam builders don't want you to know (a critique of the Three Gorges Water Control Project Feasibility Study)*, 2nd edn. London: Earthscan.

Barnaby, F. 1991: The environmental impact of the Gulf War. *The Ecologist* **21(4)**, 166–72.

Barrow, C.J. 1981: Health and resettlement consequences and opportunities created as a result of river impoundments in developing countries. *Water Supply and Management* **5(2)**, 135–50.

Barrow, C.J. 1983: The environmental consequences of water resources development in the tropics. In Ooi Jin Bee (ed.), *Natural resources in tropical countries*. Singapore: Singapore University Press for the Commonwealth Foundation and United Nations University, 439–76.

Barrow, C.J. 1987: The impact of hydroelectric development on the Amazonian environment, with particular reference to the Tucuruí Project. *Journal of Biogeography* **15(1)**, 67–78.

Beekhuis, J.V. 1981: Tourism in the Caribbean: impact on economic, social and natural environments. *Ambio* **X(6)**, 326–31.

Beenhakker, H.L. 1979: *Identification and appraisal of rural road projects*. World Bank Staff Working Paper 362. Washington DC: World Bank.

Birley, M.H. and **Peralta, G.L.** 1995: Health impacts assessment of development projects. In Vanclay, F. and Bronstein, D.A. (eds), *Environmental and social impact assessment*. Chichester: Wiley, 153–70.

Biswas, A.K. and **Quiroz, C.T.** 1996: Environmental impact of refugees. *Impact Assessment* **14(1)**, 21–39.

Black, R. 1994a: Environmental change in refugee-affected areas of the third world: the role of policy and research. *Disasters* **18(2)**, 107–16.

Black, R. 1994b: Forced migration and environmental change: the impact of refugees on host environments. *Journal of Environmental Management* **42(3)**, 261–77.

Blaikie, P., Cameron, J. and **Seddon, J.** 1977: The effects of roads in western central Nepal (Overseas Development Group, School of Development Studies). Norwich: University of East Anglia.

Bowander, B. and **Chetri, R.** 1984: The siting of a fertilizer complex: environmental and regional development issues. *Third World Planning Review* **6(2)**, 139–56.

Brokensha, D. 1980: Rural roads evaluations: social–economic indicators. Paper to Annual Meeting of the Anthropological Association, 5 December 1980, Washington. Washington DC: American Anthropological Association.

Brown, N. 1989: Climate, ecology and international security. *Survival* **31(6)**, 519–32.

Butcher, D.A.P. 1971: *An operational manual for resettlement: a systematic approach to the resettlement problem created by man-made lakes, with special relevance for West Africa*. Rome: Food and Agriculture Organization.

Butcher, E.H. and **Schofield, C.J.** 1981: Economic assault on Chagas' disease. *New Scientist* **92(1277)**, 321–5.

Carpenter, R.A. 1995: Risk assessment. *Impact Assessment* **13(2)**, 153–87.

Carpenter, T.G. 1994: *The environmental impact of railways*. Chichester: Wiley.

Cernea, M.M. 1988: *Involuntary resettlement in development projects: policy guidelines in World Bank-financed projects*. World Bank Technical Paper 180. Washington DC: World Bank.

Cernea, M.M. and **Guggenheim, S.E.** 1993: *Anthropological approaches to resettlement: policy, practice and theory*. Boulder, CO: Westview.

Chambers, R. 1969: *Resettlement schemes in tropical Africa*. New York: Praeger.

Chambers, R. (ed.) 1970: *The Volta resettlement experience*. London: Pall Mall Press.

Chambers, R., Longhurst, R. and **Pacey, A.** 1981: *Seasonal dimensions to rural poverty*. London: Frances Pinter.

Christiansson, C. and **Ashuvud, J.** 1985: Heavy industry in a rural tropical ecosystem. *Ambio* **XIV(3)**, 122–34.

Chu, C.M. 1990: The need for a social impact assessment of reproductive technology: the case of Caesarean birth in Taipei. *Environmental Impact Assessment Review* **10(1–2)**, 165–74.

Chung, M. 1988: The impacts of a road. In Overton, J. and Banks, B. (eds), *Rural Fiji*. Suva: University of the South Pacific, Institute of Pacific Studies, 97–122.

Clark, B.D, Chapman, K., Bisset, R. and **Wathern, P.** 1981: *The assessment of major industrial applications: a manual* (Department of the Environment Research Report). London: HMSO.

Clark, B.D., Gilad, A., Bisset, R. and **Tomlinson, P.** (eds) 1984: *Perspectves on environmental impact assessment*. Dordrecht: D. Reidel.

Collins, J.L. 1986: Smallholder settlement of tropical South America: the social causes of ecological destruction. *Human Organization* **45(1)**, 1–10.

Cooper Weil, D.E., Alicbusan, A.P., Wilson, J.F., Reich, M.R. and **Bradley, D.J.** 1990: *The impact of development policies on health: a review of the literature*. Geneva: World Health Organization.

Danziger, R. 1994: The social impact of HIV/AIDS in developing countries. *Social Science and Medicine* **39(7)**, 905–17.

Davies, K. 1991: Health and environmental impact assessment in Canada. *Canadian Journal of Public Health* **82(1)**, 19–21.

DeKadt, E. 1979: *Tourism: passport to development?* Oxford: Oxford University Press.

Dickey, J.W. and **Miller, L.H.** 1984: *Road project appraisal for developing countries*. Chichester: Wiley.

Dooley, J.E. 1977: *Identification and measurement of transportation impacts*. University of Toronto/York University, Joint Program in Transport Working Paper 7. Toronto: University of Toronto.

Dooley, J.E. and **Newkirk, R.T.** 1976: A planning system to minimize environmental impact applied to route selection. *Kybernetes* **15**, 213–20.

Doornkamp, J.C. 1982: The physical basis for route planning in the third world. *Third World Planning Review* **4(3)**, 224–30.

Doornkamp, J.C. 1985: Route planning. In Doornkamp J. (ed.) *The earth sciences and physical planning in the third world*. Liverpool: Liverpool University Press, 82–7.

Edington, J.M. and **Edington, M.A.** 1986: *Ecology, recreation and tourism*. Cambridge: Cambridge University Press.

Edmunds, S. 1989: Health impact statements: challenges ahead? *Community Health Studies* **13(4)**, 448–55.

Epp, H.T. 1995: Application of science to environmental impact assessment in boreal forest management: the Saskatchewan example. *Water, Air, and Soil Pollution* **82(1–2)**, 179–88.

Farrington, J.H. and **Ord, D.M.** 1988: Bure Valley Railway: an environmental impact assessment. *Project Appraisal* **3(4)**, 210–18.

Fearnside, P. 1989: Brazil's Balbina dam – environment versus the legacy of the Pharaohs in Amazonia. *Environmental Management* **13(4)**, 401–24.

Fearnside, P. 1993: Resettlement plans for China's Three Gorges Dam. In Barber, M. and Ryder, G. (eds), *Damming the Three Gorges: what dam builders don't want you to know (a critique of the Three Gorges Water Control Project Feasibility Study)*, 2nd edn. London: Earthscan, 34–58.

Fearnside, P. 1994: The Canadian feasibility study of the Three Gorges dam proposal for China's Yangzi River: a grave embarrassment to the impact assessment profession. *Impact Assessment* **12(1)**, 21–57.

Finney, C., Brabben, T., and **Adams, F.** 1988: The environmental impact of river basin development – 3. *Proceedings of the Institution of Civil Engineers, Pt. 1 – Design and Construction* **84**, 1099–102.

Gilad, A. 1984: The health component of the environmental impact assessment process. In Bisset, R. and Tomlinson, P. (eds), *Perspectives on Environmental Impact Assessment*. Dordrecht: D. Reidel, 93–103.

Gill, G.J. 1991: *Seasonality and agriculture in the developing world: a problem for the poor and the powerless*. Cambridge: Cambridge University Press.

Gilpin, A. 1995: *Environmental impact assessment (EIA): cutting edge of the twenty-first century*. Cambridge: Cambridge University Press.

Giroult, E. 1988: WHO interest in environmental impact assessment. In Wathern P. (ed.) *Environmental impact assessment: theory and practice*. London: Unwin Hyman, 257–71.

Goldsmith, E. and **Hildyard, N.** (eds) 1984: *The social and environmental effects of large dams*, vol. 1: *Overview*. Camelford: Wadebridge Ecology Centre.

Goldsmith, E. and **Hildyard, N.** (eds) 1986a: *The social and environmental effects of large dams*, vol. 2: *Case studies*. Camelford: Wadebridge Ecology Centre.

Goldsmith, E. and **Hildyard, N.** (eds) 1986b: *The social and environmental effects of large dams*, vol. 3: *Bibliography*. Camelford: Wadebridge Ecology Centre.

Goodland, R.J.A. 1978a: *Environmental assessment of the Tucuruí Hydroproject, Amazonia*. Brasília: Eletronorte SA.

Goodland, R.J.A. 1978b: Environmental assessment of the Tucuruí Hydroelectric Project, Rio Tocantins, Amazonia. *Survival International Review* **3(2)**, 11–14.

Goodland, R.J.A. 1989: *The World Bank's new policy on the environmental aspects of dam and reservoir projects*. Paper presented to a Congress on Research Needs and Strategy for the Self-sustaining Development of the Amazon, Manaus, 28–31 August 1989. Washington DC: World Bank.

Goodland, R.J.A. 1990: World Bank's new environmental policy for dams and reservoirs. *Water Resources Development* **6(4)**, 226–39.

Goodland, R.J.A. and **Irwin, H.S.** 1974: An ecological discussion of the environmental impact of the highway construction programme in the Amazon basin. *Landscape Planning* **1(1)**, 123–254.

Gourlay, K.A. 1992: *World of waste: dilemmas of industrial development*. London: Zed.

Grossman, D. and **Derman, D.** 1989: *The impact of regional road construction on land use in the West Bank*. Jerusalem: West Bank Data Base Project.

Gunnerson, C.G. 1978: *Environmental impacts of international civil engineering*. Washington DC: American Society of Civil Engineers.

Gutman, P.S. 1994: Involuntary resettlement in hydropower projects. *Annual Review of Energy and the Environment* **19**, 189–210.

Haines, A. 1990: The implications for health. In Leggett, J. (ed.), *Global warming: the Greenpeace report*. Oxford: Oxford University Press, 149–62.

Hall, A. 1994: Grassroots action for resettlement planning: Brazil and beyond. *World Development* **22(12)**, 1793–810.

Hansen, A. and **Oliver-Smith, A.** (eds) 1982: *Involuntary migration and resettlement: the problems and responses of dislocated people*. Boulder, CO: Westview.

Harvey, P.D. 1990: Educated guesses: health risk assessment in environmental impact statements. *American Journal of Law and Medicine* **16(3)**, 399–427.

Herren, G.B., Hansen, B.K. and **Wandesforde-Smith, G.** 1982: *Environmental impact assessment in the tropics: guidelines for application to river basin development*. Davis, CA: University of California, Center for Environmental and Energy Policy Research.

HMSO 1956a: *Volta River Project, I: Report of the Preparatory Commission* (Government of UK and Gold coast). London: HMSO.

HMSO 1956b: *Volta River Project, II: Appendices to reports of the Preparatory Commission* (Government of UK and Gold Coast). London: HMSO.

Hopkinson, P.G., Nash, C.A. and **Sheehy, N.** 1992: How much do people value the environment?: a method to identify how people conceptualize and value the costs and benefits of new road schemes. *Transportation* **19(2)**, 97–115.

Howe, G.M. and **Loraine, J.A.** 1980: *Environmental medicine*. London: Heinemann.

Howe, J. and **Richards, P.** (eds) 1984: *Rural roads and poverty alleviation*. London: Intermediate Technology Publications; Boulder, CO: Westview.

Interim Committee for Coordination of Investigations of the Lower Mekong Basin 1982: *Environmental impact assessment; guidelines for application to tropical river basin development*. Bangkok: UN Economic and Social Commission for Asia and the Pacific.

Ives, J.H. (ed.) 1985: *The export of hazard*. Boston: Routledge & Kegan Paul.

Jefferies, B. 1982: Sargamatha National Park: the impact of tourism in the Himalayas. *Ambio* **XI(4)**, 326–31.

Johnson, B.L. 1994: Is health risk assessment unethical? *Environmental Impact Assessment Review* **14(5–6)**, 377–84.

Kennedy, W. 1983: The EIA of highways. In University of Aberdeen, Project and Development Control Unit (ed.), *Environmental impact assessment*. The Hague: Martinus Nijhoff, 309–20.

Keogh, P., Allen, S., Almedal, C. and **Temahagili, B.** 1994: The social impact of HIV infection on women in Kiguli, Rwanda: a prospective study. *Journal of Science and Medicine* **38(8)**, 1047–53.

Kiravanich, P.E., Uyasation, U. and **Evans, J.W.** 1980: Environmental impact assessment and environmental planning for a highway project in southern Thailand: a case study. Bangkok: ASEAN/UNEP Workshop on EIA for Decision-makers, June, 1980.

Kordagui, H. and **Alajmi, D.** 1993: Environmental impact of the Gulf War: an integrated preliminary assessment. *Environmental Management* **17(4)**, 557–62.

Kreutzmann, H. 1991: The Karakoram Highway: the impact of road construction on mountain societies. *Modern Asian Studies* **25(4)**, 711–36.

La Bounty, J.F. 1982: Assessment of the environmental effects of constructing the Three Gorges project on the Yangtze River. In Yuan, S.W. (ed.), *Energy resources and the environment*. Oxford: Pergamon.

Long Li 1990: Major impacts of the Three Gorges Project on the Yangtze, China. *Water Resources Development* **6(1)**, 63–70.

McDowall, M. 1986: *The identification of man-made environmental hazards to health*. London: Macmillan.

McGregor, J. 1993: Refugees and the environment. In Black, R. and Robinson, V. (eds), *Geography and refugees patterns and processes of change*. London: Belhaven, 157–70.

McGregor, J. 1994: Climate-change and involuntary migration: implications for food security. *Food Policy* **19(2)**, 120–32.

McLaren, D.E. 1993: Environmental considerations and public involvement in the assessment of the impacts of tourism in the third world. *Impact Assessment* **11(2)**, 175–202.

McTagert, W.D. 1980: Tourism and tradition in Bali. *World Development* **8(4)**, 457.

Marchand, M. and **Toornstra, F.H.** 1986 *Ecological guidelines for river basin development*. CML Report 28. Leiden: University of Leiden, Centre for Environmental Studies.

Martens, W.J.M., Niessen, L.W., Rotmans, J., Jetten, T.H. and **McMichael, A.J.** 1995: Potential impact of global climate change on malaria risk. *Environmental Health Perspectives* **103(5)**, 458–64.

Martin, J.E. 1986: Environmental health impact assessment: methods and sources. *Environmental Impact Assessment Review* **6(1)**, 7–48.

Martin, S.H. 1992: Refugees and the environment. In Berthiaume, C. and del Mundo, F. (eds), *Refugees*. Special report on the environment, issue **89**. Geneva: United Nations High Commission for Refugees, 4–13.

Mathieson, A.R. 1982: *Tourism: economic, physical and social impacts*. London: Longman.

Mieczkowski, Z. 1995: *Environmental issues of tourism and recreation*. London: Oryx Press.

Mitchell, F.H. and **Mitchell, C.C.** 1979: Social impact assessment for health care programs: steps toward a research protocol. *Social Impact Assessment* **40–41**, 3–6.

Moran, E.F. 1988: Following the Amazon highways. In Denslow, J.S. and Padoch, C. (eds), *People of the tropical rain forest*. Berkeley: University of California Press.

Myers, N. 1993: Environmental refugees in a globally warmed world: estimating the scope of what could well become a prominent international phenomenon. *BioScience* **43(11)**, 752–61.

NAO 1992: *Overseas aid: water and the environment*. Report by the Comptroller and Auditor General, National Audit Office. London: HMSO.

Newland, K. 1994: Refugees: the rising flood. *World Watch* **7(3)**, 10–20.

OAS 1978: *Environmental quality and river basin development: a model of integrated analysis and planning*. (Government of Argentina, Organization of American States and UNEP.) Washington DC: Secretary General, Organization of American States.

Obeng, L.E. 1978: Environmental impacts of four African impoundments. In ASCE, *Environmental impacts of international civil engineering projects and practices*. Proceedings of a session sponsored by the Research Council on Environmental Impact Assessment, ASCE Technical Council on Research at the ASCE Convention, San Francisco, 17–21 October 1977. New York: American Society of Civil Engineers.

OECD 1980: *The impacts of tourism on the environment*. Paris: Organisation for Economic Co-operation and Development.

Otway, H. and **Peltu, M.** (eds) 1985: *Regulating industrial risks*. Sevenoaks: Butterworths.

Palmer, G. 1974: The ecology of settlement schemes. *Human Organization* **33(3)**, 239–50.

Palumb, D.R. 1993: The social impact of AIDS: facing the challenge. *American Journal of Pharmacy and the Sciences Supporting Public Health* **165(1)**, 29–34.

Pearce, F. 1992a: British aid: a hindrance as much as a help. *New Scientist* **134(1822)**, 12–13.

Pearce, F. 1992b: *The dammed: rivers and the coming world water crisis*. London: Bodley Head.

Pernea, E.M. and **Pernea, J.M.** 1986: An economic and social impact analysis of small industry promotion: a Philippine experience. *World Development* **14(5)**, 637–51.

Pichon, F.J. 1992: Agricultural settlement and ecological crisis in the Ecuadorian Amazon frontier: a discussion of the policy environment. *Policy Studies Journal* **20(4)**, 662–78.

Porter, G. 1995: The impact of road construction on women's trade in rural Nigeria. *Journal of Transport Geography* **3(1)**, 3–14.

Prothero, R.M. 1994: Forced migration of population and health hazards in tropical Africa. *International Journal of Epidemiology* **23(4)**, 657–64.

Rees, C.P. 1980: Environmental impact assessment of pipelining. *Pipes and Pipelines International* **25(1)**, 15–20.

Rees, C.P. 1981: Guidelines for environmental impacts assessment of dams and reservoir projects. *Water Science and Technology* **13(6)**, 57–71.

Restrepo, H.E. and **Rozenthal, M.** 1994: The social impact of aging populations: some major issues. *Social Science and Medicine* **39(9)**, 1323–38.

Rosenberg, D.M., **Bodaly, R.A.** and **Usher, P.J.** 1995: Environmental and social impacts of large scale hydroelectric development: who is listening? *Global Environmental Change* **5(2)**, 127–45.

Scholtz, U. 1992: Transmigrasi – ein Desaster? Problemme und Chancen des indonesischen Umsiedlungsprogramms. *Geographische Rundschau* **44(1)**, 33.

Scudder, T. 1993: Development-induced relocation and refugee studies: 37 years of change and continuity among Zambia's Gwembe Tonga. *Journal of Refugee Studies* **6(2)**, 123–52.

Shami, S. 1993: The social implications of population displacement and resettlement: an overview with a focus on the Arab Middle East. *International Migration Review* **XXVII(1)**, 4–33.

Simpson, R. 1990: Health impact assessment: some problems in air pollution management. *Environmental Impact Assessment Review* **10(1–2)**, 156–63.

Small, R.D. 1991: Environmental impact of fires in Kuwait. *Nature* **350(6313)**, 11–12.

Standing Advisory Committee on Trunk Road Assessment 1992: *Assessing the environmental impact of road schemes*. London: HMSO.

Stanley, N.F. and **Alpers, P.** (eds), 1975: *Man-made lakes and human health*. London: Academic Press.

Suman, D.O. 1987: Socioeconomic impacts of Panama's trans-isthmian oil pipeline. *Environmental Impact Assessment Review* **7(3)**, 227–46.

Sutcliffe, J. 1995 Environmental impact assessment: a healthy outcome? *Project Appraisal* **10(2)**, 113–24.

Thapa, G.B. 1993: Impacts of emigration on mountain watersheds: the Upper Pokhara Valley, Nepal. *Asian and Pacific Migration Journal* **24(4)**, 417–38.

Treweek, J.R., **Thompson, S.**, **Veitch, N.** and **Japp, C.** 1993: Ecological assessment of proposed road developments: a review of environmental statements. *Journal of Environmental Planning and Management* **36(3)**, 295–307.

Turnbull, R.G.H. 1992: *Environmental and health impact assessment of development projects*. London: Elsevier.

UNECE 1987: *Application of environmental impact assessment: highways and dams*. UN Economic Commission for Europe ENV/50. New York: United Nations.

UNEP 1980: *Guidelines for assessing industrial environmental impacts and environmental criteria for the siting of industry*. Paris: UN Environmental Programme.

UNEP 1982: *Guidelines in risk management and accident prevention in the chemical industry*. Nairobi: UN Environment Programme.

Unruh, J.D. 1993: Refugee resettlement in the Horn of Africa: the integration of host and refugee landuse patterns. *Land Use Policy* **10(1)**, 49–66.

Watkins, L.H. 1981: *Environmental impact of roads and traffic*. London: Applied Science Publishers.

Wellner, P. 1994: A pipeline killing field: exploitation of Burma's natural gas. *The Ecologist* **24(5)**, 189–93.

Westing, A.H. 1984: The remnants of war. *Ambio* **XXIII(1)**, 14–17.

Westing, A.H. 1992: Environmental refugees: a growing category of displaced persons. *Environmental Conservation* **19(3)**, 201–7.

Westmore, A. 1991: Health and environmental impact assessment. *Search* **22(6)**, 197.

White, G.F. 1972: Organizing scientific investigation to deal with environmental impacts. In Farvar, M.T. and Milton, J.P. (eds), *The careless technology: ecology and international development*. New York: Natural History Press (Doubleday), 914–26.

WHO 1979: *Environmental health impact assessment*. EURO Reports and Studies 7. Copenhagen: World Health Organization, Regional Office for Europe.

WHO 1983: *Environmental health impact assessment of irrigated agriculture development projects (guidelines and recommendations: final report)*. Copenhagen: World Health Organization, Regional Office for Europe.

WHO 1987: *Health and safety components of environmental impact assessment*. Copenhagen: World Health Organization, Regional Office for Europe.

WHO 1989: *Environmental impact assessment: an assessment of methodological and substantive issues affecting human health considerations*. WHO Report 41. London: University of London.

Widgren, J. 1990: International migration and regional stability. *International Affairs* **66(4)**, 749–66.

Wilks, A. and **Hildyard, J.** 1994: Evicted! The World Bank and forced resettlement. *The Ecologist* **24(6)**, 225–9.

Wilson, F.R. and **Stonehouse, D.L.** 1983: Environmental impact assessment: highway location. *Journal of Transportation Engineering – ASCE* **109(6)**, 759–68.

Woehlcke, M. 1992: Environmental refugees. *Aussenpolitik* **III/92**, 287–96.

World Bank 1974: *Environmental health, and human ecological considerations in economic development*. Washington DC: World Bank.

World Bank 1978: *Environmental considerations for the industrial development sector*. Washington DC: World Bank.

World Bank 1980: *Social issues associated with involuntary resettlement in Bank-financed projects*. Operational Manual Statement 2–33. Washington DC: World Bank.

World Bank 1985: *Manual of industrial hazard assessment techniques*. Washington DC: World Bank and Technica Ltd.

World Bank 1994: *Resettlement and development: the Bankwide review of projects involving involuntary resettlement 1986–1993*. Washington DC: World Bank, Environment Department.

10

PROBLEMS, CHALLENGES AND THE FUTURE OF IMPACT ASSESSMENT

Impact assessment has made huge progress since 1970: methods have improved and there has been a great deal of learning from experience. However, although many people accept the value of impact assessment, there are still serious faults to correct and new challenges to meet. By modifying and adding to the subheadings used in a review of impact assessment by Ortolano and Shepherd (1995: 13–25), I prepared the following list of problems and challenges. It represents a reasonable, if not complete, stocktaking of the current situation. The following are the perennial problems in implementing environmental impact assessment:

- The process is often avoided.
- It is often not carefully integrated into planning.
- It fails to ensure that developments are environmentally sound.
- Cumulative impacts are not assessed adequately.
- Public participation in environmental impact assessment is often inadequate.
- Proposed mitigation measures may not be implemented.
- An assessment is rarely conducted after the process has been completed.
- Assessment of risks and social impacts is often omitted from environmental impact assessment.

The new challenges are:

- problems concerning the 'global commons';
- environmental impact assessment in relation to development assistance;
- environmental impact assessment and international trade;
- support for sustainable development.

Most recent books on impact assessment outline the ingredients required for successful impact assessment. Typically the list includes:

- satisfactory methods;
- effective procedures;
- supportive legislation;
- sufficient data;
- funds to pay for assessment;
- clear indication of what is subject to impact assessment;
- competent impact assessors;
- freedom to assess effectively;
- effective co-ordination, auditing and monitoring;
- investigation of assessment shortcomings and means of appeal against them;
- adequate public participation;
- integrity of those commissioning, conducting and using impact assessment.

In the real world, very few, if any, assessments have all these, and anyway, impact assessment, like cooking, demands more than just ingredients.

RECOGNIZING AND OVERCOMING THE DEFICIENCIES OF IMPACT ASSESSMENT

Clearly, there is a need for considerable improvement of methods of impact assessment and in the procedures by which it is applied. Too often it is a 'bolt-on' addition to planning or even a public relations approach. Further shortcomings have been flagged, in the chapters preceding this, including the fact that application often takes place late in the development planning process and has too narrow a temporal, and often too restricted a spatial, focus; it takes what I have termed a 'snapshot' view.

While it may be possible to ensure that impact assessment is undertaken earlier in the development process, it will often still have to be practised against strict deadlines, and frequently with financial as well as time constraints. These difficulties, together with deficiencies of data, plus procedural and methodological problems, mean that impact assessment can never be conducted with the precision of a scientific experiment. It is thus subjective, contains unexamined and unexplained value assumptions, and is, to some degree, affected by politics and special interests. But even if impact assessment is subjective, this is no reason for its rejection; it seeks to predict potential impacts, not test and refine explanations. Nevertheless, there are dangers in subjectivity that mean that it must be accepted and handled well (Beattie, 1995: 110). Subjectivity must be worked with, and assessors must clearly publish the assumptions they make and try to keep methods 'transparent' (Mostert, 1996: 191).

In modern democracies, professionals and academics strive to establish and maintain standards and independence. In pre-modern times, as in some societies today, guilds and similar organizations pursued similar goals. However, when development is undertaken, those involved often lack independence and may maintain poor professional standards. The media, academic journal articles and books on impact assessment repeatedly note, first, that those conducting assessments are too closely associated with 'developers', and so are not free to express their concerns openly; and second, that there are seldom sufficiently impartial reviewers of impact assessment. Indeed, impact assessment is likely to be part of evaluation and design with the implicit aim of supporting proposed development. It is hardly surprising that many researchers mistrust impact assessment and dismiss it as ineffective. To overcome these problems, assessment and the review process must be given more independence. Simply requiring a developer to buy-in consultants is no solution as they are very likely to tone down or conceal unwanted findings to please their paymasters.

For impact assessment to have better regulatory effect, the assessors and the reviewers must be divorced from the developer. An independent agency, possibly a professional body or even a non-governmental organization, must assume the role of broker and intermediary, and in order to have sufficient 'clout' it will need to have an international standing. In effect this has happened for Antarctica, where multinational bodies have an overseeing role and, sometimes in parallel with non-governmental bodies such as Friends of the Earth, have carried out impact assessments (*see* Chapter 6). The medical profession has long maintained standards and independence with the help of professional bodies. Eco-auditing and environmental management have moved toward accreditation of practitioners and overseeing of standards and training by independent professional bodies. There may be ways of accelerating these sorts of independent co-ordination and review, even if laws are slow to change. For example, if insurance companies and professional bodies worked together, indemnity insurance could be withheld or be made too expensive for non-accredited assessors. Developers would soon hesitate to engage non-accredited assessors for fear of loss of face, and those denied accreditation, or those who had lost accreditation through malpractice, would find it difficult to stay in business.

The problems of lack of skills or data should diminish with time. This is because hindsight knowledge has been accumulating and there are improvements in its dissemination. Also, improvements in computers and software have already helped raise the standards of data-gathering and processing. Independent accreditation bodies could levy dues to invest in data collections, research and training aimed at improving impact assessment. Expert systems (*see* Chapter 2) have potential for quickly giving basic training or for modifying assessors' skills, and might also be controlled by professional bodies.

There are some improvements that could be more immediately achieved. Impact assessment is sometimes poorly presented, making it easy for developers or special-interest groups to accidentally or deliberately manipulate things, and difficult for the public and reviewers to prevent such activities.

Assessments must be clear, concise and transparent. Unfortunately, most of the literature and impact statements are an 'alphabet soup' crammed with vague and confused acronyms (Mayda, 1996: 39). If it had not been for the criticisms of a watchful reviewer I would probably have fallen in with the practice and sprinkled this book with acronyms and abbreviations! Impact assessment will become more precise and will gain respect if the terminology is standardized and clear agreed concepts replace rhetoric. Again, professional bodies could encourage this.

Three rapidly evolving fields promise to help resolve some of the weaknesses of impact assessment: strategic environmental assessment (*see* Chapter 3); cumulative environmental assessment (*see* Chapter 5); and adaptive enviromental assessment and management (*see* Chapter 3). The first of these offers a means for expanding impact assessment to cover tiers of planning above the project level and from local up to global scale. Many researchers regard it as a way of ensuring that policy-making takes account of sustainable development principles (Partidário, 1996: 40). There are interesting possibilities for global information systems to link beneficially with strategic environmental assessment (Benoît, 1995: 423). Cumulative environmental assessment broadens the focus of impact assessment and is likely to be important in the quest for sustainable development, as many of the world's problems have been caused by and will continue to be caused by a number of causes, often unrelated and indirect ones. Adaptive environmental assessment and management offers possibilities for making impact assessment more flexible and able to cope with data-poor and rapidly changing situations.

Particularly since the publication of the 'Brundtland Report' (WCED, 1987), the world's decision-makers have started to shift from reliance solely on economic criteria to giving consideration to environmental and social issues too. Unfortunately, support for impact assessment was not as strong as it might have been at the UN Conference on Environment and Development (the Rio de Janeiro 'Earth Summit'). New activities have appeared in the past half-century that have the potential to cause impacts that are irreversible and life-threatening, including, for example, nuclear technology, genetic engineering and new chemical compounds that are persistent and toxic, and affect human health or reproduction. Reliable and, as far as possible, universal impact assessment is vital. The strengths of impact assessment meet the aforementioned challenges: it is forward-looking, gives priority to the precautionary principle (i.e. attempts to avoid, rather than react to, problems), and adopts a multidisciplinary approach. (Virtually everybody concerned with environment and development issues faces the problem of how to conduct effective multidisciplinary study; it is a problem not confined to impact assessment.) Fields such as participatory rural appraisal (*see* Chapter 2) have made some progress in improving multidisciplinary study, and this should improve impact assessment.

Mayda (1996: 37) made no exaggeration when he argued that impact assessment is potentially 'one of the key mechanisms for turning the world towards sustainable development'.

REFERENCES

Beattie, R.B. 1995: Everything you already know about EIA (but often don't admit). *Environmental Impact Assessment Review* **15(2)**, 109–14.

Benoît, J. 1995: Current and future directions for structured impact assessments. *Impact Assessment* **13(4)**, 403–32.

Mayda, J. 1996: Reforming impact assessment: key issues emerging from recent practice. *Impact Assessment* **14(1)**, 87–96.

Mostert, E. 1996: Subjective environmental impact assessment: causes, problems, solutions. *Impact Assessment* **14(2)**, 191–213.

Ortolano, L. and **Shepherd, A.** 1995: Environmental impact assessment: challenges and opportunities. *Impact Assessment* **13(1)**, 3–30.

Partidário, M.R. 1996: Strategic environmental assessment: key issues emerging from recent practice. *Environmental Impact Assessment Review* **16(1)**, 31–55.

WCED 1987: *Our common future* (the 'Brundtland Report'). Oxford: Oxford University Press.

GLOSSARY

Note: many impact assessment terms are 'quasi-technical' and have no single, firm definition.

assessment: an evaluation in as objective a manner as possible but not necessarily quantitative or verified by experiment.

audit: a comparison of impacts predicted by an assessment with those that actually happened, consideration of costs of assessment, effectiveness, etc., carried out to establish how satisfactory the exercise was. Audit may also mean the gathering of environmental, social and economic information and its presentation as a sort of stock-taking to see whether a company or authority is environmentally sound, when it is termed an **eco-audit**. **Environmental audit** may refer to an account drawn up for a nation or region showing the state of things.

baseline: a description of conditions existing at a point in time against which subsequent changes can be detected through **monitoring**. A baseline study is also required in order to establish what the conditions would be if development were not to take place. Conditions may not be stable even in the absence of development: there may be decline, improvement or cyclic conditions. Baseline data may be expressed as an environmental inventory or as a report.

BS 7750: the UK's environmental management system standard, issued by the British Standards Institution.

carrying capacity: definitions vary and can be imprecise. They include the following: 'the maximum number of individuals that can be supported in a given environment'; 'the amount of biological matter a given system can yield, for consumption by organisms, over a given period of time without impairing the ability to keep producing'; 'the maximum population of a given species that can be supported indefinitely in a particular region by a system, allowing for seasonal and random changes, without any degradation of the natural resource base'. For human society there can be a maximum and an optimum carrying capacity, because many people can be supported at survival level, but fewer supported at a good 'quality of life'.

contingent valuation: a monetary value for a good or amenity established by asking people what they would be willing to pay for it.

cost–benefit analysis: a method of assessing a project that studies its costs and benefits to society as well as the revenue it generates, or is expected to generate. Strategic advantages or net social value may outweigh monetary commercial costs; for example, a railway may fail to make a profit but be vital for regional communications.

cumulative effects assessment (or cumulative environmental assessment): a type of assessment that seeks to identify and communicate the consequences of more than one impact from a single development or combination of impacts from a develop-

ment interacting with impacts from other developments in the world at large. Cumulative effects assessment is difficult and complicated, with many of the impacts being indirect in time and/or space, but is likely to become more important in the future.
cumulative impact: the consequence of more than one direct and/or indirect impact acting together.
direct impact: the first effect in a chain of causation.
EC Directive 85/337: the 'EIA Directive' of 1985 which required all countries of the European Community (now the **European Union**) to bring **environmental impact assessment** procedures into force.
eco-auditing: a multidisciplinary methodology used to assess periodically and objectively the environmental performance of an organization, authority or, in some instances, a region.
eco-label: a marker on goods indicating that those goods themselves are 'environmentally friendly'. The scheme has been promoted in the **European Union** since 1992, although Germany introduced it in the late 1970s.
EEC: the European Economic Community, one of the communities (along with the European Coal and Steel Community, etc.) making up the European Community, now the **European Union.**
EEC Fifth Environmental Action Programme 1992: a programme designed to protect and enhance the quality of environment in **European Union** countries, in force from 1993 to 2000. A strong theme is **sustainable development**, and the programme has powers to initiate actions to pursue environmental quality and sustainability.
EIA: environmental impact assessment.
EIS: environmental impact statement.
EMAS: Environmental Management and Audit Scheme. The **European Union's** systematic approach to standardized **eco-auditing.**
empowerment: the improving of people's ability to secure their own survival and development, and to increase their ability to participate in and exercise influence over crucial decisions affecting their survival. In effect, helping people to achieve their own purpose by increasing their confidence and capacity.
environment: 'parameters of life' – the sum total of conditions within which organisms may live, interacting with those conditions.
environmental assessment: may mean a number of things. In the USA an environmental assessment is a document presenting enough information to allow a decision to be made on whether a development merits a full **environmental impact assessment** that will lead to the production of an **environmental impact statement**. (If it does not, a **FONSI** is released, in effect an initial environmental examination at an early stage of environmental impact assessment.) In the UK and Canada the term is a synonym for environmental impact assessment.
environmental impact: an alteration in the **environment** or any component of the environment, including economic, social or cultural aspects.
environmental impact assessment: accurate, critical and, it is to be hoped, objective assessment of the likely effects of a development (project, programme, policy, plan, social or economic change, environmental change, etc.). It should, in addition to identifying impacts, clarify what the situation would be if no development or change were to occur and what the impacts are for various possible development options. Impacts that are irreversible or that threaten organisms, environmental quality and **sustainable development** should be highlighted.
environmental impact report: another term for **environmental impact statement,** used in California.

environmental impact statement: the communication of the results of an **environmental impact assessment** or **social impact assessment**, usually a documentary report and/or diagrammatic presentation. It should describe the pre-development situation and likely trends were no development to occur; the proposed development(s), taking care to outline possible options; identified impacts that should be assessed for importance and magnitude; likelihood of occurrence; and reliability of prediction. It is increasingly likely that the public will be presented with an environmental impact statement, and it is the channel through which the public are informed and can respond to proposals. The environmental impact statement (possibly a more detailed version than that released to the piblic) is usually required by a government in order to determine whether to allow a development proposal to proceed. (In the UK, called an 'environment statement' by the authorities.)
environmental inventory: a description of the **environment**, usually of a site prior to a development. It has much overlap with **baseline** study.
environmental management: the process of allocating natural and artificial resources so as to make optimum use of the environment in satisfying basic human needs, if possible for an indefinite period, and with minimal adverse impacts.
environmental review: a term used in the USA to refer to a full **environmental impact assessment** process.
environmental statement: a term used in the UK to mean **environmental impact statement**.
EPA: Environmental Protection Agency. A body established in 1970 (not under **NEPA**) to act as the US regulatory agency; among its duties are the checking of **environmental impact statements** produced by other bodies.
EU: European Union.
European Union: a group of 15 countries (at present) that operate as a single market and to some extent have pooled their sovereignty. Current members are Austria, Belgium, Denmark, Finland, France, Germany, Greece, the Republic of Ireland, Italy, Luxembourg, the Netherlands, Portugal, Spain, Sweden and the UK. Formerly known as the European Community.
feedback: 1. A process within a system that accelerates or enhances something that is under way (positive feedback) or counters a change that is under way (negative feedback). 2. The presentation of the findings of an assessment, **monitoring** process, review or evaluation to people involved.
first-order impact: *see* **direct impact**.
FONSI: Finding of No Significant Impact. In the USA, if an initial environmental assessment, preliminary to an **environmental impact assessment**, conducted by a federal agency indicates that there are too few impacts to merit a full environmental impact assessment a 'finding of no significant impact' is issued – a concise statement presenting the reasons why a full environmental impact assessment is not required and an abbreviated report of the findings.

There are situations where a FONSI is issued if mitigating measures are undertaken; if this is the case, the federal agency issues a *Mitigated FONSI*.
forecast: extrapolation forwards of the trends presently evident, to assess what will probably occur.
GIS: geographical information system. A computerized data systems that stores, retrieves, manipulates, and displays spatial information.
Holdridge life zone model: a widely used ecoclimatic classification system, based on the relationship of current vegetation biomes (a biome is the ecological community of a particular large area, such as grassland or tropical rain forest) to three

general climatic parameters: annual temperature, annual precipitation and an established potential evapotranspiration. The model predicts ecoclimatic areas; it does not directly model actual vegetation or land cover distribution.

holistic: a researcher adopting a holistic approach seeks to study the whole, which may be 'more than the sum of its parts'. The study extends to examining linkages and interactions. This contrasts with a single-discipline, reductionist focus.

human ecology: an important area of application of ecosystem analysis that seeks to study how people fit into and affect the environment.

impact: a change resulting from some development – often synonymous with 'effect', however, some have suggested that 'effect' should imply a natural cause and 'impact' human-induced. Manifest through an observable change in some parameter(s).

indicators: a means whereby various **thresholds** are translated into measurable terms (i.e. basic measures of performance). Indicators help describe environmental quality and allow the measurement of progress. They can also help provide a summary of information. Some indicators are 'composite' i.e. based on several parameters, an example being a 'quality of life index'. It is possible to recognize *process indicators*, which show if a process or activity is being carried-out, and impact indicators which show the effect of a development.

indirect impact: second, third or subsequent **impact** in a chain of causation, possibly indirect in space and/or time.

initial environmental assessment: a preliminary assessment carried out in order to determine whether a full **environmental impact assessment** is required.

interval scale: a scale showing the degree of difference between objects, e.g. 10°C, 20°C, 30°C, etc.

ISO 14000: the International Standards Organization's standard for environmental management and **eco-auditing**, roughly equivalent to **EMAS** or **BS 7750**.

MASAQHE: major action significantly affecting the quality of the human **environment**. This is the label applied, where appropriate, to federal activities in the USA.

method: a technique used to provide information or assess that information.

methodology: a structure for organizing a **process**.

model: a caricature or simplification of reality; often a set of equations used to predict the behaviour of variables in a particular field.

monitoring: a combination of observation and measurement of a development. It may be continuous or periodic, and may comprise data collected by observers or data collected by remote instrumentation (including satellite remote sensing or aircraft-mounted instruments). Monitoring may be aimed at detecting new development trends; at maintaining an ongoing check on performance or compliance with laws or conditions laid down when the development was approved; or at providing warning of changes (environmental, social, economic) that might affect a development.

multiplier effect: the idea that a development may generate more benefits than obvious direct benefits; for example, a dam leads to irrigation which improves income which leads to better services and regional development.

NEPA: the National Environmental Policy Act 1969, passed by the US Congress in January 1970.

NIMBY: not in my backyard. An attitude seen as a syndrome; it refers to people's opposition to a development in their locality, even if they would be willing to accept it if it were sited elsewhere.

nominal scale: a qualitative scale, such as hot, warm, cold.

normative: applied to a process or procedure designed to rank or choose 'what ought to be' according to prescribed goals or evaluative criteria.
objective: the use of verifiable facts with (in theory) no personal reflection, feeling or prejudice involved. It may be possible to make more than one objective judgement on case data.
opportunity cost: the cost of sacrificing alternative activities or earnings.
ordinal scale: a scale that classifies by order, e.g. 1st, 2nd, 3rd, etc.
PADC: Project Appraisal for Development Control (Unit), University of Aberdeen, UK. Became CEMP, the Centre for Environmental Planning and Management. One of the UK's foremost **environmental impact assessment** development and training units.
Pareto optimum: the point at which all possible gains from a voluntary exchange have been exhausted, both parties feeling that they have obtained the maximum possible benefit from the situation. To improve one party's position further means damaging the other's.
plan: a set of co-ordinated and timed **objectives** for implementing a **policy**.
policy: the inspiration and guidance for action.
prediction: a possibly imprecise assumption of trends open to discussion and modification.
procedure: steps or responsibilities required or suggested for implementation of an environmental impact assessment process, determining when it is performed, how and by whom, etc. The steps are suggested or enforced by law or by an overseeing body.
process: the way in which an environmental impact assessment is performed; a system of conduct or administration, or a series of steps.
programme: a set of **projects** in a particular field.
project: an individual development or scheme, as opposed to a group or a strategy. 'Subproject' implies a single site within or a component of a project.
project appraisal: an exercise designed to predict the financial and economic rate of return of a proposed or completed **project**, taking into account technical, administrative, social and environmental implications. It may be conducted during implementation to assess progress and identify problems for correction.
projection: a precise form of **forecast**.
ratio scale: a scale that classifies in terms of a quantitative degree of difference between objects in relation to some starting-point, e.g. degrees above boiling point in °C.
scenario: an attempt, less precise than a forecast, to assess 'what would happen if...?' ('simulation' is roughly synonymous).
scoping: the establishing of terms of reference for an **environmental impact assessment**: what to measure, who will do what jobs, what the budget will be, what merits special attention, how long the work will take, whether the public is to be involved, and if so, how. Should also ensure the results are presented in the best manner.
screening: selection of programmes or policies or developments to subject to **environmental impact assessment**. Less frequently, the process of eliminating alternatives for development by comparing the relative generation of **impacts**.
SEA: strategic environmental assessment.
shadow price: the real value of goods, services, etc., reflecting **opportunity costs**. An aggregated 'willingness to pay' for something.
significant: has the same meaning as in everyday speech, but a cause of early troubles for **environmental impact assessment**. Still often poorly defined and

reflecting a **subjective** viewpoint. The concept of what is significant can change over time as knowledge progresses or attitudes alter.
social impact assessment: a subdivision of environmental impact assessment or, alternatively, the opposite end of the same spectrum of activities, which is concerned with the impact of development on people, whether individually or as groups – ranging from households up to global society. It is also concerned with the impact of people on developments and the environment. Social changes may affect development and generate **feedback** that alters society.
standard: an accepted or approved example of something against which others are measured. Standards allow meaningful evaluation, negotiation, lawmaking and comparison site to site, country to country and year to year. They are usually agreed, reliable and measurable (using common units) levels; for example, the level of lead in drinking water that is considered toxic, or the level of ozone in a city street regarded as harmful. For enforcement purposes a 'standard' means a limit, the maximum level permitted.
strategic environmental assesment: the formalized, systematic and comprehensive process of evaluating the environmental effects of a policy, plan or programme and its alternatives, including the preparation of a written report on the findings of that evaluation, and use of the findings in publicly accountable decision-making.
structural adjustment loans: large loans made by funding bodies such as the World Bank, which have the objective of bringing about an economy-wide reform in the recipient country. Typically the lender insists on a programme of import restrictions, 'free-market' policies and relaxed state control on the economy. To qualify for these loans, countries may have to cut expenditure, often on social services, and may have to devalue their currency. The result has usually been increased unemployment, a fall in real wages and increased cost of imported goods.
subjective: consciously or unwittingly affected by the observer's feelings and/or capacities; not **objective**.
sustainable development: an imprecise term, but can be taken as meaning development that promotes economic, social and environmental benefits in the long term. Pursuit of sustainable development involves careful management of human and natural resources to ensure that their quality is maintained or improved from generation to generation.
synergistic: applied to cases in which the sum total impact of a number of impacts exceeds the effect of the sum of the individual components ('the whole is greater than the sum of the parts').
system: a set of logically interconnected **models**.
threshold: the point after which an environment, organism, society, economy is clearly affected and degrades. Ideally, impact assessment should prevent development from reaching and exceeding thresholds.
tied aid: aid can be 'tied' by source (meaning that loans or grants have to be spent on the donor's goods and/or services) or by project (meaning that funding must be spent on a specific project or programme). Although this practice may make it easier to grant aid, it may bind recipients to purchasing expensive or inferior goods and dependency on the donor for spare parts. Tied aid may lead to less than optimum planning decisions, as was illustrated by the Pergau Dam scandal (in which the UK was the donor and Malaysia was the recipient).
tiered environmental/social impact assessment: a sequential process with assessment at **policy**, **programme** and **project** levels (representing vertical tiers within a sector) or multisectoral assessment (representing horizontal tiers).

time bomb effects: a term that refers to situations where a chemical or biological material accumulates without at first causing a problem or even being apparent. Problems arise when a threshold is suddenly reached, either through continued accumulation or because some environmental or socio-economic change(or changes) triggers it. For example, a pollutant might accumulate in the soil, remaining chemically bound to clay minerals, but then a change of agricultural practice or acid deposition suddenly releases pollution.

terms of reference: the limits for a study, **project**, **programme**, etc., and the statement of the task to be undertaken and the guidelines.

USAID: the US Agency for International Development, a US agency responsible for bilateral development and aid.

utility: a measure of the satisfaction individuals receive from the consumption of goods and services. Its measurement involves value judgements.

INDEX